Current Topics in Microbiology 212 and Immunology

Springer
Berlin
Heidelberg
New York
Barcelona
Budapest
Hong Kong
London
Milan
Paris
Santa Clara
Singapore
Tokyo

Immunology and Developmental Biology of the Chicken

Edited by O. Vainio and B.A. Imhof

With 43 Figures

Springer

Olli Vainio, M.D.
Beat A. Imhof, Ph.D.
Basel Institute for Immunology
Grenzacherstr. 487
4005 Basel
Switzerland

Cover illustration: A montage showing a 3.5-day-old embryo and two chickens one day after hatching. At this stage of embryonic development, the first hemopoietic progenitor cells emerge in intra-aortic clusters. The embryo is an adapted computer design of a drawing made by Mathias Duval in 1889 in Paris. (Designed by Andre Traunecker, Basel Institute for Immunology.)

Cover design: Künkel+Lopka, Ilvesheim

ISSN 0070-217X
ISBN-13:978-3-642-80059-7 e-ISBN-13:978-3-642-80057-3
DOI: 10.1007/978-3-642-80057-3

Softcover reprint of the hardcover 1st edition 1996

Library of Congress Catalog Card Number 15-12910

Typesetting: Thomson Press (India) Ltd., Madras
SPIN: 10495370 27/3020/SPS – 5 4 3 2 1 0 – Printed on acid-free paper

Preface

The chicken is a useful model for research into the basic features of immunology because its immune system functions in a way similar to that of human beings and because the chicken embryo is easily accessed experimentally. This book is a collection of comprehensive articles updating chicken embryogenesis and immunology.

It is now clear that it is possible to produce transgenic chickens by embryonic stem cell manipulation and transfer into the oocyte. Alternatively, the chick offers us the possibility of producing somatic chimeras by manipulating embryonic cells at the primitive streak or earlier stages. Working with later embryos, we can learn the principles involved in the positioning of organ anlagen such as the limb bud. It is shown in this book that these structures are formed by appropriate cell and tissue arrangements driven by molecular factors.

Work on the chicken has contributed substantially to our understanding of basic immunology, including the graft-vs.-host reaction, the clonal reactivity of immunocompetent cells, separation of the T and B cell lineages, somatic diversification of the B cell repertoire by immunoglobulin (Ig) class switching, antibody diversification, Ig gene conversion, and the origin of the hematopoietic stem cell precursor of lymphoid and myeloid cells. Hematopoietic stem cells emerge successively from several embryonic sites such as the yolk sac, the aorta, and the mesentery near the aorta. The simplicity of the genetic loci and the vast collection of antibodies and molecular tools that have been produced recently enables one to exploit more fully all the experimental advantages of the chicken.

In brief, we hope that this book will convince researchers that the chicken is an important animal model with which one can address developmental and hematopoietic problems that are difficult to follow in human beings and mice. In the Appendix at the end of this volume, a list of monoclonal antibodies against chicken leukocyte antigens, arranged alphabetically according to the contributing scientists, and an overview of marker antibodies are presented.

Basel

OLLI VAINIO and BEAT A. IMHOF

List of Contents

List of Contributors

(Their addresses can be found at the beginning of their respective chapters.)

Why I Think the Chicken Should Survive in Immunology and Developmental Biology

M. SIMONSEN

I propose to begin with a very unusual bird (Fig. 1). This is no chicken. Nor is it a recent Biotech scandal. It is a Tyrolean allegory of the theme of self-recognition made some 200 years ago, which can be seen today in a folk-art museum in Innsbruck.

I use it here as a picturesque example of man's fascination with birds, which, so the scholars say, goes back into prehistory and supposedly lies at the root of the domestication of avian species (WOOD-GUSH 1959; ZEUNER 1963; CRAWFORD 1990a).

Fig. 1. Self-recognition. (Courtesy of Tiroler Volkskunstmuseum, Innsbruck)

Thorshavnsgade 28, 2300 Copenhagen S, Denmark

Unlike most domesticated mammalian species, which served from the beginning as a handy source of food, domesticated birds were not originally made for eating but served cultural purposes that were hard to separate in those days from religion, whoever were the gods and spirits.

The reluctance to eat domesticated birds is on record. Caesar noted in his book on the Gallic wars that the Celts kept chickens in the British Isles before the Roman legions arrived but they would not eat them. The Romans would, but back in Rome even they held the geese to be sacred. Also, let us not forget that the Oracle in Delphi would read the fate of Greek city states and Persian empires in the entrails of chickens.

Is that why I think the chicken should survive in immunology and developmental biology? Not quite yet, I will take it stepwise.

The scholars disagree on who was first to tame the chicken. The short answer is that nobody knows, but many have voiced their opinion. The majority view seems to be that chickens were domesticated in the Indus valley some 4000 years ago and that they spread from there in different tempi to the rest of the globe.

However, recent archaeological finds in China suggest that chickens were there in domesticated form already some 8000 years ago. If so, they may have been domesticated through about as many generations as *Homo sapiens* has domesticated him/herself. Moreover, fowl have the additional appeal to genetics, in that the wild forms from which it came are probably still extant, hence amenable to genetic analysis and experiments with evolution.

That brings me to the question of who laid the eggs from which domesticated chicks were to hatch?

The prevailing opinion since DARWIN's *Variation in Animals and Plants under Domestication* (1868) has been that the Red Jungle fowl provided them, but there are additional forms of jungle fowl still around.

According to CRAWFORD (1990a) there are four extant species, including *Gallus gallus*, which is by far the most widespread and can be subdivided into five subspecies (Table 1).

It is obviously pertinent to ask how related are these wild forms of chickens

Table 1. Wild forms of *Gallus* still alive

Species		
Gallus varius	Green Jungle fowl	Java
G. sonnerati	Grey Jungle fowl	West and South India
G. lafayettei	Ceylon Jungle fowl	Sri Lanka
G. gallus	Red Jungle fowl	Southeast Asia
Subspecies of *G. gallus*		
G. gallus gallus	Red Jungle fowl	Cochin-China
G. gallus spadiceus	Red Jungle fowl	Burma
G. gallus jabouillei	Red Jungle fowl	Tonkin
G. gallus murghi	Red Jungle fowl	India
G. gallus bankiva	Red Jungle fowl	Java

Adapted from CRAWFORD (1990a), who also states: "Only in the case of hybridizing between Red Jungle Fowl and domestic Fowl does the literature reveal complete compatability. All of the other combinations warrant further study."

mutually and to domesticated forms in terms of their ability to make fertile offspring in cross-matings. That point was made very succinctly by Punnett (1923) in his book *Heredity in Poultry* (cited from Crawford 1990a); "until this has been done there can be no certainty as to the manner in which our various races of poultry have arisen, and any further discussion as to their origin is merely waste of time." In the more than 70 years that followed, a lot of time has in fact been wasted.

Today it could be argued that molecular biology offers more precise and faster ways to determining the degree of genetic relationship, in particular through sequencing of mitochondrial DNA.

Even so, I will argue that there is still a lot of scientific potential to be explored here by the well-proven method of sexual hybridization.

In all the trendy talk today on the importance of preserving genetic diversity, I think that far too little emphasis is put on the need for not just preserving, but also exploring, genetic diversity among the close relatives in the wild to the fairly restricted number of domesticated species which mankind has made itself so highly dependent upon. This seems such an elementary piece of wisdom, which plant geneticists have acted upon for many years, and I see no excuse for the continued neglect of it in animal genetics.

I think we must assume that the genetic polymorphism brought into a stock of domesticated animals (or plants) comprises only a part (perhaps even a minor one) of the pool that existed among the ancestral wild population. Hence, genes we select for consciously and systematically several thousand years later would not be found were they not included in the domesticated gene pool in the first place and had they not survived later losses by genetic drift, unless of course the wanted genes actually arose by mutation during later rounds of adaptation. The latter looks to me a distinct possibility very worthy of investigation.

At any rate, the chicken certainly has proved to be adaptable and that relates both to climate (from the jungle to Iceland) and to man's demands upon it. The latter falls into three main categories: good egg-laying, good meat production, and good fights in cock-fighting. On top of that comes religion, as mentioned, where I am less familiar with the selection criteria.

In all of their many thousand years of development, domesticated chickens have never been subject to harder or more methodical selection as has occurred since the World War II. This is the era of food production, with big multinational industries employing experts in quantitative genetics. The European landscape is therefore no longer populated by the variety of chickens familiar to Darwin and his time. I take pleasure in reproducing the three pictorial examples given in Darwin's book (1868). Figure 2a shows the stately Spanish cock, Fig. 2b the Hamburgh fowl with a seemingly self-composed, bourgeois disposition, and Fig.2c the Polish fowl, which looks indomitably independent.

Today we have an egg-layer of the industrialized world belonging to the breed of White Leghorn, which supposely stems from a single ship load of Italian chickens that left the harbor of Livorno in the 1830s for the destination of Philadelphia. After many generations of increasingly scientific selection they spread their genes to the point of almost total monopoly in egg-laying, until very

Fig. 2. a Spanish cock, **b** Hamburgh fowl, **c** Polish fowl. From DARWIN (1868)

recently, when "green" feelings among consumers paved a commercial road for layers of brown eggs, felt to be closer to Nature. The latter is clearly nonsense, but at least it has preserved a different set of genes.

The production of chicken meat used to be a by-product from the production of egg-layers, but not so any longer. It is now a big industry of its own, based on quite different breeds of meat-type chickens, which grow so fast that they can be marketed for consumption at about 5 weeks of age. Also, this development started in the USA, although the breeds sound very British, like Cornish and Plymouth Rock, and have spread from there to most of the industrialized world.

Because of widespread industrial monopoly and because the genetic base for industrial poultry appears to be very narrow, a contemporary authority on poultry breeding and genetics, R.D. CRAWFORD (1990a,b), has argued that the need for conservation of genetic diversity may be even greater in poultry species than in domesticated mammals.

This is to my mind an important issue, going much beyond both the ethics and the pleasing aesthetics of biological diversity. There is a social dimension to it which is inexorably tied to the foreseeable demographic change in the world population. Demographers seem to agree that there is no way one wants to contemplate which can prevent the world population from reaching 10 billion before the middle of next century, stabilizing eventually around 14–15 billion.

In the formidable complex of problems which could be called "How on Earth to Feed Them," I think one may predict that *Gallus domesticus* is going to be the principal source of animal protein. Already in 1985 it accounted for about 22% of the global meat supply (YAMADA 1988). Furthermore, there is good reason to believe that it is going to increase much further, as the conversion ratio of kg feed per kg meat produced is lower for chickens than for domesticated mammals. The weight of that argument can hardly be diminishing.

There is an additional reason, which brings us back to religion in somewhat inverted fashion. The religous authorities of the present, be they cardinals, rabbis, or ayatollahs, do not seem to hold the chicken particularly in either awe or contempt. The chicken has attained ideological neutrality and is therefore no longer protected from man by his faith.

I hope to have made the point now that chickens are very important. I now want to make the point that there is only one scientific attitude appropriate to such importance: that is, to study that remarkable species in depth.

Is that what agricultural biologists are doing? Some of them are clearly trying to, but their main research tradition is different. It is rather a production-oriented application of statistical methods to deal with multigenic hereditary traits. With such methods, breeding strategies have been improved from an economic point of view in one species after the other without much knowledge of the operative genes, their location, structure, organization, and function.

However, DNA technologies are now moving into this field, too, not least by way of gene map construction. The pig seems to be leading the agricultural map race, but concerted action in both Europe and the USA is rapidly refining also the chicken map, with many new polymorphic loci, both microsatellites and expressed genes. The main goal is pinpointing the QTLs (quantitative trait loci) and using these in marker-assisted selection in order to further maximize production per animal and profits in the food industry.

This is all very well for some people, but we know already that practically all the new billions of extra people are going to be born in what are now classified as the developing countries. It is to be hoped that these will largely grow their own food and their own chickens, and it is far from granted that their need for scientific insight is going to be quite the same as ours. Their needs could, for example, be much more oriented towards disease prevention and avoidance of different kinds of stress.

Be this as it may, there is bound to be much new interplay developing between basic biology studied by means of chickens and the production- and disease-oriented research into the same species. Not many of the research problems discussed in this volume will find their ultimate solution by gene maps, but the maps may be useful and we are indirectly contributing to their construction and usefulness by each new functional molecule we define. Transgenic chickens form another area for future research, in which basic and applied research interests will undoubtedly meet.

I think that networking between the two camps of biologists working on chickens is likely to increase and to benefit both. With luck, it will lead to the creation of a base of scientific knowledge which will become invaluable also to those who are going to deal with the global household problems of the next century, as I have lightly and superficially touched upon.

Let me finish with the message which I could imagine the Oracle coming up with today, if we could still solicit her opinion: *Homo sapiens would be insapient not to cultivate her friendship with Gallus gallus.*

Oracular pronouncements are in need of interpretation, but that is not so hard in this particular case. The fowl should be studied in depth because of its actual and foreseeable importance to humans and that pertains to genetics and phylogeny, reproductive biology, embryonic development, growth and sexual development, physiology and nutrition, immunology and infectious diseases, and etiology.

So that is why I think the chicken should survive in immunology and developmental biology.

References

Crawford RD (1990a) Origin and history of poultry species, (Chap 1.) In: Crawford RD (ed) Poultry breeding and genetics. Elsevier, Amsterdam, pp 1–41

Crawford RD (1990b) Experience with *in situ* preservation of poultry breeds. In: Wiener G (ed) Animal genetic resources. A global programme for sustainable development. Proceedings of an FAO expert consultation, Rome, Italy, Sept 1989. FAO, Rome, pp 142–150

Darwin C (1868) The variation of animals and plants under domestication, 1, chap 7. Murray, London, pp 225–275

Wood-Gush DGM (1959) A history of the domestic chicken from antiquity to the 19th century. Poultry Sci 38: 321–326

Yamada Y (1988) The contribution of poultry science to society. Poultry Sci 44: 172–178

Zeuner FE (1963) A history of domesticated animals. Hutschinson, London

Postrearrangement Diversification Processes in Gut-Associated Lymphoid Tissues

C.-A. Reynaud and J.-C. Weill

1 Introduction

Rearrangement of the antigen receptor coding elements is the key molecular mechanism implied in the differentiation of the lymphoid lineage. Why this mechanism has been so stringently conserved remains an intriguing question. The most plausible explanation to date is that it controls allelic exclusion, thus avoiding the dilution of surface receptors and the simultaneous presence of harmful and useful molecules on the surface of the same cell.

In some species, such as mice and humans, the rearrangement process is used to generate a diversified antibody repertoire. Accordingly, these species have evolved immunoglobulin (Ig) loci composed of many functional V encoding elements which can each rearrange, be expressed and therefore become part of the repertoire produced. For mice and humans, such rearrangement takes place in bone marrow as a life-long process.

In some other species, rearrangement is used for a short period of time during embryonic development. Once this event is performed, a fixed amount of progenitors with surface IgM molecules is produced, and the repertoire can then be generated by postrearrangement diversification processes. These postrearrangement molecular events take place in gut-associated lymphoid tissues (GALT) among a rapidly proliferating and expanding B cell population. What appears to be specific to the development of B cells, compared to T cells, is

INSERM U373, Institut Necker, Faculté de médecine Necker-Enfants malades, Université Paris 5, 156 rue de Vaugirard, 75730 Paris Cedex 15, France

therefore the wide variety of cellular and molecular strategies that can be found depending upon the species studied.

In three species studied so far, chicken, rabbit and sheep, postrearrangement mechanisms generate a diversified B cell compartment in GALT. Among them, however, still different molecular mechanisms are used, and the developmental and functional characteristics of the tissue in which such primary repertoires are generated differ to a considerable extent.

2 Gene Conversion in the Chicken Bursa

In chicken, there is only one functional heavy and light chain V gene which is productively rearranged in all B cells of the animal (reviewed in REYNAUD et al. 1994). Upstream of these genes, there is a large family of highly homologous pseudogenes, 25 for the light chain, 80–100 for the heavy chain. These pseudogenes, which cannot be rearranged or transcribed, are closely linked (with an average 0.8 kb spacing) and present in both transcriptional orientations. Diversity is generated by transfering DNA segments from the pseudogenes into the unique functional V gene, which behaves, once rearranged, as an acceptor site. Moreover, pseudogenes at the heavy chain locus are fused V-D elements, allowing gene conversion to proceed through the D region as well. When compared to the functional gene, the pseudogenes are much more homologous in framework than in complementarity-determining regions (CDRs). Selection has therefore maintained during evolution sufficient homology in the framework regions (FRs) to allow for the recombination event to occur during the gene conversion process. It has maintained sufficient diversity at the CDR level to allow for the generation of a large repertoire. In this sense, these pseudogenes have the hallmarks of "functional" V coding segments. Accordingly, although the converted DNA segments have no precise borders, gene conversion taking place in the bursa appears as a CDR replacement process, when the amino acid sequence of the modified V gene is considered (WEILL and REYNAUD 1987).

The bursa of Fabricius is where diversification of the repertoire occurs. Ig gene rearrangement is initiated very early during ontogeny and does not appear to require a specific lymphoid structure (REYNAUD et al. 1992). B cell progenitors seed various embryonic sites, but only those migrating in the bursal microenvironment receive signals for further differentiation (MCCORMACK et al. 1989; REYNAUD et al. 1992). Considerable B cell expansion takes place in bursal lymphoid follicles: such a follicular structure, seeded by a few rearranged B cell progenitors, is shared by the ileal Peyer's patches in the sheep. Although oligoclonal follicular structures may also exist in secondary lymphoid tissues, e.g., splenic germinal centers or jejunal Peyer's patches, they show specific differences: the T cell content, the presence of non-IgM isotypes at the surface of proliferating B cells, and the role of external antigens (see HEIN et al. 1989). The chicken bursa and the sheep ileal

	Chicken	Rabbit	Sheep
Ig loci			
Light chain genes	1 Vλ, 1 Jλ, 25 ψV (no κ known)	2 κ loci (κ2 minor), many Vκ (> 50) λ minor	80-100 Vλ κ locus not described (25% of the repertoire)
Heavy chain genes	1 V_H, 16 D, 1 J_H, 80-100 ψV_H	100-200 V_H (50% ψV)	not described
Light chain rearrangement	one unique Vλ-Jλ	many Vκ-Jκ	many Vλ–Jλ (some major events)
Heavy chain rearrangement	one unique V_H-D-J_H	one major V_H-D-J_H (80-90%) (three others described)	not described
Mechanisms of diversification	gene conversion	gene conversion (and somatic mutation for D regions) described for the heavy chain	somatic mutation for λ light chains
		light chain diversification not described	heavy chain diversification not described
GALT involved in Ig repertoire formation	bursa of Fabricius	appendix (and sacculus rotundus ?)	ileal Peyer's patches
Development before birth	yes (embryogenesis: 21 days)	no	yes (total gestation: 150 days)
Involution at adult stage	yes	no	yes
Function	primary organ	primary, then/and secondary?	primary
Presence of T cells	< 1% [a]	increases with age (many at adult stage)	< 1% [b]
Non-IgM isotypes	few	IgA (from 4 weeks after birth on) numerous at adult stage	few
Role of exogenous antigens	stimulation of growth after birth	necessary for development after birth	stimulation of growth after birth (no effect on diversification)

Ig, immunoglobulin; GALT, gut-associated lymphoid tissues. [a]CHEN et al. (1984). [b]HEIN et al. (1989). See text for other references.

Peyer's patches are very similar, whereas the rabbit appendix displays an intermediate phenotype (Table 1 and Chap. 3).

3 Gene Conversion (and Somatic Mutation) in the Rabbit Appendix

The rabbit heavy chain locus comprises a large family of functional and nonfunctional V genes closely linked and very homologous to each other (100–200 V genes, with probably half of them functional) (reviewed in KNIGHT and CRANE 1994). However, despite the presence of many functional V_H genes, the most proximal one (V_H1) is functionally rearranged in 80%–90% of rabbit B cells (KNIGHT and BECKER 1990). Three other V_H genes, which have not been mapped precisely, account for most rearrangements not involving V_H1 (FRIEDMAN et al. 1994). After rearrangement has been performed, the V_H gene is further diversified by a gene conversion process using as donors the upstream V_H genes, functional or not (BECKER and KNIGHT 1990). Here again, all donor genes belong, as in the chicken, to the same V_{HIII} subgroup, and have been kept by evolution rather homologous in their framework regions but much more diversified in their CDRs. What is specific to the rabbit is the occurrence on the diversified V_H gene of multiple point mutations in the D region which cannot be explained by the gene conversion process, since no donors for the D region exist in the V_H pool (SHORT et al. 1991). Untemplated mutations have been observed on chicken bursal diversified sequences, but not to such an extent (REYNAUD et al. 1987). Moreover, the rabbit V_H pool has a very "classic" gene organization, with neither the alternate orientations nor the extreme clustering of chicken V genes.

Diversification of light chain genes has not been described in the rabbit, although it is likely that both conversion and mutation contribute to the formation of the light chain repertoire. The rearrangement pattern appears more complex at the κ locus than at the heavy chain, with several rearrangement events of similar frequency discernable among appendix B cells (our unpublished data). From the Vκ hybridizing pattern of the rabbit genome, a minimum of 50 Vκ genes can be estimated, and cosmid cloning has suggested a 6–7 kb spacing (HEIDMANN and ROUGEON 1984). The two known κ loci, κ1 and κ2, have been linked within 1 megabase, the κ1 locus being mainly used (HOLE et al. 1991). A λ locus comprising several Cλ genes has also been described (DUVOISIN et al. 1988), λ chains representing 5%–10% of the expressed Ig.

The rabbit appendix was proposed 30 years ago as a putative mammalian bursa equivalent (ARCHER et al. 1963; COOPER et al. 1968). Recently, WEINSTEIN and collaborators reported data strongly suggesting that diversification by gene conversion indeed takes place in the young rabbit appendix (WEINSTEIN et al. 1994): V_H sequences with an identical VDJ junction and differing by stretches of gene conversion were found within a single follicle. Development of the rabbit appendix appears somehow paradoxical in view of its proposed role in the formation of the rabbit primary repertoire: seeding of lymphoid cells only takes place after birth (around 8–10 days of life); IgA-positive B cells appear early in life within appendix follicles (as soon as 4–5 weeks), and can reach large proportions upon oral immunization with cholera toxin (M. Kosco, C. Mackay, C.-A. Reynaud and J.-C. Weill, unpublished data); the T cell content increases with age, reaching at adult stage values similar to those found in secondary lymphoid structures. Such a shift in function is unique to the appendix, which, as opposed to the chicken bursa and the sheep ileal Peyer's patches, does not regress with age. It has thus been proposed (see discussion by Knight and Crane 1994) that, during fetal and early neonatal life, B cells would be produced in fetal liver and then bone marrow. The appendix, and possibly other rabbit GALT as well (e.g., sacculus rotundus and/or Peyer's patches), could assume the transient functions of diversification and amplification of the B cell compartment, while progressively assuming a secondary immune function (Weinstein et al. 1994). At the adult stage, no active lymphopoiesis would go on in the bone marrow, as suggested by the absence of detectable undiversified VH sequences in this organ (Knight and Crane 1994), the B cell pool being thus essentially self-renewing.

Many points need to be clarified in this system. What is the role of external antigens in the development of the B cell population? Is it the necessary mitotic stimulus for B cell expansion to occur, as was shown for the sheep GALT (Reynolds and Morris 1983b; Reynaud et al. 1995)? Or is there a more specific immune function? Does the appendix maintain a dual function during later development, superimposing primary follicles generating B cells (by gene conversion) and secondary follicles involved in an immune response (by hypermutation)?

4 Somatic Mutation in Sheep Ileal Peyer's Patches

In the sheep, ileal Peyer's patches (IPPs) as opposed to jejunal Peyer's patches, possess many histological and physiological similarities with the avian bursa. They are organized in segregated follicles which are colonized by progenitors around 110 days of fetal life (birth is at 150 days). These progenitors thereafter divide extensively within the follicles and generate a large population of surface IgM B cells. This active lymphopoiesis (more than 10^9 B cells produced per hour) goes on until 3 months after birth, at which time IPP, in contrast to jejunal Peyer's patches, start to involute and by 18 months have disappeared. During this massive proliferation, most cells die in situ and a small proportion migrate to the periphery and generate the B cell compartment of the animal. Based on these features, and on the fact that the sheep bone marrow is not a site of B lymphopoiesis, Reynolds and colleagues proposed that sheep IPPs have the property of a primary B cell organ and could be therefore considered as a mammalian bursa equivalent (Reynolds and Morris, 1983a; Reynolds et al. 1987).

The major light chain isotype of the sheep is λ. The λ light chain locus is composed of a J-C unit which is mainly used (other Cλ-hybridizing elements have been detected, but their use is probably minor, if any). A family of approximately 80–100 Vλ genes exist in the sheep genome, among which about one fifth are pseudogenes. This genomic Vλ pool may not be functionally as large, because about one third of the genes have variant recombination signals and may not rearrange efficiently and also because many of them share identical CDRs (Reynaud et al. 1991, 1995). A κ isotype has also been described in approximately 25% of sheep B cells (Griebel et al. 1992). A complete μ chain cDNA sequence has been reported (Patri and Nau 1992), but the heavy chain gene organization has not been described so far.

Multiple light chain rearrangements can be found in IPP B cells with two major events, a pattern that shows however considerable variations among animals (our unpublished data). The existence of dominant rearrangement events in some animals was nevertheless instrumental in allowing us to follow the diversification of defined germline genes. Rearranged Vλ genes indeed accumulate modifications with time in IPPs. There is, on the average one modification per sequence at birth, four to five at 5 weeks of age, seven to eight at 8 weeks and 11 at 4 months. These modifications are mainly point mutations, but also occur in doublets. No correlation of these mutations with sequences from the pool of the Vλ genes could be made, suggesting that an untemplated hypermutation process was responsible for the modifications observed (Reynaud et al. 1991).

Postrearrangement diversification by somatic mutation represents a priori a paradox: how can a random mutation mechanism generate a primary repertoire, i.e., without a selection for "good" mutants exerted by antigens? Accordingly, diversified light chain sequences in IPPs have the hallmarks of "selected" sequences: clustering of mutations in CDRs and a replacement over silent substitution (R/S) ratio biased toward replacement mutations in CDRs and silent mutations in framework regions.

One first possibility could be that antigens present in the gut (bacterial or food-derived) could exert such selection, driving the proliferation of mutant clones. This is however not the case. We studied Ig diversification in sheep IPPs in conditions of drastically reduced antigen load: in germ-free animals and in ileal segments isolated from the intestinal tract before birth, which have therefore never been in contact with antigens present in the gut. In such antigen-depleted conditions, Ig diversification takes place at the same rate as in normal animals, and the pattern of mutation is similar (clustering in CDRs, biased R/S ratios) (Reynaud et al. 1995).

The apparent selected pattern of diversified Ig sequences in IPPs appears to have another explanation: it is achieved by a coadjustment of the V gene structure and the mutation process. Mutations are biased towards purines (2.2 times more frequent than pyrimidines, mutations being referred to the coding strand), and codons with high purine content in their first two positions and low content at the third one have been selected in CDRs of the Vλ genes analyzed. Such codons have thus a high intrinsic R/S mutation ratio (the codon AGT present frequently in CDRs has an R/S of 8 if mutations are random; if purines are mutated 2.2 times more frequently than pyrimidines, its intrinsic R/S ratio is 15.2 (see REYNAUD et al. 1995)). Moreover, as apparent from the rate of silent mutation, mutations are three times more frequent in CDRs than in FRs, and defined hotspots of mutations are clustered at the same positions in CDRs, irrespectively of the presence of external antigens, indicating a preferential targeting of the mutations in CDRs. This clustering is further enhanced by the counterselection of replacement mutations in framework regions, B cells in IPPs probably requiring IgM at the surface for continuous proliferation (REYNAUD et al. 1995).

Data obtained in transgenic mice with passenger Ig transgenes acting as mutation substrates have shown that the mutation process taking place in the course of an immune response has the same intrinsic properties as the one we have described in the sheep: bias for purines, clustering in CDRs, intrinsic hotspots (BETZ et al. 1993a,b).

The parallel evolution of the mutation mechanism and of its template V elements thus results in sheep in a primary repertoire largely dictated by the nature of the germline encoded elements, in a way not so different, in spite of the difference in the mechanisms, from the germline encoded pool of donors used for gene conversion.

5 Mutation or Conversion?

The results obtained with these different models raise several comments. They at first show that the hypermutation mechanism is not strictly linked to a germinal center microenvironment generated by a T-dependent immune response, thus implying that the process can be triggered by other stimuli. They also imply that, although the gene conversion process has been maintained during evolution in the mamalian immune system, other molecular strategies can be used in GALT. One should underline again that point mutations are always present at different levels during gene conversion events, and it remains possible that both processes are effectively linked, being for example initiated by the same molecular event (a break in the DNA) or sharing factors and/or enzymatic activities.

Along with this argument, gene conversion has never been documented among mutations arising in germinal centers in the mouse (see, for example, FORD et al. 1994). However, in a specific transgenic configuration, with two V_H genes closely linked, highly homologous, and encoding the same hapten specificity, recombination along with multiple point mutations occurs between these two genes after immunization (XU and SELSING 1994). It could thus happen that, in each species, a specific balance between regulatory factors (e.g., factors in charge of the chromatin organization of the V gene pool) and a particular gene organization (harboring or not a critical level/length of identity, and allowing or not for homology scanning over appropriate distances) would shift the molecular process toward either templated or untemplated DNA modifications. Only a description of both mechanisms at the molecular level will allow a better understanding of how different diversification mechanisms become activated in different species.

References

Archer OK, Sutherland DER, Good RA (1963) Appendix of the rabbit: a homologue of the bursa in the chicken. Nature 200: 337–339

Becker RS, Knight KL (1990) Somatic diversification of immunoglobulin heavy chain VDJ genes: evidence for somatic gene conversion in rabbits. Cell 63: 987–997

Betz AG, Rada C, Pannell R, Milstein C, Neuberger MS (1993a) Passenger transgenes reveal intrinsic specificity of the antibody hypermutation mechanism: clustering, polarity, and specific hot spots. Proc Natl Acad Sci USA 90: 2385–2388

Betz AG, Neuberger M, Milstein C (1993b) Discriminating intrinsic and antigen-selected mutational hotspots in immunoglobulin V genes. Immunol Today 14: 105–411

Chen CLH, Chahn TC, Cooper MD (1984) Chicken thymocyte-specific antigen identified by monoclonal antibodies: ontogeny, tissue distribution and biochemical characterization. Eur J Immunol 14: 385–391

Cooper MD, Perey DY, Gabrielsen AE, Sutherland DER, McKneally MF, Good RA (1968) Production of an antibody deficiency syndrome in rabbits by neonatal removal of organized intestinal lymphoid tissues. Int Arch Allergy 33: 65–88

Duvoisin RM, Hayzer DJ, Belin D, Jaton JC (1988) A rabbit Ig λ L chain region gene encoding C21 allotopes. J Immunol 141: 1596–1601

Ford JE, McHeyzer-Williams MG, Lieber MR (1994) Analysis of individual immunoglobulin λ light chain genes amplified from single cells is inconsistent with variable region gene conversion in germinal-center B cell somatic mutation. Eur J Immunol 24: 1816–1822

Friedman M, Tunyaplin C, Zhai SK, Knight KL (1994) Neonatal V_H, D, and J_H gene usage in rabbit B lineage cells. J Immunol 152: 632–641

Griebel PJ, Kennedy L, Graham T, Davis WC, Reynolds JD (1992) Characterization of B-cell phenotypic changes during ileal and jejunal Peyer's patch development in sheep. Immunology 77: 564–570

Heidmann O, Rougeon F (1984) Immunoglobulin κ light-chain diversity in rabbit is based on the 3' length heterogeneity of germ-line variable genes. Nature 311: 74–76

Hein WR, Dudler L, Mackay CR (1989) Surface expression of differentiation antigens on lymphocytes in the ileal and jejunal Peyer's patches of lambs. Immunology 68: 365–370

Hole NJK, Young-Cooper GO, Mage RG (1991) Mapping of the duplicated rabbit immunoglobulin κ light chain locus. Eur J Immunol 21: 403–409

Knight KL, Becker RS (1990) Molecular basis of the allelic inheritance of rabbit immunoglobulin V_H allotypes: implications for the generation of antibody diversity. Cell 60: 963–970

Knight KL, Crane MA (1994) Generating the antibody repertoire in rabbit. Adv Immunol 56: 179–218

McCormack WT, Tjoelker LW, Barth CF, Carlson LM, Petryniak B, Humphries EH, Thompson CB (1989) Selection for B cells with productive IgL gene rearrangements occurs in the bursa of Fabricius during chicken embryonic development. Genes Dev 3: 838–847

Patri S, Nau F (1992) Isolation and sequence of a cDNA coding for the immunoglobulin μ chain of the sheep. Mol Immunol 29: 829–836

Reynaud CA, Anquez V, Grimal H, Weill JC (1987) A hyperconversion mechanism generates the chicken light chain preimmune repertoire. Cell 48: 379–388

Reynaud CA, Mackay CR, Müller RG, Weill JC (1991) Somatic generation of diversity in a mammalian primary lymphoid organ: the sheep ileal Peyer's patches. Cell 64: 995–1005

Reynaud CA, Imhof BA, Anquez V, Weill JC (1992) Emergence of committed B lymphoid progenitors in the developing chicken embryo. EMBO J 12: 4349–4358

Reynaud CA, Bertocci B, Dahan A, Weill JC (1994) Formation of the chicken B-cell repertoire: ontogenesis, regulation of Ig gene rearrangement, and diversification by gene conversion. Adv Immunol 57: 353–378

Reynaud CA, Garcia C, Hein WR, Weill J-C (1995) Hypermutation generating the sheep immunoglobulin repertoire is an antigen-independent process. Cell 80: 115–125

Reynolds JD, Morris B (1983a) The evolution and involution of Peyer's patches in fetal and postnatal sheep. Eur J Immunol 13: 627–635

Reynolds JD, Morris B (1983b) The influence of gut function on lymphoid cell populations in the intestinal mucosa of lambs. Immunology 49: 501–509

Short JA, Sethupathi P, Zhai SK, Knight KL (1991) VDJ genes in V_{Ha2} allotype-suppressed rabbits. Limited germline gene usage and accumulation of somatic mutations in D regions. J Immunol 147: 4014–4018

Weill JC, Reynaud CA (1987) The chicken B cell compartment. Science 238: 1094–1098

Weinstein PD, Anderson AO, Mage RG (1994) Rabbit IgH sequences in appendix germinal centers: VH diversification by gene conversion-like and hypermutation mechanisms. Immunity 1: 647–659

Xu B, Selsing E (1994) Analysis of sequence transfers resembling gene conversion in a mouse antibody transgene. Science 265: 1590–1593

Current Concepts in Chicken B Cell Development

P.E. Funk[1] and C.B. Thompson[1,2]

1 Introduction

B lymphocytes produce antibody molecules that provide the basis for humoral immune responses. In addition, antibodies in secretions are critical in immune defense. Maternal transmitted antibody plays an important role in immunoprotection of newborns with developing immune systems. An effective humoral immune repertoire has been estimated to require the production of 10^9 distinct antibody molecules. However, most B cells only produce a single antibody specificity. Thus, an effective immune repertoire requires the maintenance of a vast repertoire of B cells, only a few of which will ever encounter their cognate antigen in any given time period. How such diversity is created during B cell development has been a central question in the field of immunology.

The avian immune system has provided a number of insights into vertebrate immunity and has been particularly valuable in studying B cell development. In

[1]Howard Hughes Medical Institute, Gwen Knapp Center for Lupus and Immunology Research, 924 East 57th Street, Chicago, IL 60637-5420, USA
[2]Departments of Medicine, Molecular Genetics and Cell Biology, and Committee on Immunology, The University of Chicago, Chicago, IL 60637, USA

fact, the term "B cell" was coined because avian B cells undergo part of their development in the bursa of Fabricius, a primary lymphoid organ unique to birds. The development of B cells in an organ specialized for that purpose and secluded from other developing cells provides a unique opportunity to study B cell development. This review will focus on recent advances in the study of B cell development in chickens and indicate potential avenues of future study. The discoveries discussed here demonstrate that the chicken remains a fruitful animal model in which to study B cell development and may well contribute new insights into mammalian B cell studies.

2 B Cell Development

2.1 Rearrangement of Immunoglobulin Genes

Birds and mammals create a diverse immunoglobulin repertoire in different ways. In mammals the immunoglobulin locus contains multiple functional variable (V), diversity (D), and joining (J) gene elements that can be recombined during B cell development to generate an immunoglobulin gene (TONEGAWA 1983). The combinatorial possibilities of assembling these elements into distinct V(D)J antigen binding exons creates a varied set of Ig molecules. In addition, the independent rearrangement and association of Ig heavy and light chains within a single cell further increases the spectrum of antibody specificities. Avian species, in contrast, have only limited combinatorial diversity. For example, chickens have only single functional V and J gene segments at both the heavy and light chain locus (REYNAUD et al. 1985, 1987). This means that every chicken B lymphocyte creates a nearly identical surface Ig (sIg) receptor by V(D)J recombination. In order to generate the varied repertoire necessary for an effective humoral response, chicken B cells diversify their Ig genes during development in the bursa of Fabricius through a process of somatic gene conversion (REYNAUD et al. 1987; THOMPSON and NEIMAN 1987; THOMPSON 1992). In this process portions of sequence from nonfunctional gene segments are unidirectionally copied onto the rearranged gene. Such conversion events can occur at multiple positions along the rearranged gene and recombined V gene segments can undergo repetitive gene conversion during B cell development in the bursa (for review see WEILL, this volume). Such conversion events account for nearly all of the variability found in the chicken Ig repertoire. Surprisingly, gene conversion is not restricted to birds as rabbits also appear to generate their Ig repertoire in this way (KNIGHT and CRANE 1994).

2.2 The Bursa of Fabricius

The bursa of Fabricius is a primary lymphoid organ in avian species that is required for the development of humoral immune responses. The bursa is first observed as an epithelial bud arising from the cloacal wall. The bursa is seeded by lymphoid precursors between days 8 and 14 of embryogenesis (HOUSSAINT et al. 1976). Once in the bursa these precursors begin to proliferate, forming lymphoid follicles. Each follicle contains approximately 10^5 lymphocytes that originated from only two to four initial precursor cells (PINK et al. 1985). The bursa contains roughly 10^4 such follicles, or 10^9 total cells. This massive production of B lymphocytes continues until sexual maturity when the bursa involutes.

The central role of the bursa in B cell development was shown by experiments in which the bursa was either surgically or chemically ablated. In these animals B cell numbers are drastically reduced, serum levels of IgG and IgA are lowered, and the animals are incapable of mounting specific humoral immune responses (WARNER et al. 1969; SUBBA RAO et al. 1978; VEROMAA et al. 1988; GRANFORS et al. 1982; JALKANEN et al. 1983). Paradoxically, these animals have elevated serum levels of IgM. The immunoglobulin repertoire of these animals is oligoclonal, containing only a few antibody specificities (MANSIKKA et al. 1990a). It is now apparent that the bursal microenvironment is required for full gene conversion to take place (MANSIKKA et al. 1990a; LASSILA et al. 1988). Each follicle contains cells of varied specificity, indicating that gene conversion occurs independently in each cell. LYDARD et al. (1976) reported that several individual follicles independently acquired the ability to bind sheep red blood cells, even in the absence of antigenic exposure. Using 2D gel electrophoresis as a measure of light chain diversity, LASSILA et al. (1988) found that the follicles in young birds were relatively undiversified. However, as the bird aged the light chain diversity within an individual follicle increased dramatically, eventually mirroring the diversity found in the peripheral B cell population.

2.3 The Bursal Stem Cell

TOIVANEN and TOIVANEN (1973) demonstrated that there are cells within the young bursa that are capable of fully reconstituting the B cell lineage in irradiated recipient birds. These cells have a finite lifetime within the bird and their numbers gradually decline during development until they are no longer detectable by about 2 weeks after hatch. These cells have been referred to as bursal stem cells. HOUSSAINT and coworkers (1976) determined that bursal stem cells originate outside of the bursa. Using cell sorting they showed that bursal stem cells are present in the embryonic spleen. They further showed that these bursal stem cells express the chB6 (BU-1) antigen, a marker often used to identify B cells in chickens (HOUSSAINT 1987; HOUSSAINT et al. 1989, 1991). Cells lacking the chB6 marker cannot reconstitute a depleted bursa. Previous experiments indicated that these bursal stem cells enter the bursa already committed to the expression of Ig molecules, suggesting that the actual rearrangement of the Ig locus occurs outside of the bursa (WEILL et al. 1986; RATCLIFFE et al. 1986; MANSIKKA et al. 1990b). This was confirmed using a PCR based assay. Two divergent PCR primers were designed such that the circular episome created when Ig genes rearrange would generate a PCR product but germline DNA would not be amplified (McCormack et al. 1989; Pickel et al. 1993). Since the episome has no origin of replication it is lost as the cell divides, therefore the PCR product can only be found in cells having recently undergone V(D)J recombination. Such (VJ) episomes are first detected in spleen at 10 days of embryonic development. Furthermore, the episome is found exclusively in the chB6+ fraction of splenocytes. Few, if any, rearranging cells are detected in the bursa. Combined, this data argues that B cells enter the bursa with their Ig loci already rearranged.

2.4 Migration of B Cell Progenitors to the Bursa

The studies described above establish the migration of B cells from the microenvironment of the spleen to the microenvironment of the bursa as a critical step in the development of a complete humoral immune system. Therefore, our laboratory began studying the homing molecules present on avian B cells. The initial interactions in the migration of mammalian cells to sites of inflammation is binding of carbohydrate structures on leukocytes to selectin molecules present on endothelial cells (Varki 1994). The selectin family of molecules is so named because its members possess a lectin-like domain that binds carbohydrate ligands. Given that B cells enter the bursa via the vasculature, their first contact would be the bursal endothelium. It therefore seemed reasonable to screen B cells for the expression of known selectins and/or selectin ligands. These studies revealed that the chB6+ cells in embryonic day 12 spleen, a population known to contain bursal stem cells, express sialyl-Lewis-x (sLe^x), a common ligand for mammalian selectins (Masteller et al. 1995).

The DT40 lymphoma cell line possesses several characteristics of bursal stem cells, including expression of sLe^x. This makes it an excellent cell in which to study adhesion in the bursa. DT40 cells adhere to frozen sections of bursa in a perifollicular pattern along the vascular regions (Masteller et al. 1995). This adhesion is found under conditions of high rotary shear that favor carbohydrate-selectin adhesion. This adhesion can be inhibited by antibodies to sLe^x, removal of sLe^x by neuraminidase treatment, or chelation of calcium, suggesting that bursal stem cells adhere to the bursal vasculature in an sLe^x dependent fashion. These experiments support the hypothesis that a selectin-carbohydrate interaction may be an important first step in the migration of B cell precursors into vascular regions of the bursa.

Another aspect of B cell entrance into the bursa is the discrimination of cells that have undergone productive Ig gene rearrangements. It is now apparent that some degree of B cell proliferation occurs in the bursa before the onset of gene conversion. This may be important because gene conversion appears to be error prone and can generate nonsense mutations, leading to the loss of Ig expression on lymphocytes. Soon after migration to the bursa of Fabricius, bursal stem cells undergo proliferative expansion. This expansion is selective for cells that have productive Ig rearrangements (McCormack et al. 1989). McCormack et al. found that up to two thirds of B cells entering the bursa before day 15 have nonproductive Ig gene rearrangements. With time the fraction of nonproductively rearranged B cells decreases, indicating that cells bearing productive rearrangements are selectively expanded.

Around day 15 of embryonic development the B cells begin to lose expression of sLe^x and express the related carbohydrate Lewis-x (Le^x). Within a few days virtually all of the B cells in the bursa become Le^x positive. This switch in carbohydrate expression coincides with another important event in B cell development, the initiation of gene conversion. After embryonic day 15 the process of gene conversion begins to diversify the B cell repertoire (McCormack et al. 1989). Single cell sorting experiments show that most gene conversion events are found in the Le^x population while the sLe^x population contains predominantly undiversified Ig genes (Masteller et al. in preparation).

These observations are consistent with a model in which B lymphocytes enter the bursa by virtue of their expression of sLe^x. These cells then migrate to the epithelium where they begin to proliferate, switch the expression of cell surface carbohydrates, and begin gene conversion. Further support for the commonality of these events is that the bursa is required for both gene conversion and the switch from sLe^{x+} to Le^{x+}. Cells coexpressing chB6 and sLe^x continue to reside in the spleen during embryonic development. However, these cells do not switch to expression of Le^x (Masteller et al. 1995). Gene conversion does not occur in bursectomized birds, implicating an interrelationship between these events (Mansikka et al. 1990a). The important events in B cell development in chicken are schematized in Fig. 1.

2.5 Cell Death in the Bursa

One of the major cell fates in the bursa is death. Many B cells in the bursa die by apoptosis, an ordered process involving degradation of DNA, cytoplasmic

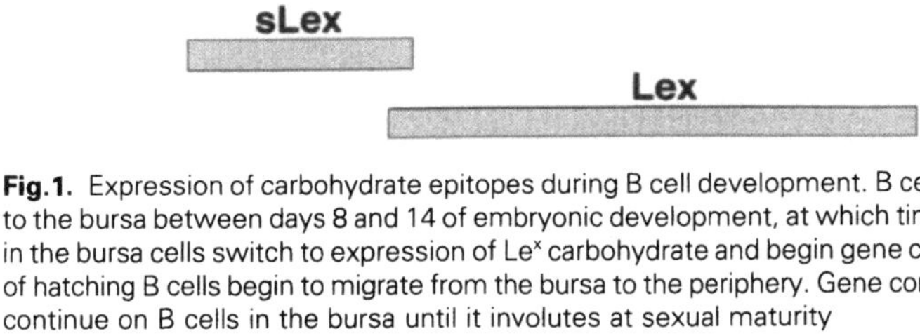

Fig. 1. Expression of carbohydrate epitopes during B cell development. B cells migrate from the spleen to the bursa between days 8 and 14 of embryonic development, at which time they express sLe^x. Once in the bursa cells switch to expression of Le^x carbohydrate and begin gene conversion. Around the time of hatching B cells begin to migrate from the bursa to the periphery. Gene conversion and Le^x expression continue on B cells in the bursa until it involutes at sexual maturity

shrinkage, and membrane blebbing of the dying cell (NEIMAN et al. 1991; MOTYKA and REYNOLDS 1991). Apoptosis is an active process by the dying cell, requiring energy and the expression of specific genes (COHEN et al. 1992). The vast majority of B cells produced in the bursa are destined to die there, only 5% migrate to the periphery (LASSILA 1989). This parallels mammalian B cell development in which many cells are lost at each stage of Ig gene rearrangement (OSMOND 1990). Remnants of these apoptotic cells are seen within macrophages in the avian bursa and the mammalian bone marrow. Such massive cell death could be due to ongoing gene conversion. With each round of gene conversion it is possible to generate stop codons or frameshift mutations in the Ig gene. The inability to express Ig would render the B cell nonfunctional. It is not clear if such a cell might still undergo conversion to rescue the nonfunctional Ig, in a manner similar to the receptor editing reported to occur in mice (GAY et al. 1993; TIEGS et al. 1993). PARAMITHIOTIS et al. (1995) have reported that chicken B cells down-regulate their surface Ig prior to undergoing apoptosis. This would support the possibility that loss of functional Ig might trigger apoptosis. Alternatively, loss of surface Ig might be a secondary event after apoptosis has begun.

The marked susceptibility of bursal cells to apoptosis was noted by NEIMAN and coworkers, who found that B cells die rapidly upon removal from the bursal microenvironment (NEIMAN et al. 1991). Also, B cells still in contact with bursal epithelium remain viable longer in culture, suggesting that the epithelium may elaborate survival factors. Subsequent reports showed that B cell survival is prolonged in the presence of phorbol myristate acetate, suggesting that protein kinase C activation can effect resistance to apoptosis (ASAKAWA et al. 1993), again implying the existence of survival signals. The balance of positive and negative signals within the bursa must be carefully poised to control lymphocyte survival. Perturbation of this balance typically leads to cell death.

Recent evidence has also suggested that cell death can occur not only from the absence of survival signals but also as a result of receptor signaling that induces cell death. The prototypic model of such a cell death receptor is the Fas antigen described in mammals (NAGATA 1994). To address the existence of cell death and cell survival signals within the bursa we set out to screen a panel of antibodies and lectins for the ability to either prevent or accelerate the death of lymphocytes in culture. One lectin, *Aleuria aurantia* agglutinin (AAA), was found that intensely stains all B cells and rapidly causes them to lose viability, as determined by uptake of the vital dye propidium iodide. These cells are undergoing apoptosis and internucleosomal DNA fragmentation is clearly visible in 1–2h. These changes in cell viability are inhibited in the presence of L-fucose, but not D-fucose (Fig. 2). Since AAA is a dimer with two binding sites for α1–6 linked fucose (DEBRAY and MONTREUIL 1989; YAMASHITA et al. 1985) the lectin may be cross-linking a cell death receptor. Addition of increasing concentrations of lectin causes less cell death, consistent with lowered cross-linking of the receptor under saturating concentrations of lectin. Cell surface carbohydrates can be placed on several different protein backbones. To examine the proteins bearing the α1–6 linked fucose, AAA was used to probe protein blots. These experiments reveal a

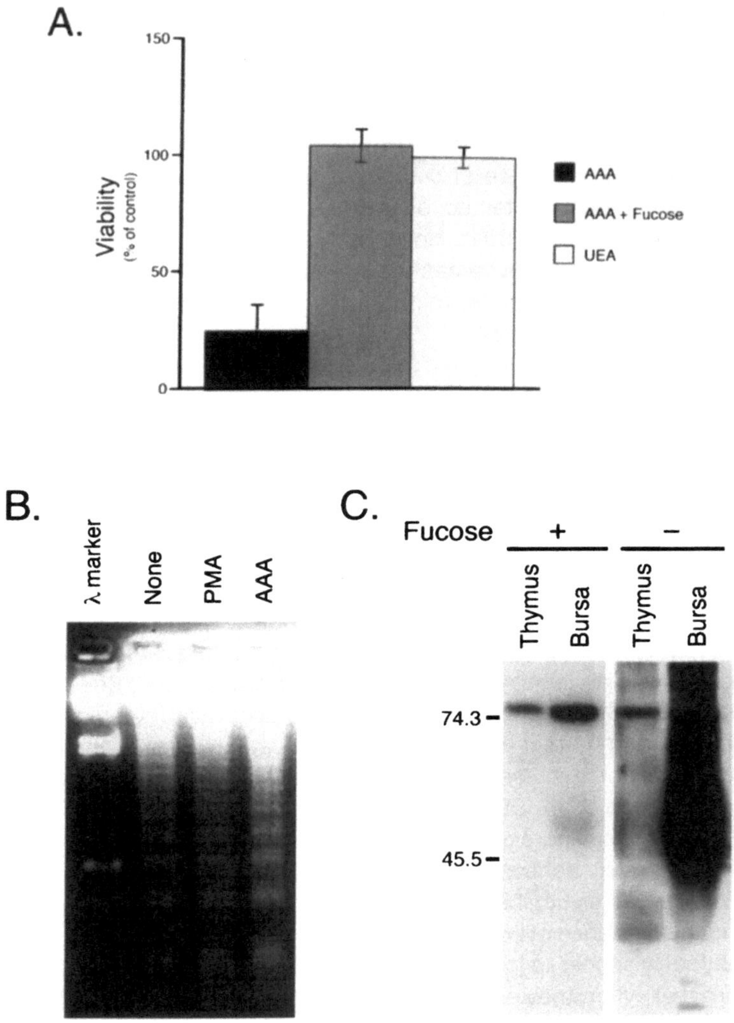

Fig. 2A–C. The lectin *Aleuria aurantia* agglutinin (*AAA*) binds a cell death inducing molecule on B cells. **A** Viability of freshly isolated B cells as measured by the exclusion of the vital dye propidium iodide. Data are expressed as a percentage of the viability of untreated B cells. Cells were incubated with 10 μg/ml AAA lectin, AAA lectin with 0.1 *M* L-fucose, or 10 μ/ml of another fucose-specific lectin, *Ulex europaeus* agglutinin (*UEA*). **B** AAA leads to DNA degradation charcteristic of apoptosis. Freshly isolated bursal lymphocytes were cultured for 2 h at 37°C with no additions, 5 ng/ml phorbol myristate acetate (*PMA*) or 10 μg/ml AAA. Cells were washed and genomic DNA isolated as described (NIEMAN et al. 1991). DNA from equivalent numbers of cells was run on a 1.8% agarose gel and stained with ethidium bromide. **C** AAA lectin binds one predominant protein on B cells. Extracts of bursa and thymus from a 5 week old chicken were electrophoresed through 10% SDS-PAGE gel, transferred to nitrocellulose sheets, and probed with biotinylated AAA lectin. Identically prepared sheets were probed in the presence of 0.1 *M* L-fucose. Bound lectin was detected with peroxidase labeled ABC reagent and visualized by chemiluminescence. Molecular weight markers are indicated in kDa. Another band is seen in both bursa and thymus (above the 74.3 kDa marker). Appearance of this band is not inhibited in the presence of fucose so it may represent an endogenously biotinylated protein

single broad band of 48–50kDa reactive with AAA. The appearance of this band is inhibited by the presence of L-fucose during incubation with the lectin (Fig. 2). The appearance of a single protein bearing this glycosylation was surprising and indicates a high degree of specificity in the glycosylation of proteins. It also suggests that this fucose group is functionally important for this molecule, further highlighting the potential importance of carbohydrates in avian B cell development. The size of the protein detected by AAA is similar to the size of the mammalian Fas antigen (NAGATA 1994). However, Fas appears to mediate the deletion of T cells predominantly, whereas the AAA binding protein is found on B cells. The AAA binding protein, then, might be a new member of a family of molecules that influence cell survival.

2.6 Other Proteins Governing Cell Survival

Other critical proteins governing cell survival are bcl-2 family members. The *bcl*-2 gene was originally described in human follicular lymphomas where a chromosomal translocation brings the gene under the control of the Ig promoter and enhancer, causing its high level expression (CLEARY et al. 1986). The bcl-2 protein is now known to protect a wide variety of cells from apoptosis and is the model member of a family of related genes that influence a cell's susceptibility to apoptosis (HOCKENBERRY et al. 1990; VAUX et al. 1988; NUNEZ et al. 1990; BOISE et al. 1993; OLTVAI et al. 1993). Since bcl-2 protects cells from apoptosis it would seem to be an ideal candidate to protect bursal B cells from death. However, the avian homologue of bcl-2 is not expressed in the bursa, although it is detected in thymus and spleen (EGUCHI et al. 1992). A related gene, called *bcl*-x, was found in the bursa (BOISE et al. 1993). There are two forms of *bcl*-x, a long form (*bcl*-x_L) and a short form (*bcl*-x_S). Bcl-x_L behaves similarly to bcl-2 in protecting cells from apoptosis. Bcl-x_S acts as an inhibitor of bcl-2, rendering cells susceptible to apoptosis even in the presence of bcl-2 (BOISE et al. 1993). The different members of the *bcl*-2 gene family appear to interact at the protein level, possibly modifying their activity from that when they are expressed individually (OLTVAI et al. 1993). The members of this gene family are highly conserved and the bursa could prove to be an excellent organ to define further the biological function of this important family of molecules.

3 Future Avenues of Investigation

3.1 Molecular Identification of B Cell Differentiation Markers

The study of B cell development in mammals has been dramatically aided by the identification of molecules to subdivide the B lineage into discrete developmental steps. Identification of these molecules has led to an understanding of some of the molecular mechanisms that control development. Similar advances have been lagging in avian immunology. Numerous antibodies have been made to chicken B cells but molecular characterization is only beginning. The evolutionary distance between mammals and birds make it possible to identify highly conserved molecules that may have escaped detection in studies using more closely related species such as mice or rats. Molecular characterization of previously identified antigens on avian B cells is likely to significantly advance our understanding of B cell development and provide invaluable insights into B cell differentiation in mammals. As well, a continued search for new antigens on chicken B cells will likely reveal new and interesting molecules.

Several antigens present on avian lymphocytes bear such investigation. Perhaps one of the best known markers of avian B cells is the chB6 (BU-1) alloantigen (GILMOUR et al. 1976; HOUSSAINT et al. 1987, 1989), which is expressed on the earliest identifiable B cell precursors in the embryonic spleen (HOUSSAINT et al. 1989) and continues to be expressed on B cells throughout ontogeny, with the exception of plasma cells in the Harderian gland (HOUSSAINT et al. 1987). Aside from B cells, chB6 is expressed on a subset on macrophages present in the bursa, liver, and intestine (HOUSSAINT et al. 1987). Since chB6 expression is rather restricted to B cells yet ubiquitously expressed on B cells, it may be an important molecule in B cell physiology. Although several laboratories have used chB6 to identify B cells, there is no information as to its function (HOUSSAINT et al. 1989). The BU-2 antigen is also expressed on 95% of bursal cells as well as B cells in peripheral blood and spleen. It is found on both Ig+ and Ig-cells, so it might be an indicator of commitment to the B lineage (HUFFNAGLE et al. 1989). The chL12 molecule is another alloantigenic marker that warrants molecular analysis; it is a 40 kDa molecule found on most hematopoietic cells in chick embryos (HOUSSAINT et al. 1991). However, after hatching, chL12 is found predominantly on T cells. It also remains on a small number of bursal cells as well as B cells that have recently emigrated from the bursa. Therefore, chL12 could be a marker for a distinct subset of B cells or a homing molecule necessary for exit from the bursa.

3.2 Regulation of Gene Conversion and B Cell Selection

B cells enter the bursa with a common Ig receptor and are mutated by gene conversion to develop unique receptors (REYNAUD et al. 1987; THOMPSON and NIEMAN 1987). Only a few cells seed each bursal follicle and produce 10^5 cells with tremendous Ig diversity (LASSILA et al. 1988). All of these cells must be screened to prevent the appearance of self-reactive cells in the periphery. The means by which the bursa selects Ig+ cells, initiates their gene conversion, and deletes nonfunctional or self-reactive cells remains a central, and difficult, question to resolve. Our laboratory has proposed a model in which the undiversified sIg on the bursal lymphocyte recognizes a "self" antigen present on the bursal epithelium (MASTELLER and THOMPSON 1994). Upon contacting this antigen the B cell is triggered to proliferate and begin gene conversion. This proliferative stimulus is responsible for the formation of the lymphoid follicles in the bursa. As the sIg is diversified it no longer binds this self antigen, it therefore ceases proliferation and gene conversion. This cell can then exit the bursa for the periphery. Cells that fail to express Ig would be selected for death. However, the putative ligand for the undiversified Ig remains unknown. Identification of this ligand would constitute a great advance in our understanding of B cell differentiation. Studies of T cell selection have indicated that several balanced signals are needed for appropriate screening of self-reactive T cells (Wallace et al. 1993). Several molecules may also be needed for B cell selection. The identification of how the process of gene conversion is turned on and off during development will greatly facilitate our understanding of this important issue.

3.3 Regulation of Cell Survival

The extreme sensitivity of chicken B cells to apoptosis makes them an excellent population to study the genes governing cell survival. Large numbers of lymphocytes die in the bursa and they are very sensitive to cell death when removed from the bursal microenvironment (Lassila 1989; Neiman et al. 1991). Yet, from only a few thousand immigrant B cell precursors, over a billion cells are produced in the bursa. Therefore, the bursa must produce inhibitors of apoptosis that can maintain B cell survival. Chicken B cells may prove to be a useful hunting ground for yet other genes controlling cell death. As noted previously, we have uncovered a cell surface molecule that can trigger rapid cell death of B cells through ligation of a unique fucose linkage. There are as yet no reports of molecules behaving this way on mammalian cells, even though we know that many cells are lost during mammalian B cell development (Osmond 1990). Once identified in the chicken it will be interesting to determine if such death-inducing receptors are also conserved in mammals.

4 Summary

The chicken has provided fundamental insights into the workings of vertebrate immunity. In particular, the development of B cells in a unique organ, the bursa of Fabricius, has provided a novel opportunity to study B cell development. Although chickens generate their Ig repertoire in a different way than mice and humans, there are many striking similarities in the developmental process. In particular, the control of lymphocyte migration and survival is key to the development of an immune system. The evolutionary distance of chickens and mammals underscore how common the problems are as well as how the solutions are often similar. Such commonalities serve to maintain the chicken as a compelling animal in which to study B cell development.

Acknowledgments. The authors wish to thank Emma Masteller, G. Todd Pharr, and Patricia Noel for critical reading of the manuscript and Therese Conway for help in preparing the manuscript.

References

Asakawa J, Tsiagbe VK, Thorbecke GJ (1993) Protection against apoptosis in chicken bursa cells by phorbol ester in vitro. Cell Immunol 147: 180–187

Boise LH, Gonzalez-Garcia M, Postema C, Ding L, Lindsten T, Turka LA, Mao X, Nunez G, Thompson CB (1993) Bcl-x, a bcl-2 related gene that functions as a dominant regulator of apoptotic cell death. Cell 74: 597–608

Cleary ML, Smith SD, Sklar J (1986) Cloning and structural analysis of cDNAs for bcl-2 and a hybrid bcl-2/immunoglobulin transcript resulting from the t(14;18) translocation. Cell 47: 19–28

Cohen JJ, Duke RC, Fadok VA, Sellins KS (1992) Apoptosis and programmed cell death in immunity. Annu Rev Immunol 10: 267–293

Debray H, Montreuil J (1989) *Aleuria aurantia* agglutinin. a new isolation procedure and further study of its specificity towards various glycopeptides and oligosaccharides. Carbohyd Res 185: 15–26

Eguchi Y, Ewert DL, Tsujimoto Y (1992) Isolation and characterization of the chicken bcl-2 gene: expression in a variety of tissues including lymphoid and neuronal organs in adult and embryo. Nucleic Acids Res 20: 4187–4192

Gay D, Saunders T, Camper S, Weigert M (1993) Receptor editing: an approach by autoreactive B cell to escape tolerance J Exp Med 177: 999–1008

Gilmour DG, Brand A, Donnelly N, Stone HA (1976) Bu-1 and Th-1, two loci determining surface antigens of B or T lymphocytes in the chicken. Immunogenetics 3: 549–563

Granfors K, Martin C, Lassila O, Suvitaival R, Toivanen A, Toivanen P (1982) Immune capacity of the chicken bursectomized at 60 hr of incubation: production of the immunoglobulins and specific antibodies. Clin Immunol Immunopathol 23: 459–469

Hockenberry D, Nunez G, Milliman C, Schreiber RD, Korsmeyer SJ (1990) Bcl-2 is an inner mitochondrial membrane protein that blocks programmed cell death. Nature 348: 334–336

Houssaint E (1987) Cell lineage segregation during bursa of Fabricius ontogeny. J Immunol 138: 3626–3634

Houssaint E, Belo M, Le Douarin NM (1976) Investigations on cell lineage and tissue interactions in the developing bursa of Fabricius through interspecific chimeras. Dev Biol 53: 250–264

Houssaint E, Diez E, Pink JRL (1987) Ontogeny and tissue distribution of the chicken BU-1a antigen. Immunology 62: 463–470

Houssaint E, Lassila O, Vainio O (1989) Bu-1 antigen expression as a marker for B cells precursors in chicken embryos. Eur J Immunol 19: 239–243

Houssaint E, Mansikka A, Vainio O (1991) Early separation of B and T lymphocyte precursors in chick embryo. J Exp Med 174: 397–406

Huffnagle GB, Ratcliffe MJH, Humphries EH (1989) Bu-2, a novel avian cell surface antigen on B cells and a population of non-lymphoid cells, is expressed homogeneously in germinal centers. Hybridoma 8: 589–604

Jalkanen S, Granfors K, Jalkanen M, Toivanen P (1983) Immune capacity of the chicken bursectomized at 60 hours of incubation: failure to produce immune, natural, and autoantibodies in spite of immunoglobulin production. Cell Immunol 80: 363–373

Knight KL, Crane MA (1994) Generating the antibody repertoire in rabbit. Adv Immunol 56: 179–218

Lassila O (1989) Emigration of B cells from chicken bursa of Fabricius. Eur J Immunol 19: 955–958

Lassila O, Alanen A, Lefkovits I, Cooper MD, Pink JR (1988) Immunoglobulin diversification in embryonic chicken bursae and in individual follicles. Eur J Immunol 18: 943–949

Lydard PM, Grossi CE, Cooper MD (1976) Ontogeny of B cells in the chicken: I.sequential development of clonal diversity in the bursa. J Exp Med 144: 79–97

Mansikka A, Jalkanen S, Sandberg M, Granfors K, Lassila O, Toivanen P (1990a) Bursectomy of chicken embryos at 60 hours of incubation leads to an oligoclonal B cell compartment and restricted Ig diversity. J Immunol 145: 3601–3609

Mansikka A, Sandberg M, Lassila O, Toivanen P (1990b) Rearrangement of immunoglobulin light chain genes in the chicken occurs prior to colonization of the embryonic bursa of Fabricius. Proc Natl Acad Sci USA 87: 9416–9420

Masteller EL, Thompson CB (1994) B cell development in the chicken. Poultry Sci 73: 998–1011
Masteller EL, Larsen RD, Carlson LM, Pickel JM, Nickoloff B, Lowe J, Thompson CB, Lee KP (1995) Chicken B cells undergo discrete developmental changes in surface carbohydrate structure that appear to play a role in directing lymphocyte migration during embryogenesis. Development 12: 1657–1667
McCormack WT, Tjoelker LW, Barth CF, Carlson LM, Petryniak B, Humphries EH, Thompson CB (1989) Selection of B cells with productive IgL gene rearrangements occurs in the bursa of Fabricius during chicken embryonic development. Genes Dev 3: 838–847
Motyka B, Reynolds JD (1991) Apoptosis is associated with extensive B cell death in the sheep ileal Peyer' patch and the chicken bursa of Fabricius: a possible role in B cell selection. Eur J Immunol 21:1951–1958
Nagata S (1994) Apoptosis-mediating Fas antigen and its natural mutation. In:Tomei LD, Cope FO (eds) Apoptosis II. The molecular basis of apoptosis in disease. Cold Spring Harbor Press, Plainview, NY, pp 313–326
Neiman PE, Thomas SJ, Loring G (1991) Induction of apoptosis during normal and neoplastic B-cell development in the bursa of Fabricius. Proc Natl Acad Sci USA 88: 5857–5861
Nunez G, London L, Hockenberry D, Alexander M, McKearn JP, Korsmeyer SJ (1990) Deregulated bcl-2 gene expression selectively prolongs survival of growth factor-deprived hemopoietic cell lines. J Immunol 144: 3602–3610
Oltvai ZN, Milliman CL, Korsmeyer SJ (1993) Bcl-2 heterodimerizes in vivo with a conserved homolog, Bax, that accelerates programmed cell death. Cell 74: 609–619
Osmond DG (1990) B cell development in the bone marrow. Semin Immunol 2: 173–180
Paramithiotis E, Jacobsen KA, Ratcliffe MJH (1995) Loss of cell surface immunoglobulin expression precedes B cell death by apoptosis in the bursa of Fabricius. J Exp Med 181:105–113
Pickel JM, McCormack WT, Chen CH, Cooper MD, Thompson CB (1993) Differential regulation of V (D) J recombination during development of avian B and T cells. Int Immunol 5: 919–927
Pink JR, Vainio O, Rijnbeek A (1985) Clones of B lymphocytes in individual follicles of the bursa of Fabricius. Eur J Immunol 15: 83–87
Ratcliffe MJH, Lassila O, Pink JRL, Vainio O (1986) Avian B cell precursors: surface immunoglobulin expression is an early, possibly bursa-independent event. Eur J Immunol 16:129–133
Reynaud CA, Anquez V, Dahan A, Weill JC (1985) A single rearrangement event generates the chicken light chain preimmune repertoire Cell 40: 283–291
Reynaud CA, Anquez V, Grimal H, Weill JC (1987) A hyperconversion mechanism generates chicken light chain preimmune repertoire. Cell 48: 379–388
Subba Rao, DSV, McDuffie FC, Glick B (1978) The regulation of IgM production in the chick: roles of the bursa of Fabricius, environmental antigens, and plasma IgG. J Immunol 120:783–787
Thompson CB (1992) Creation of immunglobulin diversity by intrachromosomal gene conversion. Trends Genet 8: 416–422
Thompson CB, Neiman PE (1987) Somatic diversification of the chicken immunoglobulin light chain gene is limited to the rearranged variable gene segment. Cell 48: 369–378
Tiegs SL, Russel DM, Nemazee D (1993) Receptor editing in self-reactive bone marrow B cells. J Exp Med 177: 1009–1020
Toivanen P, Toivanen A (1973) Bursal and post-bursal stem cells in the chicken. Functional characteristics. Eur J Immunol 3: 585–595
Tonegawa S (1983) Somatic generation of antibody diversity Nature 302: 575–581
Varki J (1994) Selection ligands. Proc Natl Acad Sci USA 91: 7390–7397
Vaux DL, Cory S, Adams JM (1988) Bcl-2 gene promotes haemopoietic cell survival and cooperates with c-myc to immortalize pre-B cells. Nature 335: 440–442
Veromaa T, Vainio O, Jalkanen S, Eerola E, Granfors K, Toivanen P (1988) Expression of B-L and Bu-1 antigens in chickens bursectomized at 60H of incubation. Eur J Immunol 18: 225–230
Wallace VA, Penninger J, Mak TW (1993) CD4, CD8, and tyrosine kinases in thymic selection. Curr Opin Immunol 5: 235–240
Warner NL, Uhr JW, Thorbecke GJ, Ovary Z (1969) Immunoglobulins, antibodies and the bursa of Fabricius: induction of agammaglobulinemia and the loss of all antibody-forming capacity by hormonal bursectomy. J Immunol 103: 1317–1329
Weill JC, Reynaud CA, Lassila O, Pink JRL (1986) Rearrangement of chicken immunoglobulin genes is not an ongoing process in the embryonic bursa of Fabricius. Proc Natl Acad Sci USA 83: 3336–3340
Yamashita K, Kochibe N, Ohkura T, Ueda I, Kobata A (1985) Fractionation of L-fucose-containing oligosaccharides on immobilized Aleuria aurantia lectin. J Biol Chem 260: 4688–4693

Evidence for Phenotypic Heterogeneity Among B Cells Emigrating from the Bursa of Fabricius: A Reflection of Functional Diversity?

E. PARAMITHIOTIS[1] and M.J.H. RATCLIFFE[2]

1 Introduction

The bursa of Fabricius is a primary lymphoid organ with a central role in the establishment of the chicken B cell compartment (for review see RATCLIFFE 1989a). The bursa is organized as folds of tissue containing about 10^4 discrete follicles surrounding a central lumen. Each follicle is divided into an outer cortex and an inner medulla, separated by a complex nucleated cortico-medullary boundary. The bursal lumen is connected by the bursal duct to the large intestine and gut-derived antigens in the chicken rectum can be transported into the bursal lumen (SOVARI and SOVARI 1977). Indeed, since the terminal portion of the chicken large intestine is capable of reverse peristaltic contractions, this transport can also result in the bursal lumen containing a variety of external antigens.

Antigens in the bursal lumen are pinocytosed by epithelial cells which form the epithelial tufts connecting the bursal lumen with the follicular medulla (BOCKMAN and COOPER 1973). Following pinocytosis, lumen derived antigens are actively transported into the medulla of bursal follicles. This is a rapid process, material applied to the chicken cloaca being observed within the bursal follicles within 30 min (SOVARI and SOVARI 1977). There has been some debate as to

[1]Division of Developmental and Clinical Immunology, University of Alabama at Birmingham, 378 Wallace Tumor Institute, Birmingham, AL 35294-3300, USA
[2]Department of Microbiology and Immunology, McGill University, 3775 University Street, Montreal, Quebec, Canada H3A 2B4

whether gut-derived antigens transported into the follicular medulla can be subsequently found in the cortex of bursal follicles. Although the vast majority of carbon black particles introduced into the bursal lumen are found in the follicular medulla, a few can be found in the cortex. To date, however, no evidence for the presence of gut-derived protein antigens in the follicular cortex has been presented although protein antigens are effectively transported to the follicular medulla. Thus the cortico-medullary boundary, at a minimum, substantially impedes the diffusion of external antigen from the medulla to the cortex of a typical bursal follicle. As a result, within each bursal follicle two potentially different antigenic environments exist, distinguished by the presence or absence of exogenous antigen.

Cortical and medullary lymphocytes themselves show considerable physiological differences. The former tend to be smaller, more densely packed and more rapidly dividing than the latter medullary population. The establishment of bursal follicles experimentally reconstituted from single precursors demonstrates that a single cell can give rise to both cortical and medullary populations (Pink et al. 1985; Ratcliffe 1989b). As follicles develop in the embryo bursa, dividing B cells cluster between the bursal epithelial layer and the basement membrane. This region is ultimately destined to become the follicular medulla. After hatch, cells migrate across the basement membrane to form the cortex of bursal follicles, the basement membrane becoming the cortico-medullary junction. Once the cortex medullar dichotomy is established however, very little is known about lineage relationships among cortical and medullary lymphocytes in the juvenile bursa.

Recent evidence has suggested that peripheral blood B cells are heterogeneous with regard to physiological criteria. In particular, chicken peripheral blood lymphocyte (PBL) B cells can be divided into three populations (Paramithiotis and Ratcliffe 1993). Population 1 B cells (~ 60% of PBL B cells in 3 week old chickens) are short lived in the periphery (2–3 days), nondividing and emigrate from the cortex of bursal follicles (Paramithiotis and Ratcliffe 1994). Population 2 B cells (~ 35% of PBL B cells in 3 week old chickens) are longer lived cells (2–3 weeks) and are also nondividing in the periphery following their emigration from the bursa. The anatomical origin of these cells within the bursa is currently unclear. The third population of PBL B cells (~ 5% of PBL B cells in 3 week old chickens) are short lived cells, generated in post bursal sites of B cell division (Paramithiotis and Ratcliffe 1993).

The physiological distinction between population 1 and population 2 B cells in the peripheral blood could have two causes. Either these B cells might emigrate from the bursa as two discrete populations, as a reflection of events occurring within the bursa, or alternatively, a single bursal emigrant population could give rise to the two peripheral B cell populations as a consequence of events occurring within the periphery. Here we identify a monoclonal antibody, LT2, which distinguishes population 1 and population 2 B cells and demonstrate that bursal emigrants are heterogeneous with regard to expression of the antigen recognized by LT2 (LT2ag). The relative proportion of LT2ag$^+$ and LT2ag$^-$ bursal emigrants are consistent with the expected relative proportions of population 1 and population 2 emigrants.

2 Results

2.1 The LT2 Antibody Detects an Antigen Heterogeneously Expressed on B Cells

The LT2 antibody was isolated from a fusion using spleen cells from mice immunized with chicken bursal cells and was selected for further analysis on the basis that the staining of juvenile bursal cells was heterogeneous. Among embryo bursal cells (day 16–18e) essentially all bursal B cells express uniformly high levels of LT2ag (data not shown). Staining of bursal cells from juvenile chickens revealed, however, that, while the majority of cells express high levels of the LT2ag, there were clear populations of $LT2^{lo}$ and $LT2ag^{-}$ cells (see below). The proportion of $LT2ag^{-}$ bursal cells in the bursa increased with age.

Among PBL B cells, the proportion of $LT2ag^{+}$ B cells was about 50% in 3 week old chickens. This proportion decreased with time such that by 5 months of age fewer than 4% of PBL B cells are $LT2ag^{+}$ (Fig. 1a–c).

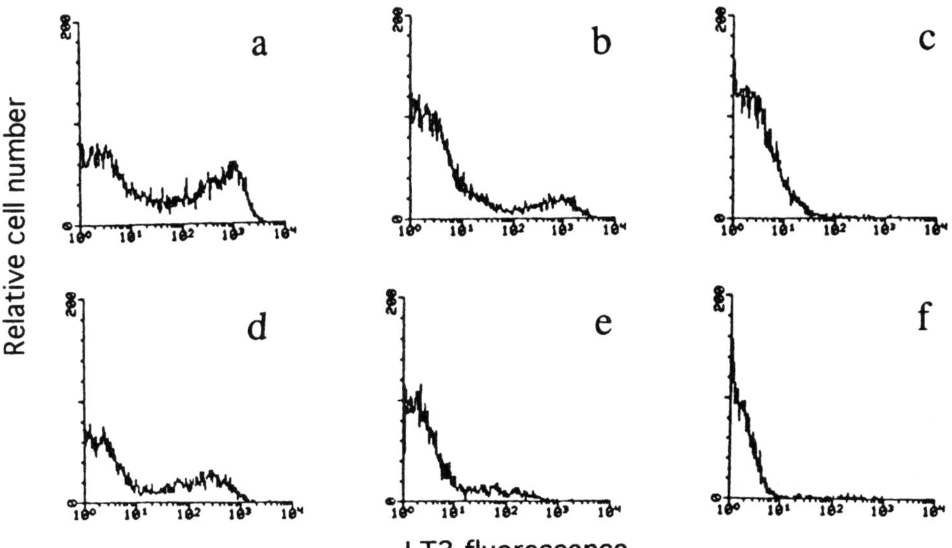

Fig. 1a–f. Heterogeneity of LT2 antigen (LT2ag) expression on peripheral blood B lymphocytes (PBL). PBLs from 3 week old (**a, d–f**), 7 week old (**b**) or 22 week old (**c**) normal chickens or 3 week old chickens 1 (**d**), 2 (**e**) or 3 (**f**) days after surgical bursectomy were stained with LT2 and anti-chB6 (Bu-1) (VEROMAA et al. 1988) monoclonal antibodies, developed with FITC conjugated anti-μ and PE-conjugated anti-γ1 antibodies respectively. The LT2ag (FL1) profiles of those cells stained with anti-chB6 are shown

2.2 LT2 Antigen Is Expressed on Short Lived Bursal Emigrants

Following surgical removal of the bursa, the proportion of $LT2ag^+$ B cells decreased rapidly in the periphery (Fig. 1a, d–f). Three days after bursectomy fewer than 4% of PBL B cells remained $LT2ag^+$. The intensity of LT2ag expression among $LT2ag^+$ B cells 1 and 2 days after bursectomy remained indistinguishable from that seen in normal controls. Consequently the reduction in $LT2ag^+$ cells following bursectomy suggests that $LT2ag^+$ cells are removed from circulation rather than remaining in circulation while undergoing a progressive loss of LT2ag expression. The rate of disappearance of $LT2ag^+$ cells was equivalent to the rate of loss of B cells for the first 3 days after bursectomy, during which time about half the peripheral blood B cells in 3 week old chicks are eliminated (Paramithiotis and Ratcliffe 1993). We conclude therefore that the LT2ag is expressed on population 1 B cells, i.e., those PBL B cells which following their emigration from the bursa have a lifespan of a few days.

2.3 LT2 Antigen Is Expressed Heterogeneously on Bursal Emigrants

To determine whether population 1 (short lived) and population 2 (longer lived) PBL B cells emigrate from the bursa as two distinct populations, bursal emigrants were analyzed for their expression of the LT2ag. Bursal emigrants are detected following intrabursal labeling with fluorescein isothiocyanate (FITC) which labels bursal cells in situ (Paramithiotis and Ratcliffe 1993). FITC-labeled PBL B cells accumulate at a rate of about 1% of the PBL B cell pool per hour corresponding to the rate of B cell emigration from the bursa. The LT2ag was expressed on about 90%–95% of FITC-labelled PBL B cells (Fig. 2), the remaining cells lacking expression of this antigen. In contrast, 100% of bursal emigrant cells expressed high levels of MHC class II and the chL12 antigen (Paramithiotis and Ratcliffe 1993). Thus bursal emigrants are heterogeneous with regard to the expression of the LT2ag. We conclude therefore that B cells emigrate from the bursa as two phenotypically distinct populations which correspond to the functionally distinct B cell populations observed in peripheral blood.

3 Discussion

The results presented here suggest that bursal cells emigrate to the periphery as two distinct populations. This conclusion is based on the expression of a cell surface antigen expressed heterogeneously on bursal and peripheral B cells. LT2ag expression is not restricted to B cells. The LT2 antibody stains at least some $CD8^+$ peripheral T cells with low intensity. At a molecular level this antigen

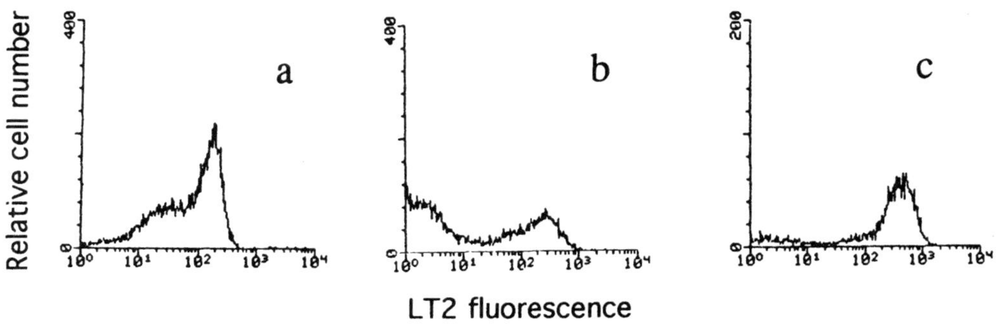

Fig. 2a–c. Heterogeneity of LT2 antigen (LT2ag) expression on bursal emigrants. Three week old chickens were labeled intrabursally with FITC (PARAMITHIOTIS and RATCLIFFE 1993). Six hours later peripheral blood lymphocytes (PBLs) and bursa were harvested and stained with LT2, developed with PE-conjugated anti-μ, and anti-ChB6, developed with biotinylated anti-γ1 followed with streptavidin-phycoerythrin-Texas red. **a** LT2ag expression of bursal cells from one such chicken: **b** LT2ag expression of PBL B cells, excluding those cells which emigrated from the bursa within the previous 6 h (stained with anti-chB6, but FITC); **c** LT2ag expression of those PBL B cells which have emigrated from the bursa within the previous 6 h (stained with anti-chB6 and FITC$^+$)

has not been characterized. The antibody is an IgM reagent and to data we have been unable to immunoprecipitate (a) band(s) from surface labeled bursal cells. It is possible that this antibody detects a carbohydrate determinant.

We have previously demonstrated that the short lived population 1 B cells emigrate directly from the bursal cortex without any distinguishable transit time within the bursa following their final intrabursal cell division (PARAMITHIOTIS and RATCLIFFE 1994). Preliminary results (not shown) suggest that the cells of the follicular cortex express the highest levels of the LT2ag in juvenile bursal follicles. In addition the follicular cortex contains cells which express high levels of both chL12 and MHC class II, a phenotype associated with immediately pre-emigrant bursal cells (PARAMITHIOTIS and RATCLIFFE 1993). These observations are consistent with the expression of LT2ag being restricted in the peripheral blood to short lived population 1B cells which have emigrated from the bursal cortex.

The bursal cortex is a site of rapid B cell division (REYNOLDS 1987; EKINO 1993; PARAMITHIOTIS and RATCLIFFE 1994) and is also the presumptive site of immunoglobulin V gene diversification by gene conversion. We have previously shown that maintenance of surface Ig (sIg) expression is required for the progression of normal B cell development in the bursa and that there is a significant amount of cell death in the cortex of bursal follicles (PARAMITHIOTIS et al. 1995). Consequently there are several potential fates for the progeny of dividing cortical lymphocytes. Further cell division is required to maintain the cellularity of the cortical population, cell death and emigration. Emigration from the bursal cortex could either be to the periphery, to give rise to the LT2ag$^+$ population 1 B cells or could potentially be to the follicular medulla. Among medullary bursal cells there are also cells expressing high levels of both chL12 and MHC class II, again suggesting that the medulla of bursal follicles contains pre-emigrant bursal cells. In addition, from the detection of apoptotic cells by electron microscopy, it is clear that a substantial proportion of medullary cells die in situ (Paramithiotis et al. 1995). Since few medullary cells undergo in situ cell division (Reynolds 1987), the apparent fates of medullary bursal cells are restricted to cell death or emigration to the periphery.

3.1 A Model for the Generation of Peripheral B Cell Repertoires in Bursal Follicles

We have developed a model for B cell development in the bursa to account for the heterogeneity of functionally and phenotypically distinct populations of bursal emigrants (Fig. 3). Among cortical lymphocytes there is clear evidence for rapid cell division, cell death and emigration to the periphery. However, although at around the time of hatch there is migration of cells from the follicular medulla to establish the follicular cortex, there is little direct evidence as to traffic between the cortical and medullary compartments within a bursal follicle. It has however been clearly established that there is no significant traffic of cells from one follicle to another. Indirect evidence suggests however that there is a flow of cells from

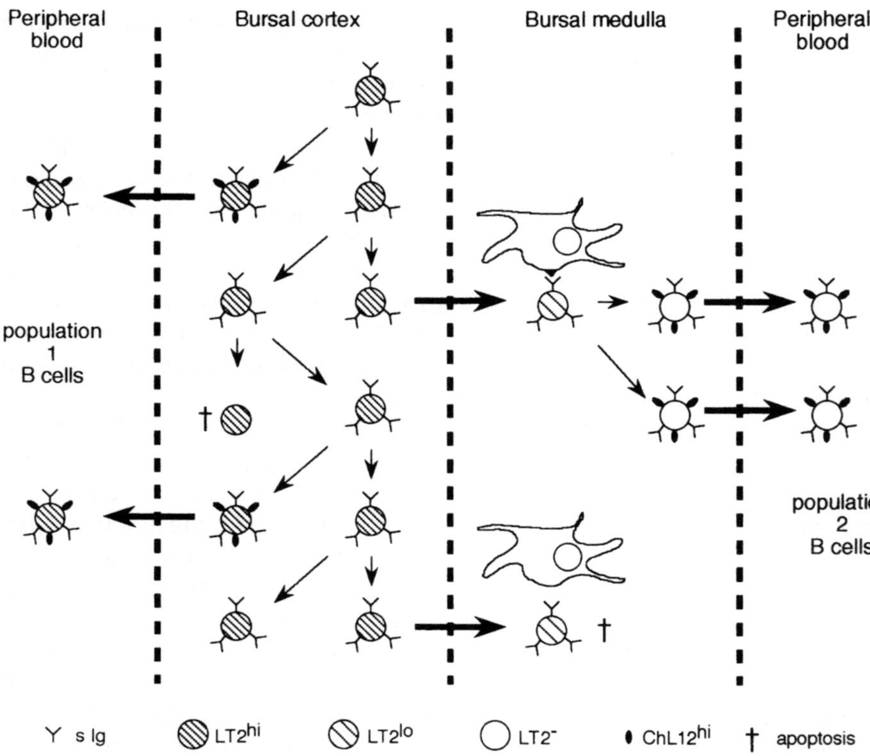

Fig. 3. B cell development in the bursa. This scheme is explained in detail in the text; it has been developed to account for the emigration of two phenotypically and functionally distinct populations of B cells from the bursa

the follicular cortex to the follicular medulla. Following injection of 3 week old chickens with vincristine sulfate (which blocks cell division in metaphase), there is a rapid accumulation of metaphase arrested cells in the follicular cortex. However, there is a very little evidence of metaphase arrested cells in the follicular medulla suggesting that medullary cells are essentially nondividing (REYNOLDS 1987). Similarly, if 3 week old chickens are injected with bromodeoxyuridine (BrdUrd), which becomes incorporated into the DNA of cells during S phase of the cell cycle without inhibiting subsequent cell division, there is rapid labeling of cortical cells in bursal follicles consistent with results obtained following vincristine treatment. However, from about 6 h after BrdUrd labelling there is a rapid accumulation of BrdUrd-labelled cells in the follicular medulla. This labeling increases with time such that, by about 16 h after the start of labeling, the great majority of medullary cells are labeled (EKINO 1993; PARAMITHIOTIS and RATCLIFFE 1994). This apparent conflict with results obtained following vincristine sulfate injection can be resolved by postulating that there is a postmitotic flow of cells from the bursal cortex to the medulla which would be inhibited by metaphase blockers such as vincristine but not by BrdUrd.

Therefore a cell in the bursal cortex might have four potential fates: further round(s) of cell division, elimination by apoptotic cell death, emigration from the bursa as a population 1 (short lived) $LT2ag^{+}$ B cell, or trafficking from the cortex to the medulla. The signals involved in making the determination between these potential fates are currently unclear.

Antigens have been shown to be transported from the bursal lumen to the follicular medulla and to be associated with dendritic cells, which may function analogously to the follicular dendritic cells of mammalian germinal centers in presenting antigen to B cells (EKINO, this volume). In addition, it has been shown that intrabursal priming of chickens with antigen enhances subsequently induced responses to the same antigen injected peripherally (EKINO et al. 1979). We propose therefore that the difference between life and death of bursal cells in the follicular medulla is a reflection of whether the bursal cell interacts with its cognate antigen. In the absence of such contact, the cell would die in situ; following cognate recognition of antigen, the medullary bursal cell would become a long lived (population 2) emigrant. Whether medullary cells are stimulated to undergo further cell division as indicated in Fig. 3 is currently speculative but is consistent with the very low levels of cell division observed in the follicular medulla (REYNOLDS 1987).

The bursal emigrant population would therefore contain two superimposed repertoires of B cell specificities: a naive repertoire which is highly diverse, has not been selected for recognition of any particular specificities, and rapidly turns over as a consequence of the short lifespan of cells within this population and a second repertoire of specificities which has been selected on the basis of antigen recognition within the bursal medulla. This latter repertoire would be a reflection of the external antigenic environment of the individual and would therefore specifically enhance the capacity to respond to potentially pathogenic entities in the environment. This repertoire would also have the potential to gradually change with the changing antigenic environment, since the lifespan of population 2 B cells is in the order of a few weeks.

Acknowledgments. This work was supported by the Medical Research Council of Canada. EP is a Fellow of the Irvington Institute for Medical Research, MJHR is a Senior Chercheur-boursier of the fonds de la Recherche en Santé du Québec.

References

Bockman DE, Cooper MD (1973) Pinocytosis by epithelium associated with lymphoid follicles in the bursa of Fabricius, appendix and Peyer's patches. Am J Anat 136: 455–477

Ekino S (1993) Role of environmental antigens in B cell proliferation in the bursa of Fabricius at neonatal stage. Eur J Immunol 23: 772–775

Ekino S, Matsumo K, Harada S, Fujii H, Nawa Y, Kotani M (1979) Amplification of plaque-forming cells in the spleen after intracloacal antigen stimulation in neonatal chicken. Immunology 37: 811–815

Paramithiotis E, Ratcliffe MJH (1993) Bursa dependent subpopulations of peripheral B lymphocytes in chicken blood. Eur J Immunol 23: 96–102

Paramithiotis E, Ratcliffe MJH (1994) B cell emigration directly from the cortex of lymphoid follicles in the bursa of Fabricius. Eur J Immunol 24: 458–463

Paramithiotis E, Jacobsen KA, Ratcliffe MJH (1995) Loss of surface immunoglobulin expression precedes B cell death by apoptosis in the bursa of Fabricius. J Exp Med 181: 105–113

Pink JRL, Vainio O, Rijnbeek A-M (1985) Clones of B lymphocytes in individual follicles of the bursa of Fabricius. Eur J Immunol 15: 83–87

Ratcliffe MJH (1989a) Development of the avian B lymphocyte lineage. CRC Crit Rev Poult Biol 2: 207–234

Ratcliffe MJH (1989b) Generation of immunoglobulin heavy chain diversity subsequent to cell surface immunoglobulin expression in the avian bursa of Fabricius. J Exp Med 170: 1165–1173

Reynolds J (1987) Mitotic rate maturation in the Peyer's patches of fetal sheep and in the bursa of the chick embryo. Eur J Immunol 17: 503–507

Sovari R, Sovari TE (1977) Bursa Fabricii as a peripheral lymphoid organ. Transport of various materials from the anal lips to the bursal lymphoid follicles with reference to its immunological importance. Immunology 32: 499–505

Veromaa T, Vainio O, Eerola E, Toivanen P (1988) Monoclonal antibodies against chicken Bu–1a and Bu-1b alloantigens. Hybridoma 7: 41–48

T Cell Receptors and T Cell Development

C.H. CHEN,[1] A. SIX,[2] T. KUBOTA,[1,3] S. TSUJI,[1,4] F.-K. KONG,[1] T.W.F. GÖBEL,[1,5] and M.D. COOPER[1,2]

1 Introduction

Birds provide a valuable model system for the study of early development. Studies of immune system development and function in chickens have contributed to some of the central tenets in immunology. These include:(1) allorecognition by lymphocytes in graft-vs-host (GVH) reactions (SIMONSEN 1985), (2) clonal reactivity of immunocompetent lymphocytes (BURNET 1957), (3) the separate differentiation pathways of T and B cells which cooperate in immune responses (COOPER et al. 1965), (4) hematopoietic stem cell origin of both lymphoid and

[1] Division of Developmental and Clinical Immunology, Department of Microbiology, University of Alabama at Birmingham, 378 Wallace Tumor Institute, Birmingham, AL 35294-3300, USA
[2] Howard Hughes Medical Institute, University of Alabama at Birmingham, 378 Wallace Tumor Institute, Birmingham, AL 35294-3300, USA
[3] Laboratory of Molecular Immunology, National Institute of Animal Health, Tsukuba Science City, Ibaraki, 305 Japan
[4] Department of Veterinary Medicine, University of Osaka, Osaka, Japan
[5] Basel Institute for Immunology, Grenzacherstrasse 487, 4005 Basel, Switzerland

myeloid lineages (Moore and Owen 1965), (5) the developmental migratory waves of thymocyte progenitors (Le Douarin 1978), (6) immunoglobulin (Ig) isotype switching by IgM-bearing B lymphocytes (Kincade et al. 1970; Kincade and Cooper 1973), and (7) somatic diversification of the B cell repertoire by gene conversion (Reynaud et al. 1987; Thompson and Neiman 1987).

Information on T cell development has been derived mainly from studies in mammals. More recently, the production of the monoclonal antibodies reactive with avian T cell receptors (TCRs) and T cell-specific differentiation antigens has allowed a detailed analysis of T cell development in the chicken. These studies reveal that T cell differentiation and function are remarkably similar in birds and mammals (Cooper et al. 1991; Chen et al. 1990, 1994). One hallmark of this similarity is the development of T cells along two discrete pathways characterized by the expression of αβ or γδ TCR, thus indicating a relatively ancient separation of these T cell lineages during vertebrate evolution.

Although the selection, differentiation and function of αβ T cells have been well studied in mammals, the earliest events in TCR rearrangement, the mechanism for divergence of αβ and γδ T cell lineages, the function of γδ T cells, their interaction with αβ T cells, and the possible extrathymic pathway for T cell development are still incompletely understood. The chicken provides a very attractive model to explore these important issues because of the advantages of embryo accessibility, easily definable thymic waves of progenitors, thymocyte progeny and daughter T cells, and the relative abundance of both embryonic thymocytes and γδ T cells.

2 The T Cell Receptor/CD3 Complex

The chicken TCRs were initially defined by specific monoclonal antibodies. Because of the evolutionary distance between mammals and birds, it proved relatively easy to produce mouse monoclonal antibodies against avian cell surface antigens including the TCR molecules (Chen et al. 1991). Both TCR αβ and γδ receptor molecules are disulfide-linked heterodimers of 50 kDa and 40 kDa glycopeptide chains that are noncovalently associated with a CD3 complex of proteins to form a signal transduction unit (Sowder et al. 1988; Chen et al. 1988, 1989; Char et al. 1990).

The cloning of the chicken TCR genes proved to be more difficult because the mammalian TCR genes are not very homologous to their avian counterparts. The specificity and diversity of T cells are determined by the TCR genes, which in mammals are composed of multiple variable (V), diversity (D), joining (J) and constant (C) gene segments (Davis and Bjorkman 1988). During T lymphocyte development the repertoire of TCRs are generated by recombination of individual V, (D), J and C segments. Additional diversity is created at the junctions of the V, D and J elements. It was uncertain whether this would be the case in birds, because in avian species antibody diversity is generated primarily by gene conversion (REYNAUD et al. 1987; THOMPSON and NEIMAN 1987). It was therefore of great interest to determine whether birds use gene conversion or somatic recombination to generate their TCR diversity.

2.1 The T Cell Receptor β Genes

In 1990, Tjoelker and colleagues identified the TCRβ genes in the chicken, as the first TCR genes successfully cloned in a nonmammalian species. The cloning of the chicken TCRβ gene was achieved by cross-hybridization of a splenic cDNA library with a pool of short (~100 bp) fragments of a mixture of mammalian TCRβ DNA under conditions of low stringency (TJOELKER et al. 1990). The full-length (1.3 kb) chicken TCRβ cDNA encodes a protein of approximately 300 amino acids including leader (L), V, D, J and C regions. Like its mammalian counterpart, the chicken TCRβ locus undergoes VDJ gene rearrangement and multiple distinct recombination events are observed during thymocyte development to achieve combinatorial diversity. However, only two Vβ gene families exist in the chicken. The sequences of the two Vβ families have low homology, but each Vβ family contains several members with highly homologous sequences. The TCRβ genomic organization is not yet complete, but six Vβ1 segments, four Vβ2 segments, one D element, four Jβ segments and one Cβ region have been mapped (TJOELKER et al. 1990; COOPER et al. 1991; L. HOOD, personal communication). One of the Vβ2 gene is located 3' to the Cβ gene, and this Vβ gene is in reverse transcriptional orientation relative to the other genes in the TCRβ locus. No Vβ pseudogenes have been identified. The four Jβ sequences are very similar to each other. The TCRβ locus has been mapped to chromosome 1 (N. BUMSTEAD et al., unpublished observation). Because of the limited heterogeneity among germline V, D and Jβ elements, TCRβ diversity is largely contributed by the variable N nucleotide addition to the coding joints of V-D and D-V recombinations (TJOELKER et al. 1990; MCCORMACK et al. 1991).

2.2 The T Cell Receptor α Genes

The chicken TCRα genes were cloned by affinity chromatographic isolation of the TCRα proteins, determination of partial amino acid sequences and screening of cDNA library with degenerate oligonucleotide probes. A TCR α cDNA (1.7 kb) was obtained which encodes a 275 amino acid protein that contains Vα, Jα and Cα regions (GÖBEL et al. 1994a). Genomic analysis indicates a single Cα gene and multiple Jα segments (~20). The 2.9 kb Cα gene segment, although shorter than its human and mouse counterparts (4.7 kb and 3.6 kb, respectively), contains four exons and the exon/intron organization is similar to that in mammals (T.W.F. GÖBEL et al., unpublished observation). The chromosome location of the TCRα locus has not yet been determined. The size of the introns, however, suggests that the TCRα gene may also be localized on a macrochromosome.

The Vα1 gene family contains multiple Vα genes(~25) with relatively high homology (GÖBEL et al. 1994a). To identify additional Vα families, Cα and Vα1 cDNAs were employed as probes to screen a cDNA library. Vα1⁻Ca⁺ clones were isolated and those containing inserts of > 1 kb were sequenced to identify a second Vα family (Vα2) that exhibits only around 24% identity with Vα1 (KUBOTA et al. 1995). Our analysis thus far suggests that the Vα1⁺ and Vα2⁺ genes represent the sole Vα gene families expressed in the spleen. Members of the Vα1 or Vα2 gene families are also used by the dozen T cell lines examined to date. These results suggest that the chicken has only two Vα gene families.

2.3 The T Cell Receptor γ Genes

Yet a different strategy was used in the recent identification of chicken TCRγ genes. Rast and Litman designed short, minimally degenerate oligonucleotide primers complementing conserved V region segments of TCR and used these to amplify TCR-like products from genomic DNA of lower vertebrate representatives by the polymerase chain reaction (PCR) (RAST and LITMAN 1994). This approach yielded two PCR products through the amplification of chicken genomic DNA. Sequence analysis revealed that both PCR products display characteristics of the V segment of TCR but do not belong to any of the chicken TCRVα or TCRVβ families that we had identified previously. Two cDNA clones identified by screening a splenic cDNA library with PCR products as probes were sequenced. The derived amino acid sequence of each clone (approximately 340 residues) contains L, V, J and C regions. Both clones have an identical C region that exhibits approximately 30% identity with human and mouse TCR Cγ at the amino acid level. Northern blot analysis with the C region probe reveals TCRγ transcripts (~1.9 kb) in the thymus, spleen and γδ T cell lines, but not in the bursa or liver. The two V region cDNA sequences have only 33% homology and are designated as prototypes of the Vγ1 and Vγ2 families (SIX et al. 1995).

To identify additional Vγ families, differential screening of the splenic library with Cγ and Vγ1+2 probes was performed to isolate Vγ⁻Cγ⁺ clones. One additional Vγ family, designated Vγ3, was identified in this way. It has 25%–29% identity with the Vγ1 and Vγ2 families. Three Jγ genes, which share 60%–65% identity at the amino acid level, have been identified, and these can be expressed with the different Vγ families (SIX et al. 1995). Twenty γδ T cell lines have been examined and each expresses Vγ1, Vγ2 or Vγ3 (our unpublished observation; W. MCCORMACK et al., personal communication).

Multiple Cγ genes have been identified in mammalian species. However, Southern blot analysis of chicken genomic DNA with the C region probe and multiple restriction enzymes yielded either one or two hybridizing bands of relatively small size (< 5 kb). The same bands were observed under conditions of low and high stringency. The mammalian Cγ genes exhibit 70%–75% or greater identity at the nucleotide level (RAULET 1989; TAKEUSHI et al. 1992; HEIN and DUDLER 1993; THOME et al. 1993), and our washing conditions should allow the detection of other Cγ genes with a similar degree of homology. The same Cγ region has been found in all of the full length cDNAs that have been sequenced to date as further evidence for a single Cγ gene in the chicken TCRγ locus. The chicken TCRγ locus, like the TCRα and TCRβ loci, thus appears relatively simple.

2.4 The T Cell Receptor δ Genes

The TCRδ genes are located within the TCRα locus in mammals (Davis and Bjorkman 1988; Koop and Hood 1994). Members of ten Vα families in humans and six Vα families in mice rearrange to the DJ segments of the TCRδ gene family (Raulet 1989; Genevée et al. 1994). Since our studies of the TCR genes and T cell development in the chicken indicate the conservation of many central features in birds and mammals, it seemed likely that the same TCRα/δ locus organization would be conserved in the chicken. This suggested that the isolation of $V\alpha^+C\alpha^-$ cDNA clones from a chick splenic library might lead to the identification of TCRδ genes. With this in mind, a $V\alpha1^+C\alpha^-$ clone was isolated and found to contain a Vα1 region, a relatively long joint region and a C-like region differing from chicken Cα, Cβ or Cγ. This candidate Cδ gene region consists of 144 amino acids and shares approximately 33% identity with human and mouse Cδ. Moreover, northern blot analysis using this C region probe reveals 1.7 kb transcripts in the thymus, spleen and γδ T cell lines, but none in the bursa, liver or αβ T cell lines (Kubota et al. 1995) further suggesting that this is the chicken TCRδ gene. Preliminary results indicate that both Vα1 and Vα2 gene segments can be used to encode TCRδ chains. Two Vδ families have also been identified (T. Kubota et al., unpublished observations).

2.5 The CD3 Complex

The chicken CD3 complex consists of at least three proteins with molecular masses of 20 kDa, 19 kDa and 17 kDa that can be coprecipitated with antibodies to αβ or γδ TCR when the T cell membrane is solubilized with a mild detergent (Chen et al. 1986). The gene encoding the 19 kDa protein has been cloned (Bernot and Auffray 1991) and its sequence is most homologous to mammalian CD3γ and δ (36%–40%). It is difficult to assign this chain to a single mammalian CD3 subunit, and the data raise the possibility that this polypeptide may represent an ancestral form of the γ and δ chains. The 20 kDa and 17 kDa elements of the CD3 can also be detected in the cytoplasm of avian NK cells. Although the 19 kDa protein is not detectable, its mRNA is present in the NK cells (Göbel et al. 1994b; see chapter by Göbel et al.).

2.6 Evolutionary Conservation of the T Cell Receptor Genes

As in mammals the avian TCR genes undergo tissue-specific rearrangements and the TCR repertoire is generated by combinatorial and junctional diversity. Although TCR chains in chickens and mammals display only ~30% identity in amino acid sequence, the consensus amino acids found in most mammalian TCR chains are also conserved in the chicken TCR chains. These include the residues that stabilize the structure of the heterodimers, the cysteine residues that form intra- and interdisulfide bonds, and positively charged lysine and arginine residues in the transmembrane domain that are thought to form salt bridges with negatively charged amino acid residues of CD3 molecules. This suggests that the evolutionary conservation is selected at the level of proteins. Interestingly, while the homology between chicken TCR V families is relatively low (<30%), each family has higher homology to a specific human or mouse TCR V family (>40%).

Genomic organization of the TCR loci is also similar in birds and mammals. V, D and J elements are flanked by classical heptamer/nonamer recombination signal sequences with the conserved length of the spacer between them. The exon-intron organization of the constant region is also similar to the mammalian counterparts. Interestingly, a single exon encodes L and V regions in avian TCRα and γ loci in contrast to the avian TCRβ and mammalian TCR loci (Göbel et al. 1994a; Six et al. 1995).

The TCRβ chain sequences of 20–30 Vβ families in mammals are divisible into two large subgroups, VβI and VβII, based on structural similarities (Schiffer et al. 1986; Bougueleret and Claverie 1987) that are found to be conserved in the chicken Vβ1 and Vβ2 families (Tjoelker et al 1990; Cooper et al. 1991). These include a conserved arginine at position 64 required to form a salt bridge with the aspartic acid at position 86 in the chicken Vβ1 family and the mammalian VβI group. Like the mammalian VβII subgroup the chicken Vβ2 family lacks the ability to form a salt bridge between amino acids 64 and 86. Instead, a tyrosine at position 65 is characteristic of the chicken Vβ2 family and the mammalian VβII group. Interestingly, the characteristic arginine-aspartic acid salt bridge present in genes of the Vβ1 family can be found in Vα2 sequences but not in the Vα1 gene family. A preliminary analysis of the utilization of Vα families in defined $V\beta1^+$ and $V\beta2^+$ cell lines suggests the possibility of preferential pairing: Vα1-Vβ1 and Vα2-Vβ2 (T. Kubota et al., unpublished observation). If preferential pairing of Vα and Vβ families proves to be the rule, the chicken model could be very useful in exploring the structure-function relationships of the preferred TCRα/β chain pairings.

3 Segregation of γδ and αβ T Cell Lineages

The segregation mechanism in mammals for αβ and γδ T cell lineages is not entirely clear, but it has been proposed that αβ T cells derive from T cell progenitors that fail to express the γδ TCR. Deletion of the δ locus by a novel recombination mechanism may be a prerequisite of αβ T cell development (de Villartay et al. 1988). Deletion of the mammalian TCRδ locus occurs by rearrangement between an upstream δ deleting element, δREC, and a downstream δ deleting element, ψJα. Coexistence of a TCRα and a TCRδ rearrangement on the same allele is impossible. Deletion of TCRδ genes must therefore occur on at least one allele in αβ T cells. In fact, δ deletion occurs at a high frequency in the human thymus and on both alleles in most αβ T cell lines (DE VILLARTAY et al. 1987). If δ deletion were essential for segregation of the two T cell lineages, it should also be operative in birds. We therefore examined αβ T cell lines to determine if the TCRδ genes are deleted and found the Cδ gene to be deleted on both alleles in all five αβ T cell lines (KUBOTA et al. 1995). These results further indicate conservation of the TCRα/δ locus in the chicken and reinforce the importance of δ deletion in the divergence of γδ and αβ T cell lineages.

Alternatively, it has been proposed that transcriptional silencing of the TCRγ genes is required for the development of αβ T cells (BONNEVILLE et al. 1989). This hypothesis would not predict the presence of TCRγ mRNA and protein expression in αβ T cell lines. Contrary to this expectation, three of eight αβ T cell lines transformed by Marek's disease virus were found to express TCRγ transcripts and proteins in the cytoplasm (C.H.CHEN et al., unpublished observation). This result casts doubt on the existence of an efficient role for a γ-silencer in regulating αβ and γδ lineage segregation during avian ontogeny.

4 The Distinctive Vβ1 and Vβ2 Subpopulations of Avian αβ T Cells

Chicken αβ T cells have characteristics similar to those of their mammalian counterparts (COOPER et al. 1991; CHEN et al. 1990, 1994), but can be divided into two discrete subpopulations by their reactivity with the TCR2 and TCR3 antibodies (CHEN et al. 1988; CHAR et al. 1990). The molecular basis for this division is that the TCR2 cells express Vβ1 genes and TCR3 cells express the Vβ2 genes (LAHTI et al. 1991); both subpopulations use the same Dβ and Jβ gene segments.

Thymocytes expressing Vβ1 and Vβ2 TCR exhibit similar developmental patterns, but the two αβ subpopulations differ in ontogeny, tissue distribution and apparent function. The Vβ1 and Vβ2 genes are rearranged sequentially during ontogeny (TJOELKER et. al. 1990; COOPER et al. 1991). Rearrangement of the TCR Vβ1 genes can begin either by V to D or D to Jβ joining, and the transcription of unrearranged TCRβ genes precedes V(D)J rearrangements (DUNON et al. 1995). Vβ1 TCR can be detected in the cytoplasm of thymocytes as early as embryonic day 12 (E12) (BUCY et al. 1990). Thymocytes expressing Vβ1-TCR appear around E15 and these become the predominant thymocyte subpopulation by E17-E18 (CHEN et al. 1988). The $V\beta2^{+}$ subpopulation of αβ T cells develops later (E18) and is maintained as a minor T cell subset (CHAR et al. 1990).

Migration of T cells to the periphery proceeds in the same order. $V\beta1^{+}$ T cells appear in the spleen around E19 and $V\beta2^{+}$ T cells appear around 5 days later, day 3 after hatching (CHEN et al. 1988; CHAR et al. 1990). The homing patterns of αβ and

$\gamma\delta$ T cells in periphery tissues are very different (Bucy et al. 1988; Char et al. 1990). The $\alpha\beta$ T cells home to the periarteriolar sheaths in the spleen and to the lamina propria in the intestine; some $\alpha\beta$ T cells enter the intestinal epithelium. Both $V\beta1^+$ and $V\beta2^+$ T cells are present in the spleen, but the intestinal $\alpha\beta$ T cells utilize the $V\beta1$ gene family almost exclusively (Char et al. 1990; Dunon et al. 1994). Ontogenic studies employing the ovalloantigen chimeric model indicate that the T cell progeny of each wave of thymocyte progenitors migrate to both the spleen and the intestine (Dunon et al. 1994).

The two $\alpha\beta$ T cell subpopulations that express $V\beta1$ or $V\beta2$ receptors also appear to exhibit functional differences. When $V\beta1^+$ cells are depleted by embryonic injection of anti-$V\beta1$ monoclonal antibody together with thymectomy, the chickens so treated later acquire a compensatory increase in the number of $V\beta2^+$ cells (Chen et al. 1989; Cihak et al. 1991), and this compensatory mechanism extends to the intestine (Cihak et al. 1991). These birds can respond normally to many T-dependent and T-independent antigens. They also appear to produce normal levels of IgM and IgG antibodies, but their capacity for IgA production is severely compromised (Cihak et al. 1991). Secretory IgA concentrations in bile and lung lavage fluid are reduced 1,000 to 10,000-fold, and secretory IgA antibodies are not produced in response to mucosal immunization. These results suggest that the $V\beta1^+$ T cells are essential for mucosal IgA antibody responses. Functional differences in the $V\beta1$ and $V\beta2$ cells may also be evident in their relative ability to induce GvH reactions. Although each subset is capable of GvH alloreactivity in the Simonsen assay, $V\beta1$ and $V\beta2$ T cells vary in their GvH potential depending on the donor and recipient strain combination (see below).

5 Thymus Dependence of Avian T Cell Development

Although thymus is an important source of T cells in all vertebrate species that have been studied, an extrathymic pathway of T cell development has also been proposed. The evidence for extrathymic intestinal T cell development includes the presence of T cells in congenitally athymic nude mice, evidence of T cell differentiation in thymectomized and lethally irradiated mice reconstituted with bone marrow, a unique cellular phenotype and TCR V gene usage by intestinal lymphocytes, and the suggestion of in situ rearrangement of TCR genes in the intestines (Rocha et al. 1992; Poussier and Julius 1994; Le Francois and Puddington 1995).

Intestinal epithelial lymphocytes (IELS) in the mouse display an unusual phenotype relative to most T cells in the blood, spleen and lymph nodes. The $\gamma\delta$ IEL that express CD8 α chain homodimers (CD8$\alpha\alpha$), rather than CD8$\alpha\beta$ heterodimers, are generally thought to be generated extrathymically. The unique phenotype of IELS and their biased V gene usage could also be explained by preferential homing, retention and proliferation of a particular subset of thymus-derived T cells (LIN et al. 1994; LEFRANCOIS and PUDDINGTON 1995). As a first step in examining the origin of CD8αα T cells in the chicken, the avian CD8 antigen has been characterized. The chicken CD8 molecule is a 64 kDa protein consisting of two chains of identical molecular weight (CHAN et al. 1988). However, removal of N-linked oligosaccharides reveals that the CD8 antigen is composed of two core proteins (TREGASKES et al. 1995). Cloning and sequence analysis of cDNAs for these proteins indicate that the CD8α chain has a peptide backbone of 24 kDa and is heavily glycosylated with O-linked sugars. The smaller CD8β chain contains four potential N-glycosylation sites and has a peptide backbone of 21 kDa. Sequence and protein analyses thus indicate that the CD8 antigen is conserved during vertebrate evolution and can be expressed on the cell surface either as a CD8αα homodimer or as a CD8αβ heterodimer in birds as well as in mammals (TREGASKES et al. 1995).

Immunofluorescence analysis using monoclonal antibodies that discriminate between CD8α and CD8β chains indicates that the vast majority of the $CD8^+$ cells in the thymus, spleen and blood of adult chickens express both CD8α and CD8β. In contrast, approximately half of the $CD8^+$ IELs express CD8α molecules only. Among IELs, 60% of the $CD8^+\gamma\delta$ T cells and 35% of the $CD8^+\alpha\beta$ T cells express CD8αα homodimers. In striking contrast to the γδ T cell profile in the adult spleen, a relatively large proportion of the $CD8^+$ γδ T cells in the spleen of embryos and young chicks express CD8αα homodimers (TREGASKES et al. 1995). $CD4^+$ T cells expressing relatively low levels of CD8αα molecules can also be detected in the blood, spleen and intestine. Moreover, when purified $CD8^-$ thymocytes were stimulated with an anti-TCRγδ antibody (TCR1) plus exogenous growth factors, CD8αα expression was induced (F. KONG et al., unpublished observation). While murine intestinal T cells that express the CD8αα phenotype have been proposed to be generated extrathymically (ROCHA et al. 1992; POUSSIER and JULIUS 1994; LEFRANCOIS and PUDDINGTON 1995), the relative frequency of the CD8αα subpopulation of the IELs is unaltered in chickens that have been thymectomized at hatching. Thus, in the chicken, CD8αα expression does not appear to be a phenotypic marker of T cells derived via an extrathymic pathway.

The idea of an extrathymic origin for T cells grew out of the observation that T cells can be found in the intestine of thymectomized mice, nude mice and athymic irradiation chimeras (ROCHA et al. 1992; POUSSIER and JULIUS 1994; LEFRANCOIS and PUDDINGTON 1995). However, thymus remnants in nude mice or incomplete depletion of donor T cells from hematopoietic tissue transplants to athymic irradiation chimeras supply the T cells in these animals (IKEHARA et al. 1987; LIN et al. 1994; LEFRANCOIS and PUDDINGTON 1995). Thymectomy of newly hatched chicks drastically inhibits development of the γδ T cell population in the peripheral tissues including intestine (CHEN et al. 1989; CIHAK et al. 1993; F. KONG et al., unpublished observation), and this result has been confirmed recently in mice (LIN et al. 1994). Similarly, thymectomy of sheep embryos results in a long-lasting depletion of T cells, including the γδ T cells in the intestine (HEIN et al. 1990). In an infant with complete thymic aplasia (DiGeorge syndrome), this development defect resulted in the absence of αβ and γδ T cells in the periphery (ATKINSON 1995).

While it has been proposed that an extrathymic pathway may reflect the persistence of a more primitive pathway present in lower vertebrates, removal of the thymus from 5–6 day *Xenopus laevis* larvae prevents T cell development in frogs even though they may survive beyond 2 years of age. These athymic frogs, which appear to have normal NK cells, fail to develop T cells (J.D. HORTON, M.D. COOPER et al., unpublished observations).

The issue of extrathymic T cell development has been explored further in an analysis of the origin of T cells in chick-quail chimeras. While transplants of chick embryonic thymus (after receiving the first wave of progenitor cells) generate both γδ and αβ chick T cells in quail recipients, transplantation of the embryonic spleen, bursa or intestine does not yield donor T cells in quail embryo recipients (COLTEY et al. 1989; BUCY et al. 1989). These observations indicate that neither γδ nor αβ T cells are generated outside of the thymus in avian embryos. To examine the generation of T cells beyond the embryonic stage, the origin and kinetics of intestinal and splenic colonization by γδ T cells were analyzed in chimeras of two congenic chick strains that differ in their ov alloantigen T cell marker (DUNON et al. 1993a,b). The results of these studies indicate that the intestine is populated by long-lived γδ T cell clones of thymic origin. Examination of the repertoire of the Vβ1$^+$ cells in ov alloantigen chimeras reveals that the TCRVβ repertoire derived from each migratory wave is similar in thymus, spleen and intestine. None of the Vβ1 or Jβ gene segments are used exclusively in the spleen or the intestine. A thymic origin of all of the different types of T cells is thus indicated in birds.

Studies have also been performed to determine whether rearrangement of TCR genes can take place outside of the thymus. TCR genes are formed by the combinatorial joining of individual members of the V, D and J gene families during which the joining of two coding segments in the same transcriptional orientation on the chromosome results in deletion of the intervening DNA (FUJIMOTO and TAMAGISHI 1987; OKAZAKI et al. 1987). The deleted DNA forms a circular episome which can be detected by PCR as an indication of rearrangement in a particular location (MCCORMACK et al. 1989a). Another indication of TCR rearrangement is the expression of the recombinase activating genes (RAGs), since RAG-1 and RAG-2 are both necessary for V(D)J recombination during lymphocyte development (SCHATZ et al. 1989; OETTINGER et al. 1990). V(D)J deletion circles and the expression of RAGs have been reported in mouse intestinal epithelium (GUY-GRAND et al. 1991, 1992; MAKINO et al. 1993), thus suggesting that TCR rearrangement may take place in the intestinal epithelial environment. In addition, TCR deletion circles and RAG expression have also been reported in the spleen (MAKINO et al. 1993).

Whereas TCR V-DJβ deletion circles could not be detected in chick embryo spleens (PICKEL et al. 1993), we could detect Vβ1 deletion circles using DNA purified from the lymphocytes of the adult chicken spleen and blood. This could either indicate that V-DJβ rearrangements occur outside of the thymus in older birds or that the V-DJβ deletion circles persist in recent thymic emigrants. Our preliminary data suggest the latter possibility.

While chicken thymocytes express both RAG-1 and RAG-2 mRNA, bursal lymphocytes contain RAG-2 transcripts only (CARLSON et al. 1991). Since V(D)J recombination of chicken Ig genes occurs primarily in prebursal cells (WEILL et al. 1986; McCORMACK et al. 1989b; PICKEL et al. 1993), the function of RAG-2 in the bursal cells is unknown. Intracellular staining of permeabilized cells and western blot analysis with an anti-RAG-2 antibody indicate the expression of RAG-2 proteins in thymocytes but not in bursal lymphocytes (FERGUSON et al. 1994), suggesting that the presence of RAG-2 transcripts may not coincide with the expression of the protein. The recent cloning of all of the different types of TCR genes in the chicken and the relative simplicity of the TCR α, β, γ and δ loci should allow a critical re-examination of the issue of possible TCR gene rearrangement in extrathymic tissues.

6 Functional Interaction of αβ and γδ T Cells

The biological role of γδ T cells is still a mystery, although recent studies in mice indicate that they may play an important role in protection against certain microorganisms (HAAS et al. 1993). It has been tacitly assumed that they act like αβ T cells to function in cell-mediated responses. This may be true in part, but recent evidence suggests that γδ T cells may also function as regulatory cells for αβ T cells (KASAHARA et al. 1993; RAULET1994). Studies in the chicken suggest that γδ T cells differ remarkably from the αβ T cells in their proliferative characteristics. Normal polyclonal γδ T cells cannot grow well in response to plant mitogens or specific TCR ligation, except in the presence of αβ T cells or their growth factors (ARSTILA et al. 1993; KASAHARA et al. 1993). Interestingly, only the $CD8^+$ subpopulation of γδ T cells divide in response to TCR ligation and growth factors produced by $CD4^+$ T cells. The $CD8^+$γδ T cells are relatively large and they express MHC class II, suggesting that they have been activated in vivo (EWERT et al. 1984). Activated human T cells can capture, process and present tetanus toxoid antigen to class II-restricted T cell clones (BARNABA et al. 1994). This capacity correlates with the rate of class II synthesis, B7 expression and up-regulation of adhesion molecules. On the basis of this constellation of findings we suggest that a two-way interaction between γδ and αβ T cells results in mutual regulatory roles of the two subpopulations in the immune response (KASAHARA et al. 1993; COOPER and CHEN 1993).

We have selected the GvH reaction as a model in which to examine the interaction between αβ and γδ T cells. GvH reaction, one of the most serious complications of bone marrow transplantation, is initiated when donor-derived T cells recognize an MHC-related antigenic disparity between the donor and host (STORB and THOMAS 1983; KORNGOLD and SPRENT 1991). Experiments in mice indicate that αβ T cells recognize allogeneic MHC proteins or their captive peptides to induce GvH disease (MATIS et al. 1987). Relatively little is known about the role of γδ T cells in allograft reactions, in part because of the difficulty in obtaining sufficient numbers of polyclonal γδ T cells for experimental analysis in most conventional animal models, although the numbers of γδ T cells have been found to increase in the circulation of patients receiving bone marrow transplants (VILMER et al. 1988; VAN DER HARST et al. 1991). The chick embryo provides an excellent model for the analysis of interactions between donor and host cells during the acute GvH reaction (SIMONSEN 1985, 1990). Unlike model systems that employ irradiated recipients, the chick embryo has an intact, albeit immature, lymphohematopoietic system, and host cells are engaged in the GvH response. The GvH potential of T cell subpopulations can be analyzed in detail using the Simonsen assay (SIMONSEN 1985). Purified donor T cell subpopulations are injected into the chorioallantoic vessels of allogeneic embryos, which are then examined over the succeeding days. Our results in these experiments indicate that both $V\beta1^+$ and $V\beta2^+\alpha\beta$ T cells are fully capable of inducing GvH lesions, but only the $CD4^+$ cells can induce the reaction (TSUJI et al. 1995). Conversely, the γδ T cells are incapable of initiating an acute GvH reactions.

The relative GvH potency of the $V\beta1^+$ and $V\beta2^+$ T cells varies with the donor/host strain combination, suggesting that $V\beta1^+$ and $V\beta2^+$ populations from the same donor possess a different TCR repertoire for a given set of alloantigens. Studies in mammals indicate that the TCR recognition sites for antigenic peptides within the MHC class II molecule are also involved in alloantigen recognition (SHERMAN and CHATTOPADHYAY 1993). Differences in repertoires of the CDR3 regions (that are responsible for peptide contact) are thus expected to account for the differential capacity of the Vβ1 and Vβ2 subsets of avian T cells to induce GvH reactions in the same inbred recipients (CHIEN and DAVIS 1993). An extensive analysis of the repertoire of αβ T cells in GvH foci is currently underway.

The early GvH lesions in recipient spleens consist of focal lymphocyte infiltrates and surrounding edema, and these evolve to become acute necrotic foci by 5–7 days (TSUJI et al. 1995). The GvH attack is initiated by focal accumulations of donor αβ T cells in the host spleen beginning on the second day after the intravenous infusion of $CD4^+$ αβ T cells, and an abundance of recipient γδ T cells are quickly drawn into the lesions initiated by the donor αβ T cells. Most cells in the neighborhood of these lesions are induced to express host MHC class II at relatively high levels. By the fifth day the αβ T cells surround the necrotic lesions to outnumber the γδ cells, and most of the splenic cells express MHC class II molecules.

This GvH picture, in which host γδ T cells accumulate around the alloreactive donor αβ T cells, apparently results from both recruitment of γδ T cells from other sites and local proliferation. When the splenic γδ T cells from embryos injected with $CD4^+\alpha\beta$ T cells were assayed for DNA content, most were in G0/G1 phases of the cell cycle on day 3, but by the fifth day of the GvH reaction the percentage of γδ T cells in S plus G2/M phases was higher than in control embryos. The activation status of host γδ and donor αβ cells in the GvH lesions was further assessed by determining MHC class II expression. Although donor αβ T cells did not express class II molecules before injection into allogeneic embryos, MHC class II expression was detected on both donor αβ T cells and recipient γδ T cells in the GvH lesions. Serial analysis suggests that activation of host $\gamma\delta$ T cells occurs as a consequence of their interaction with activated $CD4^+\alpha\beta$ T cells of donor origin. It is also possible that the activated $\gamma\delta$ T cells in turn influence the $\alpha\beta$ T cells, a possibility that could be tested by analyzing the GvH reaction in hosts depleted of $\gamma\delta$ T cells.

7 Conclusion

Analysis of the chicken TCR genes and their protein products reveals divergent evolution at the level of nucleotide sequence, but the conservation of important structural features of the TCR polypeptides in avian and mammalian species. The genomic organization and the mechanisms for repertoire diversification of the TCR are very similar in birds and mammals. However, the chicken α and β TCR loci are much less complex than the mammalian $\alpha\beta$ TCR loci. There are only two $V\alpha$ and $V\beta$ families in the chicken, compared to the 20–30 families of $V\alpha$ and $V\beta$ genes that exist in both mice and human (MAK 1993). To a lesser extent the avian γ and δ loci are also less complex. The chicken thus provides a good model for the analysis of rearrangement and expression of all four sets of TCR genes in parallel. This information should provide insight into the general rules concerning the coordinated pattern of TCR gene rearrangements and expression in the four loci and their role in divergence of the $\alpha\beta$ and $\gamma\delta$ T cell lineages.

The possibility that T cells, particularly the intestinal $\gamma\delta$ T cells expressing $CD8\alpha\alpha$, can be generated in extrathymic as well as intrathymic sites has important basic and clinical implications. While this hypothesis is based on considerable experimental evidence in mice, the results of studies in the chicken cast doubt on the importance of an extrathymic pathway of T cell development. A detailed ontogenetic analysis of the TCR V(D)J rearrangement process in the chicken may provide an unambiguous answer to the question of extrathymic generation in the chicken, and may stimulate interest in re-examining these issues in mammalian species.

The dependence of $\gamma\delta$ T cells on $\alpha\beta$ T cells for proliferative responses suggests the mutual regulatory roles of these two subpopulations in immune responses. The chick embryo model, employed in the Simonsen assay for the GvH reaction, provides an intact but immunologically immature recipient model for analysis of the roles of T cell subsets in allorecognition. The high frequency of $\gamma\delta$ T cells in the chicken and the ease of separating donor $\alpha\beta$ T cells on the basis of their $V\beta1$ or $V\beta2$ gene usage make the Simonsen's assay an excellent system to study the interaction of $\alpha\beta$ and $\gamma\delta$ T cells in the immune response.

Acknowledgments. We thank Ms. Ann Brookshire for preparation of the manuscript. This work was supported in part by NIH grant AI 30879.

References

Arstila TP, Toivanen P, Lassila O (1993) Helper activity of CD4⁺ αβ T cells is required for the avian γδ T cell response. Eur J Immunol 23: 2034–2037

Atkinson TP (1995) Absence of γδ T lymphocytes in a patient with complete DiGeorge syndrome. FASEB J 9: A780

Barnaba V, Watts C, de Boer M, Lane P, Lanzavecchia A (1994) Professional presentation of antigen by activated human T cells. Eur J Immunol 24: 71–75

Bernot A, Auffray C (1991) Primary structure and ontogeny of an avian CD3 transcript. Proc Natl Acad Sci USA 88: 2550–2554

Bonneville M, Ishida I, Mombaerts P, Katsuki M, Verbeek S, Berns A, Tonegawa S (1989) Blockade of αβ T cell development by TCR γδ transgenes. Nature 342: 931–932

Bougueleret L, Claverie J-M (1987) Variability analysis of the human and mouse T-cell receptor β chains. Immunogenetics 26: 304–308

Bucy RP, Chen CH, Cihak J, Lösch U, Cooper MD (1988) Avian T cells expressing γδ receptors localize in the splenic sinusoids and the intestinal epithelium. J Immunol 141: 2200–2205

Bucy RP, Coltey M, Chen CH, Char D, Le Douarin NM, Cooper MD (1989) Cytoplasmic CD3⁺ surface CD8⁺ lymphocytes develop as a thymus-independent lineage in chick-quail chimeras. Eur J Immunol 19: 1449–1445

Bucy RP, Chen CH, Cooper MD (1990) Ontogeny of T cell receptors in the chicken thymus. J Immunol 144: 1161–1168

Burnet FM (1957) A modification of Jerne's theory of antibody production using the concept of clonal selection. Aust J Sci 20: 67–69

Carlson LM, Oettinger MA, Schatz DG, Masteller EL, Hurley EA, McCormack WT, Baltimore D, Thompson CB (1991) Selecive expression of RAG-2 in chicken B cells undergoing immunoglobulin gene conversion. Cell 64: 201–208

Chan MM, Chen CH, Ager LL, Cooper MD (1988) Identification of the avian homologues of mammalian CD4 and CD8 antigens. J Immunol 140: 2133–2138

Char D, Sanchez P, Chen CH, Bucy RP, Cooper MD (1990) A third sublineage of avian T cells can be identified with a T cell receptor-3-specific antibody. J Immunol 145: 3547–3555

Chen CH, Ager LL, Gartland GL, Cooper MD (1986) Identification of a T3/T cell receptor complex in chickens. J Exp Med 164: 375–380

Chen CH, Cihak J, Lösch U, Cooper MD (1988) Differential expression of two T cell receptors, TCR 1 and TCR2, on chicken lymphocytes. Eur J Immunol 18: 539–543

Chen CH, Sowder JT, Lahti JM, Cihak J, Lösch U, Cooper MD (1989) TCR3: A third T-cell receptor in the chicken. Proc Natl Acad Sci USA 86: 2351–2355

Chen CH, Bucy RP, Cooper MD (1990) T cell differentiation in birds. Sem Immunol 2: 79–86

Chen CH, Pickel JM, Lahti JM, Cooper MD (1991) Surface markers on avian immune cells. In: Sharma JM (ed) Avian cellular immunology. CRC Press, Boca Raton, FL, pp 1–22

Chen CH, Göbel TWF, Kubota T, Cooper MD (1994) T cell development in the chicken. Poultry Sci 73: 1012–1018

Chien Y-H, Davis MM (1993) How αβ T-cell receptors 'see' peptide/MHC complexes. Immunol Today 14: 597–602

Cihak J, Hoffmann–Fezer G, Ziegler-Heitbrock HWL, Stein H, Kaspers B, Chen CH, Cooper MD, Lösch U (1991) T cells expressing the Vβ1 T-cell receptor are required for IgA production in the chicken. Proc Natl Acad Sci USA 88: 10951–10955

Cihak J, Lösch U, Hoffmann-Fezer G, Chen CH, Cooper MD, Ziegler-Heitbrock HWL (1993) In vivo depletion of chicken T-cell subsets. Scand J Immunol 38: 123–129

Coltey M, Bucy RP, Chen CH, Cihak J, Lösch U, Char D, Le Douarin NM, Cooper MD (1989) Analysis of the first two waves of thymus homing stem cells and their T cell progeny in chick quail chimeras. J Exp Med 170: 543–557

Cooper MD, Chen CH (1993) Avian T cell development and the γδ-αβ T cell connection. In: Proceedings of the Xth World Veterinary Poultry Association Congress, Sydney, pp 27–31

Cooper MD, Peterson DA, Good RA (1965) Delineation of the thymic and bursal lymphoid systems in the chicken. Nature 205: 143–146

Cooper MD, Chen CH, Bucy RP, Thompson CB (1991) Avian T cell ontogeny. Adv Immunol 50: 87–117

Davis MM, Bjorkman PJ (1988) T-cell antigen receptor genes and T-cell recognition. Nature 334: 395–401

de Vilartay J-P, Lewis D, Hockett R, Waldmann TA, Korsmeyer SJ, Cohen DJ (1987) Deletional rearrangement in the human T-Cell receptor α-chain locus. Proc Natl Acad Sci USA 84: 8608–8612
de Villartay J-P, Hockett RD, Coran D, Korsmeyer SJ, Cohen DI (1988) Deletion of the human T-cell receptor α-gene by a site-specific recombination. Nature 335: 170–174
Dunon D, Cooper MD, Imhof BA (1993a) Thymic origin and migration patterns of intestinal γ/δ T cells. J Exp Med 177: 257–263
Dunon D, Cooper MD, Imhof BA (1993b) Migration patterns of thymus-derived γδ T cells during chicken development. Eur J Immunol 23: 2545–2550
Dunon D, Schwager J, Dangy J-P, Cooper MD, Imhof BA (1994) T cell migration during development: homing is not related to TCR Vβ1 repertoire selection. EMBO J 13: 808–815
Dunon D, Schwager J, Dangy J-P, Imhof BA (1995) Ontogeny of TCR Vβ1 expression revealed novel invariant alternative transcripts. J Immunol 154: 1256–1264
Ewert DL, Munchus MS, Chen CH, Cooper MD (1984) Analysis of structural properties and cellular distribution of avian Ia antigen by using monoclonal antibodies to monomorphic determinants. J Immunol 132: 2524–2530
Ferguson SE, Accavitti MA, Wang DD, Chen CH, Thompson CB (1994) Regulation of RAG-2 protein expression in avian thymocytes. Mol Cell Biol 14: 7298–7305
Fujimoto S, Yamagishi H (1987) Isolation of an excision product of T-cell receptor α-chain gene rearrangements. Nature 327: 242–243
Genevée C, Chung V, Diu A, Hercend T, Triebel F (1994) TCR gene segments from at least one third of Vα subfamilies rearrange at the α locus. Mol Immunol 31: 109–115
Göbel TWF, Chen CH, Lahti JM, Kubota T, Kuo CL, Aebersold R, Hood L, Cooper MD (1994a) Identification of T-cell receptor TCRα-chain genes in the chicken. Proc Natl Acad Sci USA 91:1094–1098
Göbel TWF, Chen CH, Schrimpf J, Grossi CE, Bernot A, Bucy RP, Auffray C, Cooper MD (1994b) Characterization of avian NK cells and their intracellular CD3 protein complex. Eur J Immunol 24: 1685–1691
Guy-Grand D, Cerf-Bensussan N, Malissen B, Malassis-Seris B, Briottet C, Vassalli P (1991) Two gut intraepithelial $CD8^+$ lymphocyte populations with different T cell receptors: a role for the gut epithelium in T cell differentiation. J Exp Med 173: 471–481
Guy-Grand D, Vanden Broecke C, Briottet C, Malassis-Seris M, Selz F, Vassalli P (1992) Different expression of the recombination activity gene RAG-1 in various populations of thymocytes, peripheral T cells and gut thymus-independent intraepithelial lymphocytes suggests two pathways of T cell receptor rearrangement. Eur J Immunol 22: 505–510
Haas W, Pereira P, Tonegawa S (1993) Gamma/delta cells. Annu Rev Immunol 11: 637–685
Hein WR, Dudler L (1993) Divergent evolution of T cell repertoires: extensive diversity and developmentally regulated expression of the sheep γδ T cell receptor. EMBO J 12: 715–724
Hein WR, Dudler L, Morris B (1990) Differential peripheral expansion and in vivo antigen reactivity of α/β and γ/δ T cells emigrating from the early fetal lamb thymus. Eur J Immunol 20: 1805–1813
Ikehara S, Shimuzu J, Yasumizu R, Nakamura T, Inaba M, Inoue S, Oyaizu N, Sugiura K, Oo MM, Hamashima Y, Good RA (1987) Thymic rudiments are responsible for induction of functional T cells in nu/nu mice. Thymus 10: 193–205
Kasahara Y, Chen CH, Cooper MD (1993) Growth requirements for avian γδ T cells include exogenous cytokines, receptor ligation and in vivo priming. Eur J Immunol 23: 2230–2236
Kincade PW, Cooper MD (1973) Immunoglobulin A: site and sequence of expression in developing chicks. Science 179: 398–400
Kincade PW, Lawton AR, Bockman DE, Cooper MD (1970) Suppression of immunoglobulin G synthesis as a result of antibody-mediated suppression of immunoglobulin M synthesis in chickens. Proc Natl Acad Sci USA 67: 1918–1925
Koop BF, Hood L (1994) Striking sequence similarity over almost 100 kilobases of human and mouse T-cell receptor DNA. Nature Genet 7: 48–53
Korngold R, Sprent J (1991) Graft-versus-host disease in experimental allogeneic bone marrow transplantation. Proc Soc Exp Biol Med 197: 12–18
Kubota T, Chen CH, Hockett R, Göbel TWF, Cooper MD (1995) Conservation of the T cell receptor α/δ locus in chickens. FASEB J 9: 4736
Lahti JM, Chen CH, Tjoelker LW, Pickel JM, Schat KA, Calnek BW, Thompson CB, Cooper MD (1991) Two distinct αβ T-cell lineages can be distinguished by the differential usage of T-cell receptor Vβ gene segments. Proc Natl Acad Sci USA 88: 10956–10960
Le Douarin NM (1978) Ontogeny of hematopoietic organs studied in avian embryo interspecies chimeras. In: Clarkson B, Marks PA, Till JE (eds) Differentiation of normal and neoplastic hematopoietic cells. Cold Spring Harbor Laboratory, Cold Spring Harbor, New York, pp 5–31

Lefrançois L, Puddington L (1995) Extrathymic intestinal T-cell development: virtual reality? Immunol Today 16: 16–21
Lin T, Matsuzaki G, Kenai H, Nakamura T, Nomoto K (1994) Thymus influences the development of extrathymically derived intestinal intraepithelial lymphocytes. Eur J Immunol 23: 1968-1974
Mak TW (1993) The T-cell antigen receptor enters its second decade. Scand J Immunol 38: 209–211
Makino Y, Yamagata M, Sasho T, Adachi Y, Kanno R, Koseki H, Kanno M, Taniguchi M (1993) Extrathymic development of Vα 14-positive T cells. J Exp Med 177: 1399–1408
Matis LA, Sorger SB, McElligott DL, Fink PJ, Hedrick SM (1987) The molecular basis of alloreactivity in antigen-specific, major histocompatibility complex-restricted T cell clones. Cell 51: 59–69
McCormack WT, Tjoelker LW, Carlson LM, Petryniak B, Barth CF, Humphries EH, Thompson CB (1989a) Chicken IgL gene rearrangement involves deletion of a circular episome and addition of single nonrandom nucleotides to both coding segments. Cell 56: 785-791
McCormack WT, Tjoelker LW, Barath CV, Carlson LM, Petryniak B, Humphries EH, Thompson CB (1989b) Selection for B cells with productive IgL gene arrangements occurs in the bursa of Fabricius during the chicken embryonic development. Genes Dev 3: 838–847
McCormack WT, Tjoelker LW, Stella G. Postema CB, Thompson CB (1991) Chicken T-cell receptor β-chain diversity: An evolutionarily conserved Dβ-encoded glycine turn within the hypervariable CDR3 domain. Proc Natl Acad Sci USA 88: 7699–7703
Moore MAS, Owen JJT (1965) Chromosome marker studies on the development of the haemopoieitic system in the chick embryo. Nature 208: 959-990
Oettinger MA, Schatz DG, Gorka C, Baltimore D (1990) RAG-1 and RAG-2, adjacent genes that synergistically activate V(D)J recombination. Science 22: 1517–1523
Okazaki K, Davis DD, Sakano H (1987) T cell receptor β gene sequences in the circular DNA of thymocyte nuclei: direct evidence for intramolecular DNA deletion in V-D-J joining. Cell 49:477–485
Pickel JM, McCormack WT, Chen CH, Cooper MD, Thompson CB (1993) Differential regulation of V(D)J recombination during development of avian B and T cells. Int Immunol 5: 919–927
Poussier P, Julius M (1994) Thymus independent T cell development and selection in the intestinal epithelium. Anna Rev Immunol 12: 521–553
Rast JP, Litman GW (1994) T cell receptor gene homologs are present in the most primitive jawed vertebrates. Proc Natl Acad Sci USA 91: 9248–9252
Raulet Dh (1989) The structure, function, and molecular genetics of the γδ T cell receptor. Annu Rev Immunol 7: 175–207
Raulet DH (1994) How γδ T cells make a living. Curr Biol 4: 246–248
Reynaud C-A, Anquez V, Dahan A, WAeill J-C (1987) A hyperconversion mechanism generates the chicken light chain preimmune repertoire. Cell 48: 379–388
Rocha B, Vassalli P, Guy-Grand D (1992) The extrathymic T-cell development pathway. Immunol Today 13: 449–454
Schatz DG, Oettinger MA, Baltimore D (1989) The V(D)J recombination activating gene, RAG-1. Cell 59: 1035–1048
Schiffer M, Wu TT, Kabat EA (1986) Subgroups of variable region genes of β chains of T-cell receptors for antigen. Proc Natl Acad Sci USA 83: 4461–4463
Sherman LA, Chattopadhyay S (1993) The molecular basis of allorecognition. Annu Rev Immunol 11: 385–402
Simonsen M (1985) Graft-versus-host reactions: the history that never was, and the way things happened to happen. Immunol Rev 88: 5–24
Simonsen M (1990) Alloreactive T cells. Scand J Immunol 32: 565–575
Six A, Rast JP, Li Y, Cooper MD, Chen CH, Litman GW (1995) Identification of the chicken T-cell receptor γδ genes. FASEB J 9: 4737
Sowder JT, Chen CH, Ager LL, Chan MM, Cooper MD (1988) A large subpopulation of avian T cells express a homologue of the mammalian T γ/δ receptor. J Exp Med 167: 315–322
Storb R, Thomas ED (1983) Allogeneic bone-marrow transplantation. Immunol Rev 71: 77–102
Takeushi N, Ishiguro N, Shinigawa M (1992) Molecular cloning and sequence analysis of bovine T-cell receptor γ and δ chain genes. Immunogenetics 35: 89–96
Thome M, Saalmuller A, Pfaff E (1993) Molecular cloning of porcine T cell receptor α, β, γ and δ chains using polymerase chain reaction fragments of the constant regions. Eur J Immunol 23: 1005–1010
Thompson CB, Neiman PE (1987) Somatic diversification of the chicken immunoglobulin light chain gene is limited to the rearranged variable gene segment. Cell 48: 369–378
Tjoelker LW, Carlson LM, Lee K, Lahti J, McCormack WT, Leiden JM, Chen CH, Cooper MD, Thompson CB (1990) Evolutionary conservation of antigen recognition: the chicken T-cell receptor β chain. Proc Natl Acad Sci USA 87: 7856–7860

Tregaskes CA, Kong F, Paramithiotis E, Chen CH, Ratcliffe MJH, Davison TF, Young JR (1995) Identification and analysis of the expression of CD8αβ and CD8αα isoforms in chickens reveals a major TCRγδ CD8αβ subset of intestinal intraepithelial lymphocytes. J Immunol 154: 4485–4494
Tsuji S, Illges H, Kubota T, Char D, Bucy P, Simonsen M, Chen CH, Cooper MD (1995) Donor αβ T cell clones recruit host γδ T cells in acute GvH reactions. 9th international congress of immunology 1995, p 251, abstr 1486
van der Harst D, Brand A, van Luxemburg SAP, Kooij-Winkelaar YMC, Zwaan FE, Koning F (1991) Selective outgrowth of $CD45RO^+$ $V\gamma 9^+/\delta 2^+$ V-cell receptor γ/δ T cells early after bone marrow transplantation. Blood 78: 1875–1881
Vilmer E, Triebel F, David V, Rabian C, Schumpp M, Leca G, Degos L, Hercend T, Sigaux F, Bensussan A (1988) Prominent expansion of circulating lymphocytes bearing γδ T-cell receptor with preferential expression of variable γ genes after allogeneic bone marrow transplantation. Blood 72: 841–849
Weill JC, Reynaud C-A, Lassila O, Pink JRL (1986) Rearrangement of chicken immunoglobulin genes is not an ongoing process in the embryonic bursa of Fabricius. Proc Natl Acad USA 83: 3336–3340

Chicken γδ T Cells

F.J. SANCHEZ-GARCIA and W.T. MCCORMACK

1 Introduction

Functional and genetic studies of the γδ T cell receptor (TCR) have progressed in reverse order compared to the study of αβ T cells (reviewed by RAULET 1989). A wealth of information about the αβ T cell population and its function was available before the genes for the αβ TCR were cloned. Studies of the αβ TCR repertoire and gene expression then helped elucidate molecular mechanisms for generating diversity of the αβ T cell repertoire and the nature of antigen recognition by αβ T cells. However, the identification of a third TCR locus (γ), and the subsequent discovery that a fourth TCR chain (δ) was encoded by genes embedded within the TCR-α locus, led to the description of the γδ TCR expressed by a T cell subpopulation with no known function. Over the last 10 years, laboratories around the world have been trying to assign functional roles for T cells expressing the γδ TCR. Studies of mouse and human γδ T cells have revealed many interesting differences in the biology of γδ vs. αβ T cells, such as tissue distribution, repertoire restrictions, requirement for a thymus for their development, and lack of classical MHC restriction (for reviews see ALLISON 1993; HAAS et al. 1993; KRONENBERG 1994; LEFRANCOIS and PUDDINGTON 1995).

Dept. of Pathology and Laboratory Medicine, Center for Mammalian Genetics, Box 100275, Health Science Center, University of Florida College of Medicine, Gainesville, FL 32610-0275 USA

In mice and humans, γδ T cells comprise only approximately 5% of the peripheral blood and splenic T cell population, but are more prominent at several epithelial tissue surfaces (HAAS et al. 1993). The immune systems of several other animal species, such as cattle, sheep and chickens, are now under scrutiny, in part because the peripheral T cell pool in these species includes a larger proportion of γδ T cells (30%–50%) than is found in mice and humans (HEIN and MACKAY 1991; COOPER et al. 1991). Furthermore, the epithelial localization of major γδ T cell populations in all species examined, especially in the intestinal epithelium, suggests an important role for γδ T cells at mucosal surfaces. Studies of the evolutionarily conserved features of γδ T cell function and γδ TCR genetics in animal models such as ruminants and the chicken will provide insight into the function of these cells in immunosurveillance and infectious diseases. In this chapter, we review recent studies of γδ T cells using the chicken model system.

Chicken T cells have been well characterized by the use of polyclonal and monoclonal antibodies raised against a variety of functionally important T cell molecules (reviewed by COOPER et al. 1991). Most importantly, monoclonal antibodies (mAbs) define the chicken homologues for the CD3, CD4 and CD8 molecules, and three lineages of T cells, named for the order in which they appear in the embryonic thymus: TCR1 (γδ T cells), TCR2 (αβ T cells expressing Vβ1 genes) and TCR3 (αβ T cells expressing Vβ2 genes) (CHEN et al. 1988, 1989; LAHTI et al. 1991).

2 Ontogeny of Chicken γδ T Cells

Chicken T cell development occurs in three waves during embryonic development, following three brief periods during which the thymus is receptive to precursor stem cells (COOPER et al. 1991). Each of the three waves of T cell development produces all three chicken T cell subpopulations, TCR1, TCR2 and TCR3. Chicken γδ T cells ($TCR1^+$) appear before αβ T cells during each wave. For example, during the first thymocyte wave, $TCR1^+$, $TCR2^+$ and $TCR3^+$ cells reach peak frequency in the thymus at embryonic days E15, E18, and E21 (hatch), respectively.

Chicken γδ T cells ($TCR1^+$) first appear in the thymus at day 11 of embryonic development (E11), and peak in relative cell number (30% of thymocytes) at day E14–15. After hatching, ≤10% of thymocytes express TCR1. γδ T cells first appear in the embryonic intestine at day E16–17 (DUNON et al. 1993a). In contrast to αβ T cells, which appear to undergo the same positive and negative selection events in the thymus as mammalian αβ T cells, chicken γδ T cells do not appear to undergo intrathymic selection. $TCR1^+$ cells migrate quickly from the thymus without undergoing clonal expansion (COOPER et al. 1991), immediately express high levels of surface TCR molecules ($>10^4$/cells vs 10^3/cells for αβ T cells) (GEORGE and COOPER 1990), are resistant to receptor modulation by cross-linking with anti-TCR antibody (GEORGE and COOPER 1990), and are not sensitive to developmental arrest by cyclosporin A treatment (BUCY et al. 1990) . Taken together, these data suggest that chicken γδ T cells are not subject to the same selection events in the thymus as are αβ T cells.

It has been suggested that gut-associated lymphoid tissue (GALT) may have evolved prior to the thymus as a site of T cell development (FICHTELIUS et al. 1969; DUPASQUIER 1989). An extrathymic pathway of T cell development is thought to have preceded or diverged from the thymic pathway and to remain important during early ontogeny in some mammals, especially in ruminant species (HEIN and MACKAY 1991). Extrathymic development of γδ T cells has been extensively studied in the mouse (reviewed by ROCHA et al. 1992), although recent evidence suggests that the thymus still exerts some influence on "extrathymic" T cell development (LEFRANCOIS and PUDDINGTON 1995).

Early studies of chicken T cell development suggested that the intestinal epithelium might also be a site of γδ T cell development in the chicken. These studies took advantage of the antibody suppression model of CIHAK et al. (1993), whereby γδ T cells are specifically suppressed from developing by injection of TCR1 mAb in ovo, followed by posthatching thymectomy. In these experiments, sufficient numbers of $TCR2^+$ and $TCR3^+$ cells emigrate from the embryonic thymus prior to thymectomy that the peripheral αβ T cell population of these chicks appears to be normal. When chicken γδ T cell development is suppressed in this manner, a significant population of $TCR1^+$ cells remains in the intestinal epithelium, even though nearly all (95%) $TCR1^+$ cells are eliminated from peripheral blood (CHEN et al. 1989; COOPER et al. 1991; CIHAK et al. 1991, 1993), suggesting a possible extrathymic origin of some intestinal $TCR1^+$ cells. Alternatively, the $TCR1^+$ cells remaining in these experiments might have escaped suppression by the mAb treatment.

Early studies of the origin of peripheral γδ T cells suggested that they are thymus-derived. Thymectomy at the time of hatching results in dramatic reduction in the circulating pool of $TCR1^+$ cells, suggesting a requirement for sustained seeding of peripheral blood lymphocyte (PBL) γδ T cells from the thymus after hatching (CHEN et al. 1989). The recent studies of DUNON et al. (1993a) are consistent with a thymic origin for the embryonic intestinal γδ T cells as well. Using congenic chicken strains differing in the expression of the ov antigen, which is expressed only on T lineage cells and their precursors, the origin of γδ T cells colonizing the embryonic intestine was investigated. By injecting ov^- recipient embryos with lymphoid cell populations from ov^+ embryos, it was shown that γδ T cells from the first two waves of T cell development migrate into the intestinal epithelium, where they may persist for long periods of time (>75 days), suggesting that, at least for the early stages of development, the intestinal γδ T cells are of thymic rather than extrathymic origin. Similar experiments have demonstrated that γδ T cells arising from the three waves of T cell development in the thymus seed the spleen as well as the intestine (DUNON et al. 1993b).

The number of γδ T cells in the chicken GALT has been shown to vary with age, region of the gut, and genetic background (LILLEHOJ and CHUNG 1992). The ratio of γδ T cell to αβ T cells (TCR1+/ TCR2+) increases after hatching and reaches 3.4 and 4.28 in jejunal intraepithelial lymphocytes (IEL) of the SC and TK strains, respectively. The relative expansion of γδ T cells is greater for the IEL compartment (0.96 at 8 weeks of age to 4.29 at 12 weeks of age) than in the lamina propria lymphocyte (LPL) population (from 1.23 at 8 weeks of age to 2.15 at 12 weeks) of SC chickens.

An interesting interaction between γδ T cells and the endocrine system has recently been reported by ARSTILA and LASSILA (1993), who demonstrated that the peripheral blood and splenic γδ T cell populations are expanded in response to androgens at the time of sexual maturation in roosters, but not in hens. There is no corresponding expansion of the intestinal γδ T cell population in roosters. The expansion of peripheral blood and splenic γδ T cells is revealed by both an increased frequency and an increased absolute number of γδ T cells in males beginning at the age of 6 months. The dependence of this expansion on androgens was demonstrated by treating hens with testosterone, which resulted in the appearance of male secondary sex characteristics and an increase in the frequency of γδ T cells. The mechanism for this androgen-driven expansion of γδ T cells is unknown, as testosterone treatment has no apparent effect on the thymus, and avian T cells do not appear to express androgen receptors.

3 Tissue Distribution of Chicken γδ T Cells

In the periphery, the proportion of TCR1+ cells in PBLs reaches 20%–50% in adult chickens (COOPER et al. 1991). As in humans, splenic γδ T cells in chickens are localized to the sinusoids of the red pulp and remain dispersed rather than forming aggregates or nodules (BUCY et al. 1988, 1989). In contrast, chicken splenic αβ T cells are localized in the periarteriolar sheaths. In the mouse, the mammalian species that has been the most thoroughly studied for its γδ T cell repertoire, the γδ T cells found in the blood, spleen and lymph nodes express Vγ2 and Vγ1 genes with various Vδ genes and are characterized by a high amount of receptor diversity (i.e., junctional diversity) (HAAS et al. 1993).

Chicken γδ T cells are very predominant in intraepithelial sites in the villi of chicken intestinal epithelium (BUCY et al. 1988), but are minor populations or absent in Peyer's patches and cecal tonsils. Lymphocytes have been found to be present in the intestinal epithelium of all vertebrate species examined (FICHTELIUS et al 1969). The presence of γδ T cells within the IEL population of chickens (BUCY et al. 1988), mice (GOODMAN and LEFRANCOIS 1988; BONNEVILLE et al. 1988), humans (BUCY et al. 1989), and ruminants (HEIN and MACKAY 1991) suggests that the function of γδ T cells in this location may be phylogenetically conserved. γδ T cells are especially predominant in the intestinal IEL population of chickens and sheep, but are present in lower percentages in the IEL of rodents and humans (HEIN and MACKAY 1991; GUY-GRAND and VASSALLI 1993).

The localization of γδ T cells to mucosal surfaces suggests that a primary function of γδ T cells is to provide immunosurveillance at these sites. In the mouse, waves of γδ T cell development in the thymus are characterized by specific Vγ and Vδ gene rearrangements and correlate with the appearance of specific γδ T cell subsets in various epithelial locations, such as the skin (expressing Vγ3 and Vδ1), mucosal surfaces of the uterus, vagina and tongue (expressing Vγ4 and Vδ1), and the intestine (expressing Vγ5 with various Vδ) (ALLISON 1993; HAAS et al. 1993). In addition to preferential V gene utilization, the mouse γδ TCR repertoire is further restricted in junctional diversity at some of these epithelial sites, to the extreme of being monomorphic for the skin, tongue and reproductive tract γδ T cells. The mouse γδ T cell subsets appear following a programmed pattern of V gene rearrangements during development in the embryonic and adult thymus and preferential joining sites of V gene segments to generate the monomorphic γδ TCR. Coordinated differentiation of homing properties results in migration of the γδ T cell subsets into different tissue sites.

Mice and cattle, but not the other mammalian species examined, have a resident population of γδ T cells in the skin, known as dendritic epidermal cells (DECs) (HAAS et al. 1993). The mouse DEC population is characterized by expression of a monomorphic γδ T cell receptor repertoire and recognition of a keratinocyte-specific antigen (HAVRAN et al. 1991). In chickens, however, no TCR^+ cells are observed histologically in different kinds of normal skin, including feathered regions, comb, or waddle (BUCY et al. 1988). Interestingly, TCR^+ cells are capable of infiltrating chicken skin during an inflammatory autoimmune response, such as in the inherited scleroderma of the UCD-200 chicken line (GERSHWIN et al. 1981; GRUSCHWITZ et al. 1991) and delayed amelanosis of the Smyth line (G. ERF, Univ. Arkansas, personal communication), suggesting that γδ T cells may be involved in immune functions in the skin of chickens. Similarly, in certain human cutaneous diseases, such as leprosy, leishmaniasis and cutaneous lupus erythematosus, γδ T cells invade the skin and undergo localized expansion, presumably due to activation by pathogenic antigens and/or superantigens or by self stress antigens (ALAIBAC 1992). The chicken may therefore provide a good animal model for infiltration of the skin by γδ T cells in the absence of a resident γδ T cell population (i.e., DECs)

4 Cell Surface Markers on Chicken γδ T Cells

The surface markers on chicken immune cells have been reviewed by VAINIO and LASSILA (1989) and by CHEN et al. (1991). We focus here on the surface markers associated with γδ T cells.

The chicken CD3 complex, recognized by mAb CT3, is expressed on the cell surface in conjunction with the polymorphic TCR molecules (CHEN et al. 1986). The CT3 mAb immunoprecipitates a complex of at least three proteins from T cells, with Mr of 20000, 19000 and 17000. When the T cell surface molecules are solubilized with mild detergent, the TCR1, TCR2 and TCR3 heterodimers can be coprecipitated with the CT3 antibody.

The structure of one of the chicken CD3 chains has been determined (Bernot and Auffray 1991) by isolating a cDNA clone encoding a protein 175 amino acids long, for which mRNA expression is first detectable in thymus of 10 day embryos 2 days before the expression of the TCR. The chicken CD3 cDNA sequence displays 36%–39% and 39%–40% homology to the mammalian CD3 γ and δ polypeptides, respectively. Because the subunit encoded by the chicken CD3 clone could not be unambiguously identified based on sequence homology, Berno⁻ and coworkers have suggested that this polypeptide either evolved from an ancestral form of the CD3γ and δ chains that might provide the function of both the CD3 γ and δ subunits in the chicken or that gene duplications for the CD3 subunits have occurred independently in the avian vs. mammalian lineages.

Most chicken γδ T cells developing in the thymus are $CD4^-CD8^-$, although other subsets (i.e., $CD4^+CD8^+$, $CD4^-CD8^+$, and $CD4^+CD8^-$) have been reported to be detectable both in vivo and in thymus organ cultures (Davidson and Boyd 1992; Davidson et al. 1992). Chicken γδ T cells in the peripheral blood are $CD4^-CD8^-$, but most splenic γδ T cells (75%) are $CD8^+$ (Chan et al. 1988; Bucy et al. 1988; Chen et al. 1988). In contrast, chicken αβ T cells in the thymus are $CD4^+CD8^+$, and αβ T cells in the blood and spleen are either $CD4^+$ or $CD8^+$. Most γδ T cells (66%±2.9%) in the intestinal epithelium, like those in the spleen, coexpress the CD8 antigen, suggesting functional differences for γδ T cells depending on their tissue localization.

In addition to the evolutionary conservation of the CD4 and CD8 coreceptors (and other accessory molecules described below) between avian and mammalian species, the intracellular signaling pathways associated with these surface molecules and some cytokine receptors appear to be conserved as well. For example, Chow et al. (1992) have described the chicken homologue for the phosphotyrosine kinase gene *lck* involved in T cell activation. Similarly, the chicken homologue of the interleukin-2 (IL-2) receptor light chain (CD25) is identified by the mAb INN Ch16 and is expressed on all thymocyte subsets, including γδ T cells (Fedecka-Bruner et al. 1991), suggesting that the IL-2/IL-2 receptor differentiation pathway is evolutionarily conserved.

The chicken homologue of the T cell costimulatory molecule CD28 is recognized by the mAb 2–4, which was originally reported to recognize chicken CD2 (Vainio et al. 1991), and by the mAb AV7 (Young et al. 1994). The molecule recognized by these two mAbs has been cloned and expressed in COS cells, and the nucelotide sequence data are consistent with its identification as CD28 (Young et al. 1994). Chicken CD28 is expressed on virtually all thymocytes, including all thymic γδ T cells, but is absent from peripheral blood and spleen γδ T cells (Vainio et al. 1991). The costimulatory function of CD28 in chicken peripheral αβ T cells has been demonstrated by Arstila et al. (1994). However, the functional role of CD28 in chicken γδ T cells may be limited to a stage of maturation in the thymus, suggesting an important difference between mammalian and avian peripheral γδ T cells. Although the mouse has a population of CD28⁻ γδ T cells in the intestinal IEL population that does not appear to respond to TCR-CD3 triggering (OHTEKI and MACDONALD 1993), most mouse peripheral γδ T cells do express and require CD28-mediated constimulation for activation (SPERLING et al. 1993).

The chicken homologue for CD5 is identified by the mAb OC5 (KNABEL et al. 1993). CD5 is expressed on nearly all (99%) chicken thymocytes and on 55% of blood and spleen lymphocytes. All circulating chicken T cells, including γδ and αβ T cells, express CD5. Both soluble and Sepahrose-bound OC5 mAbs are mitogenic for unseparated blood mononuclear cells, and this effect was even greater in the presence of concanavalin (ConA), suggesting a role for CD5 in regulating chicken T cell proliferation.

KNABEL et al. (1993) have also described a mouse mAb (OC2) recognizing the chicken homologue of the mammalian CD2 antigen, a molecule that has a role in cell adhesion and T cell activation. Chicken CD2 was shown to be expressed on most peripheral blood αβ T cells, but not on peripheral blood γδ T cells. Whereas CD2 is expressed on subpopulations of γδ T cells in mammalian species, CD2 expression is not an absolute requirement for T cell function.

BILSLAND and SPRINGER (1994) recently cloned the chicken CD18 gene, which encodes the β2 subunit common to the leukocyte integrin family. The deduced 748 amino acid sequence reveals a transmembrane protein with 65% and 64% identity with its human and mouse homologues, respectively. When the chicken β2 cDNA is expressed in COS cells as a hybrid molecule with the human α subunit of LFA-1, the cells are able to bind to purified human ICAM-1 and ICAM-3 coated plates, and the hybrid LFA-1 molecules are as active as wild type LFA-1 in this binding assay. Differential expression of chicken β2 (CD18) on αβ and γδ T cells has not been reported.

Chicken intestinal T cells express an integrin-like antigen identified by the A19 mAb (HAURY et al. 1993). The mAb A19, raised against chicken intestinal IELs, recognizes a multimolecular complex of proteins on chicken αβ and γδ T lymphocytes, which displays a migration pattern similar to other members of the integrin family, such as human MLA (HML-1), which is found in human intestinal IELs and activated PBLs (CERF-BENSUSSAN et al. 1987). Similarly, A19 is found in relatively high intensity on most (53%) chicken intestinal IELs (αβ and γδ), but only at low levels on relatively few thymocytes (<5%), splenocytes (<5%), blood lymphocytes (<2%) and bone marrow mononuclear cells (<1%). Of the intestinal IEL γδ T cells, 83%±5% express the A19 antigen.

In functional studies Haury and coworkers (HAURY et al. 1993) demonstrated that ConA or immobilized anti-CD3 mAb, in the presence of transforming growth factor-β (TGF-β1), induces the expression of A19 antigen on both peripheral blood and splenic lymphocytes, although up-regulation of A19 following stimulation is more distinctive on splenic cells than on PBLs. Injection of anti-A19 mAb into newly hatched chicks does not induce significant modification of A19 expression by intestinal IELs, and the numbers of A19 positive cells in intestinal sections are the same in A19-treated and -untreated animals. Taken together, these results led Haury and coworkers to suggest that the A19 antigen is a member of the integrin family involved in the retention of T cells in the chicken intestinal epithelium.

5 Biological Function and Interaction with Other Cell Types

Cellular immunity has been implicated in resistance to several of the most common nonneoplastic avian viruses in the chicken, such as infectious bronchitis virus, fowlpox virus, Newcastle disease virus, and infectious laryngotracheitis virus (SHARMA et al. 1990). The experimental evidence has been primarily a lack of correlation between antibody responses in vaccinated chickens and disease resistance to subsequent challenge infections, and/or continued resistance in bursectomized vaccinated birds. T cell-mediated immune responses have been demonstrated for several neoplastic viral diseases in chickens, including Marek's disease (caused by a herpesvirus), avian sarcoma virus-induced tumors and reticuloendotheliosis (caused by avian retroviruses) (SHARMA et al. 1990).

Phenotypic characterizations of the T cells that may be responding to viral infections in chickens have recently been reported. BANBURA et al. (1991) reported that αβ T cells enlarge and proliferate in response to infection with either a virulent influenza A virus or an avirulent fowl pox virus. The γδ T cells respond to these viral infections by increasing in mean cell size, but the frequency of γδ T cells is either unchanged ($CD8^+$) or decreases ($CD8^-$). Thus, although γδ T cells are influenced by the viral infections, the antiviral immune response appears to be mediated primarily by αβ T cells. Similarly, MERKLE et al. (1992) reported that the cytotoxic T cell response to reticuloendotheliosis virus infection is both MHC-restricted and virus-specific and is mediated by αβ T cells but not by γδ T cells. The possibility that chicken γδ T cells might acquire natural killer (NK) activity following viral infections, as recently reported for influenza-infected mice (EICHELBERGER and DOHERTY 1994), has not been investigated.

Cell-mediated immune responses also appear to be important in bacterial and parasitic infections in chickens, and the GALTs appear to be important sites for these responses (LILLEHOJ 1990). The best studied example is infection of the chicken intestine by *Eimeria* sp., which are intracellular protozoan parasites. Eimerial infections cause coccidiosis, which is of considerable economic importance for the poultry industry. Several observations suggest a role for γδ T cells in the immune response to this infection.

First, specific in vitro T cell proliferative responses and cell-mediated protective immunity are observed after infection of chickens with *Eimeria* (LILLEHOJ 1986), and intestinal IELs with NK-like activity have been implicated in the response to *Eimeria* infection (CHAI and LILLEHOJ 1988; LILLEHOJ 1989). Second, phenotypic analyses of chicken intestinal IELs following challenge infection with

Eimeria revealed a significant increase in CD8+ γδ (and αβ) T cells (LILLEHOJ and BACON 1991). When chicken strains differing in susceptibility to coccidiosis were compared, both the SC and TK strains responded with an increased number of γδ and αβ IELs after primary infection, whereas γδ and αβ CD8+ IELs increased in number upon secondary infection only in the more resistant SC strain (LILLEHOJ 1994). Finally, chickens treated orally for 1 week with cyclosporin A beginning 1 day before primary infection are resistant to a secondary infection (LILLEHOJ 1987), raising the possibility that the responding T cells during the primary response are capable of clonal expansion by an IL-2-independent pathway. Such an alternative pathway appears to exist in some murine γδ T cells, which are activated by cytokines (e.g., IL-7) in the absence of TCR cross-linking and IL-2 signaling (LYNCH and SHEVACH 1992).

The accumulation of γδ T cells does not, however, prove that they are responding to the pathogen or providing a protective immune response, as they may be providing a regulatory role by interacting with αβ T cells, as discussed below, or may be bystander cells activated by the local cytokine secretion by αβ T cells. However, further evidence suggesting an important role for chicken γδ T cells in response to microbial pathogens is provided by experiments demonstrating in vitro responses to microorganisms or products derived from them, as discussed in the next section.

Chicken γδ T cells exhibit distinct growth requirements compared to αβ T cells. Early studies revealed that chicken γδ T cells respond poorly to stimulation by mitogens or cross-linkage of their TCR with anti-CD3 or anti-TCR1 mAb (SOWDER et al. 1988; COOPER et al. 1991). More recent studies confirm that, whereas chicken γδ T cells grow well in unseparated cultures of mitogen-stimulated splenocyte or PBL cultures, isolated γδ T cells are unable to grow when stimulated by either the presence of exogenous growth factors (from mitogen-stimulated splenocytes) or receptor cross-linking alone and require both types of stimulation for a strong proliferative response in vitro (KASAHARA et al. 1993; ARSTILA et al. 1993). This is in contrast to chicken αβ T cells (TCR2 or TCR3), which proliferate in vitro after TCR cross-linking alone. Similarly, γδ T cells from complete Freund's adjuvant-primed chickens proliferate in vitro only in the presence of purified protein derivative of *Mycobacterium tuberculosis* and either αβ T cells or ConA supernatant (ARSTILA et al. 1993). IL-2 appears to be an important growth factor for chicken γδ T cells, but IL-2 is produced in very small quantities by chicken γδ T cells (KASAHARA et al. 1993). Therefore, the requirement for helper activity from αβ T cells for γδ T cell growth may be due mostly to a requirement for growth factors that the γδ T cells cannot produce themselves.

Interestingly, the CD8+ subpopulation of chicken γδ T cells responds better to mitogens than the CD8− subpopulation, and CD8+ γδ T cells have the phenotype of activated cells, i.e., they are larger and express MHC class II antigens on their surface (KASAHARA et al. 1993). This suggests that the interaction between γδ and αβ T cells could have mutual effects, such that the activated γδ T cells stimulate the αβ T cells via MHC class II/αβ TCR interaction, with subsequent up-regulation or down-regulation of αβ T cell activity, depending on the second signals provided by the γδ T cells. Such interactions have been reported in the murine system. For example, modulation of murine γδ T cells with the anti-γδ TCR mAb GL3 induces IL-2 production by $CD4^+$ T lymphocytes and cytolytic activity by $CD8^+$ T cells as well as vigorous in vitro proliferative responses of spleen and lymph node αβ T cells when cultured in the absence of foreign antigens (KAUFMANN et al. 1993). In addition, murine γδ intestinal IELs have been shown to have distinct regulatory functions, as they abrogate oral tolerance, whereas αβ IELs provide B cell help (FUJIHASHI et al. 1992).

αβ and γδ T cell interactions might also occur during cell differentiation in the thymus. From in vitro mouse thymic cultures it has been observed that if αβ T cells are excluded from the initial population of double negative thymocytes, the $CD8^+$ γδ T cells do not appear in the cultures, thus raising the possibility of intrathymic αβ and γδ T cell interaction as well as a role for αβ T cells in the induction of CD8 molecules on thymic γδ T cells (SPETZ et al. 1991).

Another possible function of γδ T cells in epithelial sites is suggested by the studies of BOISMENU and HAVRAN (1994), demonstrating that mouse intraepithelial γδ T cells may modulate epithelial cell growth. Using a transwell culture system, the growth of cultured keratinocytes that require exogenous keratinocyte growth factor (KGF) was shown to be stimulated by DEC lines expressing the monomorphic γδ TCR (HAVRAN et al. 1991), but not by αβ cell lines or γδ cell lines expressing other Vγ/ Vδ genes. Furthermore, the stimulatory cell lines were shown to produce KGF. KGF is also produced by activated skin and intestine IEL γδ T cells, but not by lymphoid αβ or γδ T lymphocytes. BOISMENU and HAVRAN suggest that the ability of intraepithelial γδ T cells to recognize injured epithelial cells and provide KGF may provide a local self-regulating mechanism to maintain the integrity of epithelial surfaces.

6 Antigen Recognition by Chicken γδ T Cells

Further progress in understanding the physiological role of γδ T cells requires that the ligands recognized by γδ TCR be identified. Antigens reported to be recognized by mammalian γδ T cells include: cell surface molecules such as class I and class II MHC molecules, nonclassical class I molecules (Qa-1, TL and CD1) and immunoglobulin λ chains; autologous heat shock proteins, such as Hsp60 and Hsp27; various cell types, including keratinocytes (by mouse DECs), epithelial cells (by IELs), and tumor cells (Daudi, Burkitt's lymphomas and MOLT-4); and a peptide from a herpes simplex virus glycoprotein (reviewed by HAAS et al. 1993; KRONENBERG 1994; KIM et al. 1995). In addition, a number of microbial products may be recognized by mammalian γδ T cells, including staphylococcal enterotoxins A and E, tetanus toxoid, *M. tuberculosis* 65 KDa protein, *M. tuberculosis* 10–14 KDa protease-sensitive molecules, and *M. tuberculosis*- and *M. microti*-derived non protein molecules containing phosphate (RUST et al. 1990;

BOOM et al. 1994; CONSTANT et al. 1994; TANAKA et al. 1994; SCHOEL et al. 1994; SANCHEZ-GARCIA and COLSTON, unpublished).

Much of the data for mammalian γδ T cells are derived from experiments using T cells clones, which may not be representative of in vivo γδ T cell populations. Although antigen recognition by γδ T cells does not fit the paradigm of classical peptide/MHC/TCR interaction for αβ T cells, participation of classical and/or non-classical MHC molecules cannot be excluded. Other possibilities include cross-reactivities and multispecificities of the γδ TCR, such that a given TCR might be capable of recognizing both MHC plus peptide and a non-MHC element, e.g., superantigens (reviewed by KRONENBERG 1994). The current model is that γδ T cells recognize cell surface or secreted molecules expressed by invading pathogens and/or self stress response proteins induced upon infection. This recognition may or may not require antigen processing and presentation, by either classical or nonclassical MHC molecules.

It will be of considerable biological interest to know whether γδ T cells from phylogenetically distant species recognize similar or even identical molecules and express a similar γδ TCR repertoire in response to such antigens. For example, ARSTILA et al. (1993) have shown that chicken spleen γδ T cells respond in vitro in response to mycobacterial antigens, suggesting the possibility of an evolutionary conserved γδ TCR repertoire.

Experiments designed for the assessment of γδ TCR repertoire or antigen specificity of intestinal IEL γδ T cells will have to take into account the experimental difficulties in sustaining in vitro proliferation of intestinal IELs. VINEY and MACDONALD (1990) and SANO et al. (1993) have shown that murine intestinal IELs die rapidly in culture, with only 5% surviving for 2 days. In the chicken, HAURY et al. (1993) have reported a "relatively weak proliferative response" of IELs when cells are stimulated with ConA and supernatants of ConA-stimulated splenocytes. In the chicken model it may be possible to take advantage of in vitro retroviral transformation using the replication-defective avian reticuloendotheliosis virus strain T (REV-T), as described by MARMOR et al. (1993). Transformed γδ T cell lines (and αβ T cell lines) can be produced by in vitro transformation of bone marrow cells (MARMOR et al.1993) and intestinal IELs (ALLER and McCORMACK, unpublished) with REV-T. The bone marrow-derived γδ T cell lines are either $CD8^+$ or $CD8^-$ and are all $CD4^-$.

An interesting candidate molecule for T cell restriction and/or antigen recognition that is currently unique to avian species is B-G. B-G antigens are disulfide-linked, dimeric cell surface molecules encoded by a multigene family within the chicken MHC locus. B-G molecules are composed of a signal sequence, an Ig-like extracellular V domain, a transmembrane region, and a cytoplasmic region of heptad amino acid repeats of varying length (KAUFMAN and SALOMONSEN 1992). Originally defined as a polymorphic marker on chicken erythrocytes, B-G is now known to be expressed on many immunologically relevant cell types, including thrombocytes, B and T cells, and stromal cells of the bursa, thymus and intestinal epithelium (MILLER et al. 1990; SALOMONSEN et al. 1991).

7 Concluding Remarks

The function of γδ T cells appears to be one of surveillance, especially at epithelial surfaces, by reacting to antigens or superantigens expressed by bacterial pathogens and/or self stress response proteins expressed by infected or otherwise stressed cells. The presence of γδ T cells at epithelial sites in all species suggests that this function is evolutionarily conserved. The increased prevalence of γδ T cells in the chicken and other animal species suggests that γδ T cells may also have greater functional roles in immune responses in lymphoid organs such as the spleen. If the genetics of the chicken TCR loci prove to be simpler than those of mammalian species, as has proven to be the case for other immunologically relevant loci, the chicken will provide an ideal model in which to address many of the remaining questions about γδ T cell function, antigen recognition and the γδ TCR repertoire and to determine which of these functional aspects of γδ T cell biology are evolutionarily conserved.

References

Alaibac M (1992) γδ T lymphocytes: relevance of the current studies to dermatology. Int J Dermatol 31: 157–159

Allison JP (1993) γδ T cell development. Curr Opin Immunol 5: 241–246

Arstila TP, Lassila O (1993) Androgen-induced expansion of the peripheral blood γδ T cell population in the chicken. J Immunol 151: 6627–6633

Arstila TP, Toivanen P, Lassila O (1993) Helper activity of CD4+αβ T cells is required for the avian γδ T cell response. Eur J Immunol 23: 2034–2037

Arstila TP, Vainio O, Lassila O (1994) Evolutionarily conserved function of CD28 in αβ T cell activation. Scand J Immunol 40: 368–371

Banbura M, Webster RG, Cooper MD, Doherty PC (1991) Size and frequency characteristics of αβ and γδ T cells in the spleens of normal and cyclophosphamide-suppressed virus-infected chickens. Cell Immunol 136: 242–250

Bernot A, Auffray C (1991) Primary structure and ontogeny of an avian CD3 transcript. Proc Natl Acad Sci USA 88: 2550–2554

Bilsland CAG, Springer TA (1994) Cloning and expression of the chicken CD18 cDNA. J Leukoc Biol 55: 501–506

Boismenu R, Havran WL (1994) Modulation of epithelial cell growth by intraepithelial γδ T cells. Science 26: 1253–1255

Bonneville M, Janeway Jr CA, Ito K, Haser W, Ishida I, Nakanishi N, Tonegawa S (1988) Intestinal intraepithelial lymphocytes are a distinct set of γδ T cells. Nature 336: 479–481

Boom WH, Balaji KN, Nayak R, Tsukaguchi K, Chervenak KA (1994) Characterization of a 10–14 KDa protease-sensitive *Mycobacterium tuberculosis*-H37Ra antigen that stimulates γδ T cells. Infect Immun 62: 5511–5518

Bucy RP, Chen CH, Cihak J, Lösch U, Cooper MD (1988) Avian T cells expressing γδ receptors localize in the splenic sinusoids and the intestinal epithelium. J Immunol 141: 2200–2205

Bucy RP, Chen CH, Cooper MD (1989) Tissue localization and CD8 accessory molecule expression of T γ/δ cells in humans. J Immunol 142: 3045–3047

Bucy RP, Li J, Xu XY, Char D, Chen CH (1990) Effect of cyclosporin A on the ontogeny of different T cell sublineages in chickens. J Immunol 144: 3257–3265

Cerf-Bensussan N, Jerry A, Brousse N, Lisowska-Grospierre B, Guy-Grand D, Gricelli C (1987) A monoclonal antibody (HML-1) defining a novel membrane molecule present on human intestinal lymphocytes. Eur J Immunol 17: 1279–1285

Chai J-Y, Lillehoj HS (1988) Isolation and functional characterization of chicken intestinal intra-epithelial lymphocytes showing natural killer cell activity against tumour target cells. Immunology 63: 111–117

Chan MM, Chen CH, Ager LL. Cooper MD (1988) Identification of the avian homologues of mammalian CD4 and CD8 antigens. J Immunol 140: 2133–2138

Chen CH, Ager LL, Gartland GL, Cooper MD (1986) Identification of a T3/ T cell receptor complex in chickens. J Exp Med 164: 375–380

Chen CH, Cihak J, Lösch U, Cooper MD (1988) Diffferential expression of two T cell receptors, TCR1 and TCR2, on chicken lymphocytes. Eur J Immunol 18: 539–543

Chen CH, Sowder JT, Lahti JM, Cihak J, Lösch U, Cooper MD (1989) TCR3: a third T-cell receptor in the chicken. Proc Natl Acad Sci USA 86: 2351–2355

Chen CH, Pickel JM, Lahti JM, Cooper MD (1991) Surface markers on avian immune cells. In: Sharma JM (ed) Avian cellular immunology. CRC Press, Boca Raton, pp 2–22

Chow LM, Ratcliffe KJ, Veillette A (1992) *tkl* is the avian homolog of the mammalian *lck* tyrosine protein kinase gene. Mol Cell Biol 12: 1226–1233

Cihak J, Hoffman-Fezer G, Ziegler-Heibrock HWL, Stein H, Kaspers B, Chen CH, Cooper MD, Lösch U (1991) T cells expressing the Vβ1 T-cell receptor are required for IgA production in the chicken. Proc Natl Acad Sci USA 88: 10951–10955

Cihak J, Lösch U, Hoffmann-Fezer F, Chen CH, Cooper MD, Ziegler-Heitbrock HWL (1993) In vivo depletion of chicken T-cell subsets. Scand J Immunol 38: 123–129

Constant P, Davodeau F, Peyrat M-A, Poquet Y, Puzo G, Bonneville M, Fournie J-J (1994) Stimulation of human γδ T cells by nonpeptidic mycobacterial ligands. Science 264–270

Cooper MD, Chen CH, Bucy RP, Thompson CB (1991) Avian T cell ontogeny. Adv Immunol 50: 87–117

Davidson NJ, Boyd RL (1992) Delineation of chicken thymocytes by CD3-TCR complex, CD4 and CD8 antigen expression reveals phylogenically conserved and novel thymocyte subsets. Int Immunol 4: 1175–1182

Davidson NJ, Chen CH, Boyd RL (1992) Kinetics of chicken embryonic thymocyte development in vivo and in organ culture. Eur J Immunol 22: 1429–1435

Dunon D, Cooper MD, Imhof BA (1993a) Thymic origin of embryonic γ/δ T cells. J Exp Med 177: 257–263

Dunon D, Cooper MD, Imhof BA (1993b) Migration patterns of thymus-derived γδ T cells during chicken development. Eur J Immunol 23: 2545–2550

DuPasquier L (1989) Evolution of the immune system. In: Paul WE (ed) Fundamental immunology. Raven, New York, pp 139–165

Eichelberger M, Doherty PC (1994) γδ T cells from influenza-infected mice develop a natural killer cell phenotype following culture. Cell Immunol 159: 94–102

Fedecka-Bruner B, Penninger J, Vaigot P, Lehmann A, Martinez C, Kroemer G (1991) Developmental expression of IL-2-receptor light chain (CD25) in the chicken embryo. Dev Immunol 1: 237–242

Fichtelius KE, Finstad J, Good RA (1969) The phylogenetic occurrence of lymphocytes within the gut epithelium. Int Arch Allergy Appl Immunol 35: 119–133

Fujihashi K, Taguchi T, Aicher WK, McGhee JR, Bluestone JA, Eldridge JH, Kiyono H (1992) Immunoregulatory functions for murine intraepithelial lymphocytes; γ/δ T cell receptor-positive (TCR^+) T cells abrogate oral tolerance, while α/β TCR^+ T cells provide B cell help. J Exp Med 175: 695–707

George JF, Cooper MD (1990) γ/δ T cells and α/β T cells differ in their developmental patterns of receptor expression and modulation requirements. Eur J Immunol 20: 2177–2181

Gershwin ME, Abplanalp H, Castles JJ, Ikeda RM, van der Water J, Eklund J, Haynes D (1981) Characterization of a spontaneous disease of White Leghorn chickens resembling progressive systematic sclerosis (scleroderma). J Exp Med 153: 1640–1659

Goodman T, Lefrancois L (1988) Expression of the gamma/delta T-cell receptor on intestinal $CD8^+$ intraepithelial lymphocytes. Nature 333: 855–858

Gruschwitz MS, Moorman S, Kromer G, Sgonc R, Dietrich H, Boeck G, Gershwin ME, Boyd R, Wick G (1991) Phenotypic analysis of skin infiltrates in comparison with peripheral blood lymphocytes, spleen cells and thymocytes in early avian scleroderma. J. Autoimmun 4: 577–593

Guy-Grande D, Vassilla P (1993) Gut intraepithelial T lymphocytes. Curr Opin Immunol 5: 247–252

Haas W, Pereira P, Tonegawa S (1993) Gamma/delta cells. Annu Rev Immunol 11: 637–685

Haury M, Kasahara Y, Schaal S, Bucy RP, Cooper MD (1993) Intestinal T lymphocytes in the chicken express an integrin-like antigen. Eur J Immunol 23: 313–319

Havran WL, Chien Y-H, Allison JP (1991) Recognition of self antigens by skin-derived T cells with invariant γδ antigen receptors. Science 252: 1430–1432

Hein WR, Mackay CR (1991) Prominence of γδ T cells in the ruminant immune system. Immunol Today 12: 30–34
Kasahara Y, Chen CH, Cooper MD (1993) Growth requirements for avian γδ T cells include exogenous cytokines, receptor ligation and in vivo priming. Eur J Immunol 23: 2230–2236
Kaufman J, Salomonsen J (1992) B-G: we know what it is, but what does it do? Immunol Today 13: 1–3
Kaufmann SHE, Blum C, Yamamoto S (1993) Crosstalk between alpha/beta and γ/δ T cells in vivo: activation of α/β T cell responses after γ/δ T cell modulation with the monoclonal antibody GL3. Proc Natl Acad Sci USA 90: 9620–9624
Kim HT, Nelson EL, Clayberg C, Sanjanwala M, Sklar J, Krensky AM (1995) γδ T cell recognition of tumor Ig peptide. J Immunol 154: 1614–1623
Knabel M, Cihak J, Lösch U (1993) Characterization of new monoclonal antibodies identifying avian T lymphocyte antigens. Immunobiology 188: 415–429
Kronenberg M (1994) Antigens recognized by γδ T cells. Curr Opin Immunol 6: 64–71
Lahti JM, Chen CH, Tjoelker LW, Pickel JM, Schat KA, Calnek BW, Thompson CB, Cooper MD (1991) Two distinct αβ T-cell lineages can be distinguished by the differential usage of T-cell receptor Vβ gene segments. Proc Natl Acad Sci USA 88: 10956–10960
Lefrancois L, Puddington L (1995) Extrathymic intestinal T-cell development: virtual reality? Immunol Today 16: 16–21
Lillehoj HS (1986) Immune response during coccidiosis in SC and FP chickens. I. In vitro assessment of T cell proliferation response to stage-specific parasite antigens. Vet Immunol Immunopathol 13: 321–330
Lillehoj HS (1987) Effects of immunosuppression on avian coccidiosis: Cyclosporin A but not hormonal bursectomy abrogates host protective immunity. Infect Immun 55: 1616–1621
Lillehoj HS (1989) Intestinal intraepithelial and splenic natural killer cell responses to Eimerian infections in inbred chickens. Infect Immun 57: 1879–84
Lillehoj HS (1990) Cell-mediated immunity in parasitic and bacterial diseases. In: Sharma JM (ed) Avian cellular immunology. CRC Press, Boca Raton, pp 155–182
Lillehoj HS (1994) Analysis of *Eimeria acervulina*-induced changes in the intestinal T lymphocyte subpopulations in two chicken strains showing different levels of susceptibility to coccidiosis. Res Vet Sci 56: 1–7
Lillehoj HS, Bacon LD (1991) Increase of intestinal intraepithelial lymphocytes expressing CD8 antigen following challenge infection with *Eimeria acervulina*. Avian Dis 35: 294–301
Lillehoj HS, Chung KS (1992) Postnatal development of T-lymphocyte subpopulations in the intestinal intraepithelium and lamina propria in chickens. Vet Immunol Immunopathol 31: 347–360
Lynch F, Shevach EM (1992) Activation requirements of newborn thymic γδ T cells. J Immunol 149: 2307–2314
Marmor MD, Benatar T, Ratcliffe MJH (1993) Retroviral transformation in vitro of chicken T cells expressing either α/β or γ/δ T cell receptors by reticuloentheliosis virus strain T. J Exp Med 177: 647–656
Merkle H, Cihak J, Lösch U (1992) The cytotoxic T lymphocyte response in reticuloendotheliosis-infected chickens is mediated by αβ and not by γδ T cells. Immunobiology 186: 292–303
Miller MM, Goto R, Young S, Liu J, Hardy J (1990) Antigens similar to major histocompatibility complex B-G are expressed in the intestinal epithelium in the chicken. Immunogenetics 32: 45–50
Ohteki T, MacDonald HR (1993) Expression of the CD28 costimulatory molecule on subsets of murine intestinal intraepithelial lymphocytes correlates with lineage and responsiveness. Eur J Immunol 23: 1251–1255
Raulet DH (1989) The structure, function, and molecular genetics of the γ/δ T cell receptor. Annu Rev Immunol 7: 175–207
Rocha B, Vassalli P, Guy-Grand D (1992) The extrathymic T-cell development pathway. Immunol Today 13: 449–454
Rust CJ, Verreck F, Vietor H, Koning F (1990) Specific recognition of staphylococcal enterotoxin A by human T cells bearing receptors with the Vγ9 region. Nature 346: 572–574
Salomonsen J, Dunon D, Skjodt K, Thorpe D, Vainio O, Kaufman J (1991) Chicken major histocompatibility complex-encoded B-G antigens are found on many cell types that are important for the immune system. Proc Natl Acad Sci USA 88: 1359–1363
Sano Y, Dudley E, Carding S, Lin RH, Hayday AC, Janeway CA (1993) γδ T cells isolated from intestinal epithelium respond to B-cell lymphoma. Immunology 80: 388–394
Schoel B, Sprenger S, Kaufmann SHE (1994) Phosphate is essential for stimulation of Vγ9Vδ2 thymocytes by mycobacterial low molecular weight ligand. Eur J Immunol 24: 1886–1892

Sharma JM, Prowse SJ, York JJ (1990) Role of cellular immunity in neoplastic and nonneoplastic viral diseases. In: Sharma JM (ed) Avian cellular immunology. CRC Press, Boca Raton, pp 139–154
Sowder JT, Chen CH, Ager LL, Chan MM, Cooper MD (1988) A large subpopulation of avian T cells express a homologue of the mammalian Tγ/δ receptor. J Exp Med 167: 315–322
Sperling AI, Linsley PS, Barrett TA, Bluestone JA (1993) CD28-mediated costimulation is necessary for the activation of T cell receptor-$\gamma\delta^+$ T lymphocytes. J Immunol 151: 6043–6050
Spetz AL, Kourvilsky P, Larsson-Sciard EL (1991) Induction of CD8 molecules on thymic γ/δ T cells in vitro is dependent upon α/β T cells. Eur J Immunol 21: 2755–2759
Tanaka Y, Sano S, Nieves E, DeLibero G, Resa D, Modlin RL, Brenner MB, Bloom BR, Morita CT (1994) Non-peptide ligands for human γδ T cells. Proc Natl Acad Sci USA 91: 8175–8179
Vainio O, Lassila O (1989) Chicken T cells: differentiation antigens and cell-cell interactions. Poultry Biol 2: 97–102
Vaino O, Riwar B, Brown MH, Lassila O (1991) Characterization of the putative avian CD2 homologue. J Immunol 147: 1593–1599
Viney JL, MacDonald TT (1990) Selective death of T cell receptor γδ intraepithelial lymphocytes by apoptosis. Eur J Immunol 20: 2809–2812
Young JR, Davison TF, Tregaskes CA, Rennie MC, Vainio O (1994) Monomeric homologue of mammalian CD28 is expressed on chicken T cells. J Immunol 152: 3848–3851

T Cell Subsets and the Activation of γδ T Cells

T.P. ARSTILA

1 Introduction

The T cell subsets in chicken are, in most respects, similar to those in mouse or human (COOPER et al. 1991; ARSTILA et al. 1994a). As in mammals, mature αβ T cells express either CD4 or CD8. The $CD4^+$ αβ T cells function as helper T cells, recognizing antigens in a MHC class II-restricted way and producing cytokines upon activation (CHAN et al. 1988). The αβ T cells expressing CD8 are poor producers of cytokines but exhibit T cell receptor (TCR)-dependent cytotoxicity. The αβ T cells can also be divided into two subsets on the basis of the Vβ gene segment utilized (TJOELKER et al. 1990; LAHTI et al. 1991). There are two families of V genes in the TCR β locus, both containing several gene segments, and both subsets can be detected by a monoclonal antibody (CHEN et al. 1988; CIHAK et al. 1988; CHAR et al. 1990). Some studies have indicated differences between cells expressing Vβ1 or Vβ2 gene segments. For example, $V\beta1^+$ cells are numerous in the intestinal epithelium and can support IgA responses. In contrast, $V\beta2^+$ cells are rare in the intestinal epithelium and, either because of this or because of an unknown functional characteristic, are unable to provide help for IgA responses (CIHAK et al. 1991).

A noteworthy feature of the avian immune system is the high frequency of γδ T cells (SOWDER et al. 1988). In adult males, up to 50% of the circulating lymphocytes can express the γδ TCR (ARSTILA and LASSILA 1993). In blood, the majority of γδ T cells are CD4/CD8 double negative, but in the spleen and intestinal

Department of Medical Microbiology, Turku University, Turku, Finland epithelium two thirds of them express CD8 (Bucy et al . 1988). No $CD4^+$ γδ T cells have been described in the chicken. The preferential homing of γδ T cells to epithelia, so striking in the mouse, has not been observed in the chicken. As in all species, the functional role of γδ T cells in the chicken has remained mysterious. However, because of the high number of γδ T cells found in these animals, the chicken offers a challenging model to study these enigmatic cells.

Recent studies have highlighted the importance of collaboration between lymphocyte subsets and the crucial role of $CD4^+$ αβ T cells in the avian immune response (Table 1). In this chapter we will review data on the T cell subsets in the chicken and interactions between them.

2 Activation of γδ T Cells

In the chicken, αβ T cells can be activated by mitogens and monoclonal antibodies (mAbs) and antigen-specific αβ T cell lines have also been characterized (Vainio et al. 1988; Cooper et al. 1991). γδ T cells can also be activated in vitro and induced to proliferate by mitogens, such as concanavalin A (ConA), or by immobilized mAbs against the TCR (Sowder et al. 1988). These responses are, however, consistently weaker than those of αβ T cells induced by the same stimuli. In the mouse and human, mycobacterial antigens have been shown to be highly efficient stimulators of γδ T cells (Janis et al. 1989; Kabelitz et al. 1990; Constant et al. 1994). In the chicken, intramuscular administration of killed mycobacteria in mineral oil (complete Freund's adjuvant) induces a twofold increase in the frequency and a more than fourfold increase in the absolute numbers of peripheral blood γδ T cells (Arstila et al. 1995). Cells isolated from these chickens respond vigorously to mycobacterial antigens, and analysis of the proliferating cells show that they include significant numbers of large activated blasts expressing γδ TCR. Others have reported that stimulation with mitogens or anti-TCR mAb preferentially activates the $CD8^+$ γδ T cell subset. In contrast, two-color immunofluorescence analysis of peripheral blood lymphocytes (PBLs) from the

Table 1. Summary of avian T cell subsets

Subset	Frequency in blood (%)	Comments
$CD8^+$γδ	1–5	Predominant γδ T cell subset in spleen and intestinal epithelium; rare in blood; lack the costimulatory molecule CD28; preferentially activated by in vitro stimulation with mAb to TCR or Con A
$CD8^-$γδ	15–50	Predominant γδ T cell subset in thymus and blood; lack the costimulatory molecule CD28; γδ T cells dependent on help provided by $CD4^+$αβ T cells
$CD4^+$αβ	45	Produce cytokines, provide help for T cell-dependent antibody responses and γδ T cell responses; includes both Vβ1$^+$ and Vβ2$^+$ cells; Vβ2$^+$ cells cannot provide help for IgA responses
$CD8^+$αβ	15	Function as cytotoxic T cells; poor producers of cytokines

mycobacteria-primed chickens indicates that the number of both $CD8^-$ and $CD8^+$ γδ T cells increases. αβ T cells are also activated by mycobacterial antigens, but the increase in the absolute numbers of αβ T cells is less than twofold. These results indicate that, also in the chicken, mycobacterial components are a potent stimulator of γδ T cells.

One of the mycobacterial components which has received much attention is the 65 kDa heat shock protein (HSP65). Some reports have suggested it to be a dominant antigen for γδ T cells, while others have reported that mycobacterial components other than HSP65 are responsible for the stimulation of γδ T cells (Haregewoin et al. 1989; Kabelitz et al. 1990; Pfeffer et al. 1990; O'Brien et al. 1992; Constant et al. 1994). The role of HSP65 as a T cell stimulator has also been investigated in the chicken (Arstila et al. 1995). Both αβ and γδ T cells respond to HSP65, but there seems to be no difference between HSP65 and other mycobacterial preparations in the effect on γδ T cells. Moreover, HSP65 reactivity is observed only after in vivo priming with killed mycobacteria, indicating that in the naive γδ T cell repertoire HSP65 does not play a dominant role.

3 Interaction Between $CD4^+$ αβ T Cells and γδ T Cells

In chickens primed with killed mycobacteria, both αβ and γδ T cells respond to mycobacterial antigens. To analyze the contribution of γδ T cells to the response, αβ T cells can be removed with magnetic beads. In the absence of αβ T cells, no proliferation or γδ T cell blasts is observed (Arstila et al. 1993). A similar effect is observed if only the $CD4^+$ αβ T cells are removed. In other studies, when separated T cell subsets are stimulated with mitogens or anti-CD3 mAb, the γδ T cells proliferate poorly compared to αβ T cells (Kasahara et al. 1993). In bulk cultures, TCR-type specific mAbs, when immobilized, induce strong αβ T cell responses, but only very modest γδ T cell proliferation. These studies indicate that the presence of $CD4^+$ αβ T cells is essential for the initiation of a γδ T cell response. The effect of $CD4^+$ αβ T cells can be replaced by cytokines produced by ConA stimulated T cells. If these cytokines are added to cultures depleted of αβ T cells or stimulated with anti γδ TCR-specific mAb, a vigorous proliferation of γδ T cells is observed. Together these studies show that the response of avian γδ T cells is dependent on soluble factors produced by $CD4^+$ αβ T cells. The possibility of a cognate interaction between $CD4^+$ αβ T cells and γδ T cells cannot be excluded either. In the interaction between T cells and B cells, direct contacts between the cells are essential, and B cells can present antigens to $CD4^+$ T cells. An interesting question is whether γδ T cells might also be capable of presenting antigens. However, unlike αβ T cells, γδ T cells do not seem to up-regulate the expression of MHC class II molecules upon activation (unpublished observations).

It has been reported that the two lineages of αβ T cells expressing either Vβ1 or Vβ2 gene segments might differ from each other in some respects, such as repertoire, homing and possibly the capability to support B cell function (CIHAK et al. 1991; COOPER et al. 1991). Both subsets respond readily to stimulation with anti-TCR mAb. Their ability to provide help for γδ T cell responses to mycobacterial antigens has been tested by depleting one or the other of the subsets by using magnetic beads. However, the absence of either subset does not have a negative effect on the proliferative response or on the induction of γδ T cell blasts.

4 The Role of Costimulatory Signals in T Cell Activation

Experiments on mammalian T cells have made it evident that the signal provided by TCR alone is not sufficient to trigger a T cell response. An important costimulatory signal is provided by the interaction of the T cell molecule CD28 with its ligands, B7-1 and B7-2, expressed on antigen-presenting cells (LINSLEY and LEDBETTER 1993). In the chicken, the homologue of CD28 has been characterized and the gene encoding it cloned (VAINIO et al. 1991; YOUNG et al. 1994). The overall identity at the amino acid level to mammalian homologues is 50%, with important functional motifs involved in ligand-binding and signaling conserved. In contrast to the mammalian CD28, a homodimer linked together with a disulfide bond, the chicken CD28 is a monomeric molecule, with a crucial extracellular cysteine (Cys-140) replaced by a glutamine residue.

In the chicken, CD28 is expressed on practically all αβ T cells, but peripheral γδ T cells are CD28$^-$ (VAINIO et al. 1991). On αβ T cells CD28 is capable of delivering a potent costimulatory signal (ARSTILA et al.1994b). Stimulation with phorbol myristate acetate, or a soluble mAb against TCR is insufficient to induce significant proliferation. A strong synergistic effect is provided by a mAb against CD28, triggering strong proliferative responses. The signaling pathway employed by the chicken CD28 is at least partly distinct from that employed by TCR. Cyclosporin A efficiently inhibits TCR-mediated activation of T cells, but has only a small effect on CD28-mediated activation. Herbimycin A, an inhibitor of protein tyrosine kinase function, also suppresses the signal through CD28. These results indicate both structural and functional conservation of the CD28 costimulatory pathway between mammalian species and the chicken.

It is tempting to speculate that the lack of the CD28 costimulatory pathway on chicken γδ T cells makes them dependent on help from CD4$^+$ αβ T cells. In the mouse and human, CD28 signaling has been associated with induction and enhancement of cytokine production (LINDSTEN et al. 1989; FRASER et al. 1991). Moreover, a CD28$^-$ αβ T cell subset in the mouse has been reported to be refractory to TCR-mediated activation, unless external cytokines are provided (OHTEKI and MACDONALD 1993). Whether signaling through the γδ TCR in absence of costimulatory signals results in cell anergy or death remains to be studied. However, there is some evidence that activation renders chicken γδ T cells more susceptible to programmed cell death (ARSTILA et al. 1994c).

5 Concluding Remarks

The studies described here, as well as studies in other species, have made it evident that αβ and γδ T cells interact in the generation of the T cell response (ARSTILA et al. 1993; KASAHARA et al. 1993; KAUFMANN et al. 1993; PECHHOLF et al. 1994). In the chicken, in which γδ T cells form a major lymphocyte population, CD4⁺ αβ T cells can control γδ T cell responses by providing soluble factors which the γδ T cells apparently are unable to produce themselves (Fig.1). Whether this is due to the absence of CD28 on γδ T cells and whether the γδ T cells possess other costimulatory molecules remains to be studied. The exact nature of the interaction between the T cell subsets is also unclear. However, these observations emphasize the crucial role that CD4⁺ αβ T cells have in the initiation and regulation of the avian immune response.

Recent studies on mammalian γδ T cells have suggested interesting differences between them and αβ T cells. Several groups have reported that γδ T cells can recognize antigens other than peptides or proteins. For example, the stimulating component of mycobacteria has been shown to be a phosphate-linked thymidine compound (CONSTANT et al. 1994). These and other results have also casted doubt on the role of MHC restriction in antigen recognition by γδ T cells. There is evidence suggesting that the interaction between γδ TCR and antigen may resemble that of an immunoglobulin molecule and antigen, rather than that of αβ TCR (ROCK et al. 1994; SCHILD et al. 1994). If such features are shown to apply to γδ T cells in general, it becomes quite evident that this subset occupies a place in the immune system clearly distinct from that of αβ T cells. Mice lacking αβ T cells have provided some hints of the contribution of γδ T cells to antimicrobial defense, showing that in some instances γδ T cells have a specialized role in the immune response (MOMBAERTS et al. 1993; TSUJI et al. 1994). In the chicken, the high number of γδ T cells in itself suggests that these cells have a significant role

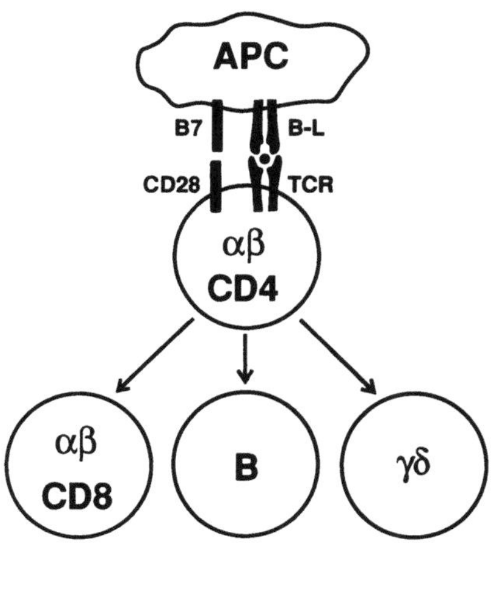

Fig. 1. Interactions between lymphocyte subsets in the avian immune response. The antigen-presenting cell (APC) presents the antigen bound to a MHC class II molecule (*B-L*) to the CD4⁺ αβ T cell and also provides a costimulatory signal, transmitted through CD28. The activated CD4⁺ αβ T cell then provides help (*arrows*) to other kinds of lymphocytes

in the avian immune system. Thus, the chicken offers an ideal experimental model to study γδ T cell immunobiology.

Acknowledgments. I would like to thank Drs. Olli Lassila and Olli Vainio for helpful discussions and constructive criticism during this work and Dr. Pia Isomäki for critical reading of the manuscript. These studies were supported by the Finnish Academy, the Cultural Fund of Southwest Finland and the Turku University Foundation.

References

Arstila TP, Lassila O (1993) Androgen-induced expansion of the peripheral blood γδ T cell population in the chicken. J Immunol 151: 6627–6633

Arstila TP, Toivanen P, Lassila O (1993) Helper activity of $CD4^+$ αβ T cells is required for the avian γδ T cell response. Eur J Immunol 23 : 2034–2037

Arstila TP, Vainio O, Lassila O (1994a) Central role of $CD4^+$ T cells in avian immune response. Poultry Sci 73: 1019–1026

Arstila TP, Vainio O, Lassila O (1994b) Evolutionarily conserved function of CD28 in αβ T cell activation. Scand. J Immunol 40: 368–371

Arstila TP, Toivanen P, Vainio O, Lassila O (1994c) γδ and αβ T cells are equally susceptible to apoptosis. Scand J Immunol 40: 209–215

Arstila TP, Toivanen P, Lassila O (1995) Primed avian γδ T cells respond to mycobacterial antigens, but show no preference for the 65-kDa heat shock protein. Cell Immunol 162: 74–79

Bucy RP, Chen CH, Cihak J, Lösch U, Cooper MD (1988) Avian T cells expressing γδ receptors localize in the splenic sinusoids and the intestinal epithelium. J Immunol 141: 2200–2205

Chan MM, Chen CH, Ager LL, Cooper MD (1988) Identification of the avian homologues of mammalian CD4 and CD8 antigens. J Immunol 140: 2133–2138

Char D, Sanchez P, Chen CH, Bucy RP, Cooper MD (1990) A third sublineage of avian T cells can be identified with a T cell receptor- 3-specific antibody. J Immunol 145: 3547–3555

Chen CH, Cihak J, Lösch U, Cooper MD (1988) Differential expression of two T cell receptors, TcR1 and TcR2, on chicken lymphocytes. Eur J Immunol 18: 539–543

Cihak J, Ziegler-Heitbrock HWL, Trainer H, Schranner I, Merkenschlager M, Lösch U (1988) Characterization and functional properties of a novel monoclonal antibody which identifies a T cell receptor in chickens. Eur J Immunol 18: 533–537

Cihak J, Hoffmann-Fezer G, Ziegler-Heibrock HWL, Stein H, Kaspers B, Chen CH, Cooper MD, Lösch U (1991) T cells expressing the Vβ1 T-cell receptor are required for IgA production in the chicken. Proc Natl Acad Sci USA 88: 10951–10955

Constant P, Davodeau F, Peyrat M-A, Poquet Y, Puzo G, Bonneville M, Fournie J-J (1994) Stimulation of human γδ T cells by nonpeptidic mycobacterial ligands. Science 264: 267–270

Cooper MD, Chen CH, Bucy RP, Thompson CB (1991) Avian T cell ontogeny. Adv Immunol 50: 87–117

Fraser JD, Irving BA, Crabtree GR, Weiss A (1991) Regulation of interleukin-2 gene enhancer activity by the T cell accessory molecule CD28. Science 251: 313–316

Haregewoin AG, Soman G, Hom CR, Finberg RW (1989) Human γ/δ T cells respond to mycobacterial heat-shock protein. Nature 340: 309–312

Janis EM, Kaufmann SHE, Schwartz RH, Pardoll DM (1989) Activation of γ/δ T cells in the primary immune response to *Mycobacterium tuberculosis*. Science 244: 713–716

Kabelitz D, Bender A, Schondelmaier S, Schoel B, Kaufmann SHE (1990) A large fraction of human peripheral blood γ/δ T cells is activated by *Mycobacterium tuberculosis* but not by its 65-kD heat shock protein. J Exp Med 171: 667–679

Kasahara Y, Chen CH, Cooper MD (1993) Growth requirements for avian γδ T cells include exogenous cytokines, receptor ligation and in vivo priming. Eur J Immunol 23: 2230–2236

Kaufmann SHE, Blum C, Yamamoto S (1993) Crosstalk between α/β T cells and γ/δ T cells in vivo: activation of α/β T cell responses after γ/δ T cell modulation with the monoclonal antibody GL3. Proc Natl Acad Sci USA 90: 9620–9624

Lahti JM, Chen CH, Tjoelker LW, Pickel JM, Schat KA, Calnek BW, Thompson CB, Cooper MD (1991) Two distinct αβ T-cell lineages can be distinguished by the differential usage of T-cell receptor Vβ gene segments. Proc Natl Acad Sci USA 88: 10956–10960

Lindsten T, June CH, Ledbetter JA, Stella G, Thompson CB (1989) Regulation of lymphokine messenger RNA stability by a surface-mediated T cell activation pathway. Science 244: 339–343

Linsley PS, Ledbetter JA (1993) The role of the CD28 receptor during T cell responses to antigen. Annu Rev Immunol 11: 191–212

Mombaerts P, Arnoldi J, Russ F, Tonegawa S, Kaufmann SHE (1993) Different roles of αβ and γδ T cells in immunity against an intracellular bacterial pathogen. Nature 365: 53–56

Ohteki T, MacDonald HR (1993) Expression of the CD28 costimulatory molecule on subsets of murine intestinal intraepithelial lymphocytes correlates withlineage and responsiveness. Eur J Immunol 23: 1251–1255

O'Brien RL, Fu Y–X, Cranfill R, Dallas A, Ellis C, Reardon C, Lang J, Carding SR, Kubo R, Born W (1992) Heat shock protein Hsp60-reactive γδ cells: a large, diversified T-lymphocyte subset with highly focused specificity. Proc Natl Acad Sci USA 89: 4348–4352

Pechhold K, Wesch D, Schondelmaier S, Kabelitz D (1994) Primary activation of Vγ9-expressing γδ T cells by Mycobacterium tuberculosis. Requirement for Th1-type CD4 T cell help and inhibition by IL-10. J Immunol 152: 4984–4992

Pfeffer K, Schoel B, Gulle H, Kaufmann SHE, Wagner H (1990) Primary responses of human T cells to mycobacteria: a frequent set of γ/δ T cells are stimulated by protease resistant ligands. Eur J Immunol 1175–1179

Rock EP, Sibbald PR, Davis MM, Chien Y-H (1994) CDR3 length in antigen-specific immune receptors. J Exp Med 179: 985–989

Schild H, Mavaddat N, Litzenberger C, Ehrich EW, Davis MM, Bluestone JA, Matis L, Draper RK, Chien Y-H (1994) The nature of major histocompatibility complex recognition by γδ T cells. Cell 76: 29–37

Sowder JT, Chen CH, Ager LL, Chan MM, Cooper MD (1988) A large subpopulation of avian T cells express a homologue of the mammalian Tγ/δ receptor. J Exp Med 167: 315–322

Tjoelker LW, Carlson LM, Lee K, Lahti J, McCormack WT, Leiden JM, Chen CH, Cooper MD, Thompson CB (1990) Evolutionary conservation of antigen recognition: the chicken T cell receptor β chain. Proc Natl Acad Sci USA. 87: 7856–7860

Tsuji M, Mombaerts P, Lefrancois L, Nussenzweig RS, Zavala F, Tonegawa S (1994) γδ T cells contribute to immunity against the liver stages of malaria in αβ T-cell-deficient mice. Proc Natl Acad Sci USA 91: 345–349

Vainio O, Veromaa T, Eerola E, Toivanen P, Ratcliffe MJH (1988) Antigen-presenting cell-T cell interaction in the chicken is MHC class II antigen restricted. J Immunol 140: 2864–2868

Vainio O, Riwar B, Brown MH, Lassila O (1991) Characterization of the putative avian CD2 homologue. J Immunol 147: 1593–1599

Young JR, Davison TF, Tregaskes CA, Rennie MC, Vainio O (1994) A monomeric homologue of mammalian CD28 is expressed on chicken T cells. J Immunol 152: 3848–3851

T Cell Migration During Ontogeny and T Cell Repertoire Generation

D. Dunon[1] and B.A. Imhof[2]

1 Introduction

Before their hematopoietic stem cell origin was recognized, chicken thymocytes were thought to be derived from epithelial precursors in the thymus that could be converted into lymphocytes under the influence of surrounding mesenchymal tissue (Auerbach 1961). Evidence obtained later indicated that embryonic yolk sac contains the precursors of lymphoid cells (Moore and Owen 1967). Yolk sac stem cells at the first and second day of embryonation (E1–2) were then shown to be derived from the embryo itself (Martin et al. 1978). Further studies indicated the presence of pluripotent stem cells in the region of the thoracic aorta as early as E4, a few days before some stem cells can be found in the spleen. These studies, conducted in chick-quail chimeras, indicated an aortic (E4) followed by a paraortic (E6) origin of precursors of thymocytes, B cells and myeloid cells. It was suggested that embryonic stem cells native to the aortic region migrate via the circulation and colonize the spleen, yolk sac, and finally the bone marrow. However, recent studies have shown that the stem cells produced in these different organs during ontogeny may correspond to different populations generated separately since they do not present identical properties (see chapter by Dieterlen).

[1] Laboratoire de Biologie Moléculaire et Cellulaire du Développement, URA-CNRS 1135, Université Pierre et Marie Curie, 75005 Paris, France
[2] Basel Institute for Immunology, Grenzacherstrasse 487, 4005 Basel, Switzerland

The chick-quail model was originally used to investigate the homing of thymocyte precursors into the thymus epithelium, derived from the third pharyngeal pouch (LE DOUARIN 1978). Thymocyte precursors enter the peripheral thymus in waves (JOTEREAU et al. 1980; LE DOUARIN et al. 1984), the first of which begins in chicken embryos on E6.5, the second on E12, and the third around E18. Each wave of influx into the thymus lasts for 1 or 2 days and is followed by the transient production of thymocyte progeny (COLTEY et al. 1987, 1989). It was found that, during the first wave of thymus colonization, most T cells probably come from the paraortic foci (COLTEY et al. 1987, 1989), whereas for the second and the third wave of colonization they originate from the bone marrow (DUNON et al. 1990a and unpublished data). The same pattern of thymus colonization has also been determined in quails (JOTEREAU and LE DOUARIN 1982), except that each colonization wave occurs 1 or 2 days earlier. Recently, congenic chicken strains allowed experiments done in chick-chick chimeras. By grafting thymic lobes from an ov^{+} chick donor into thymectomized ov^{-} recipients 2 and 9 days after hatching, it was shown that donor thymocytes were soon replaced by host ov^{-} thymocytes and their progeny, indicating that a series of waves, or more likely a continuous stream of thymocyte precursors, enters the thymus after hatching (DUNON and IMHOF 1993; DUNON et al. 1993a, b).

Comparative developmental studies of the mammalian and avian immune systems have helped to define other central features of the vertebrate immune system, and to provide insight into its evolution. Among vertebrates, birds provide several unique possibilities for studying the establishment of the T cell system. These include: (1) the dichotomy of the T and B lymphocyte system, with each population differentiating in a special central lymphoid organ, T cells in the thymus and B Cells in the bursa of Fabricius; and (2) the ready availability of a large number of exactly staged embryos, and the fact that the size of these embryos allows experimental manipulations. Although thousands of avian species exist, the domestic chicken (*Gallus gallus*) and the quail (*Cotournix cotournix japonica*) have been used in most studies of the immune system.

A renewed interest in the avian T lymphoid system has emerged with the cloning of the different T cell receptor (TCR) chains and the use of monoclonal and functional antibodies directed against T cell differentiation antigens (COOPER et al. 1991; GÖBEL et al. 1994). The analysis of T cell development has been made possible by various in vivo techniques. The pioneer work was done by the group of N. LE DOUARIN, who set up the quail-chick xenogenic chimera technique in the late 1960s. More recently, chicken strains have been used which are congenic for the ov antigen which is present on hematopoietic precursors, immature and mature T cells, and a B cell subset (HOUSSAINT et al. 1991; VAINIO et al. 1987).

In this review we show that the chicken is the best vertebrate animal to study thymus colonization by hematopoietic precursors and emigration of mature thymocytes towards the periphery.

2 Pro-T Cell Homing to the Thymus: The Chemotactic Step

Following studies performed in chickens and also in mice, a model for pro-T cell homing has emerged recently (DUNON and IMHOF 1993). In this model, pro-T cells are transported to the vicinity of the thymus via the blood circulation. Several adhesion molecules are responsible for the attachment of pro-T cells to thymic endothelium. Extravasation and migration of pro-T cells in the perivascular space through the perithymic mesenchyme towards the thymic epithelium requires interaction with extracellular matrix molecules and is driven by chemotactic molecules. Finally, a second adhesion step between the pro-T cell and the thymic epithelium is thought to occur before the pro-T cell enters the thymus itself.

As early as 1978, quail-chick chimera studies led to the hypothesis that a main step in the homing of pro-T cells was the production of chemoattractants in a periodic fashion by the thymic epithelium, generating peaks of receptivity for the blood-borne stem cells (LE DOUARIN 1978). Evidence for the chemotaxis of quail hematopoietic precursors isolated from thymus or bone marrow was initially obtained by observing the migration of individual cells in a modified Zigmond chamber (BEN SLIMANE et al. 1983; CHAMPION et al. 1986; ZIGMOND 1978). Bone marrow cells from 11.5 day old quail embryos migrated up a gradient of soluble molecules secreted by a chicken embryonic thymus during the first period of thymus colonization. When bone marrow cells were exposed to medium conditioned by a receptive thymus, two parameters of their locomotion were modified: their migration was oriented (chemotaxis) and their speed was increased (chemokinesis). The chemotactic migration was lost when the rudiment placed in the Zigmond chamber was taken from an embryo that was not at the stage of being colonized in vivo. However, putative receptors for chemotactic factors must be expressed on T cell progenitors throughout embryogenesis, since colonization of a grafted thymus which is at a receptive stage can occur during a refractory period when no colonization was taking place in the original embryo (JOTEREAU et al. 1980).

Chemotactic peptides have been partially purified from medium conditioned by thymic epithelial cells taken from avian embryos (Table 1) (BEN SLIMANE et al. 1983; CHAMPION et al. 1986). To date however, the best characterized chemotactic molecule involved in thymus homing is β_2-microglobulin (β_2m), which was originally purified from medium conditioned by a rat thymic epithelial cell line (DARGEMONT et al. 1989; DEUGNIER et al. 1989; IMHOF et al. 1988). Chicken β_2m has been shown in Boyden chamber assays to attract bone marrow cells of 13 day old chicken embryos (second wave of colonization) (DUNON et al. 1990a). When the congenic chicken strains for the ov antigen were used, the cells responsive to β_2m colonized a 13 day old thymus in vivo. Thymus colonization in vivo by injected bone marrow cells could be partly inhibited by a simultaneous injection of β_2m or of anti-β_2m-specific antibody. Moreover, during chicken embryogenesis, peaks of β_2m RNA transcripts and of free β_2m protein synthesis were only detected in the

Table 1. Thymic factors involved in chemotactic migration of hemopoietic precursors

Chemotactic factors	Species	Molecular weight (kDa)	Producers	Migrating cells
Quail peptides	Quail	>5 3<< 5 <3	Thymus during second colonization period	Hematopoietic precursors from bone marrow [1]
Chicken peptides	Chicken	>50 <12	Embryonic thymus Thymus during first colonization period	Hematopoietic precursors from bone marrow [2]
β2-Microglobulin	Chicken	11.5	Thymus during second coloniza-tion period	Lymphoid pre-cursors from bone marrow [3]

[1] CHAMPION et al. 1986; [2] BEN SLIMANE et al. 1983; [3] DUNON et al. 1990a.

thymus. The peak of free β_2m protein synthesis in the thymus and the increased number of bone marrow cells responding to β_2m occur concomitantly with the second wave of colonization (Fig. 1) (DUNON et al. 1990b). Free β_2 m was also detected in the thymus of 7 day old embryos during the first wave of thymus colonization (unpublished results). In addition, it is interesting to note that the second wave of thymus colonization occurs concomitantly with MHC class I expression (Fig. 2). The fact that β_2m is the common subunit of MHC class I antigen (KLEIN 1986) indicates that β_2m synthesized during the second wave of thymus colonization is involved in two different pathways. A first pool is produced as free β_2m, which exerts its chemotactic activity, whereas it is the second pool which is used for MHC class I expression. These sequential phenomena explain the generally high level and the peak of β_2m transcripts observed in thymus during this period, whereas MHC class I heavy chain transcripts were present at low levels and no peak was detected (DUNON et al. 1990a).

These data suggest that β_2m participates in the process of thymus colonization in the chick. It is unclear whether β_2m also plays a chemotactic role in adult birds and mammals, in which β_2m is expressed on most cells as a subunit of MHC class I antigens and is found in body fluids at high concentration. Thymus colonization in adult β_2m-deficient mice is normal (ZIJLSTRA et al. 1990), which suggests that when one molecular entity with chemotactic properties is absent, others can compensate for it.

Although α6 integrins are involved in thymus homing in mice (DUNON and IMHOF 1993; IMHOF et al. 1991), in birds such adhesion molecules involved in pro-T cell interaction with thymic endothelium have not yet been identified. However, in the presence of thymic chemotactic factors including β_2m, quail hematopoietic precursors were able to traverse a human amniotic basement membrane (SAVAGNER et al. 1986). The inhibition of this process by fibronectin-specific antibodies or by synthetic peptides containing RGDS, an integrin binding sequence of fibronectin, suggests that T cell precursors interact with fibronectin during migration in the perivascular space. It has also been shown that laminin-specific antibodies inhibit this invasive process and that migrating precursors express β_1 integrins.

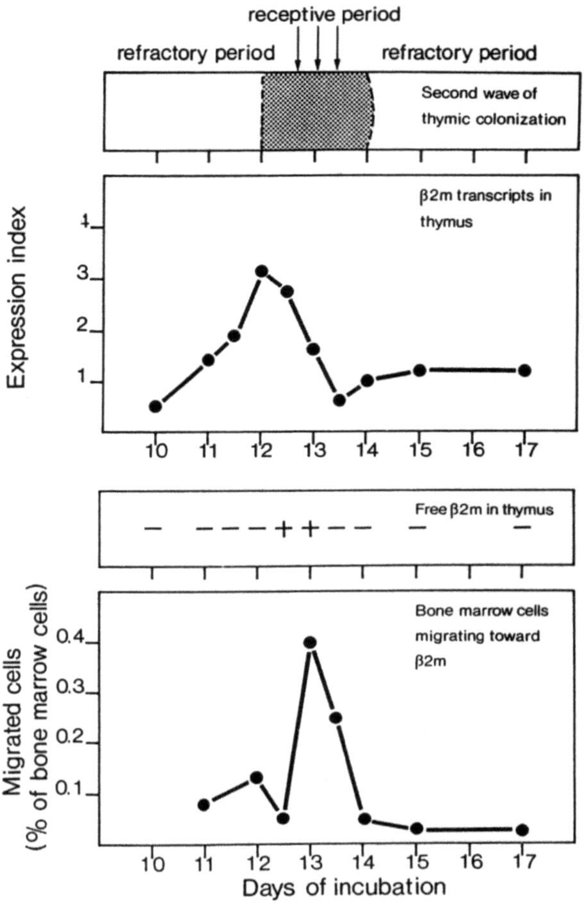

Fig. 1. β Microglobulin (β_2m) expression and β_2m chemotactic activity during the second wave of thymus colonization. Quantification of 1.3 kb β_2m transcript in the thymus during embryogenesis was performed by scanning of northern blot autoradiography after standardization (DUNON et al. 1990b). Detection of free β_2m on thymus section was obtained by peroxidase immunostaining using an antibody which detects only β_2m in its free form. The percentage of bone marrow cells migrating toward β_2m at a concentration of around 3×10^{-11} *M* in a chemotactic assay (DUNON et al. 1990b) was calculated for each different stage of embryogenesis

3 Embryonic Waves of Thymocyte Development

The availability of both an antibody which recognizes TCR γδ and two that recognize forms of TCR αβ containing a Vβ segment of the Vβ1 or Vβ2 family allowed the study of TCR ontogeny (CHAR et al. 1990; CHEN et al. 1988, 1989; CIHAK et al. 1988; LAHTI et al. 1991; SOWDER et al. 1988). On E12, about 5 days after the initial influx of thymocyte precursors, a few chick thymocytes begin to express the TCR γδ/CD3 complex on their surface (SOWDER et al. 1988). The TCR γδ/CD3$^+$ cells increase numerically to reach peak frequency levels by E15, when more than 30% of the thymocytes express TCR γδ. These constitute virtually all of the surface CD3$^+$ thymocytes until around E15 (BUCY et al. 1990). The cytoplasmic expression of the Vβ1 segment can be seen as early as E12 in a subpopulation of thymocytes (BUCY et al. 1990); however, Vβ1$^+$ thymocytes do not appear until E15

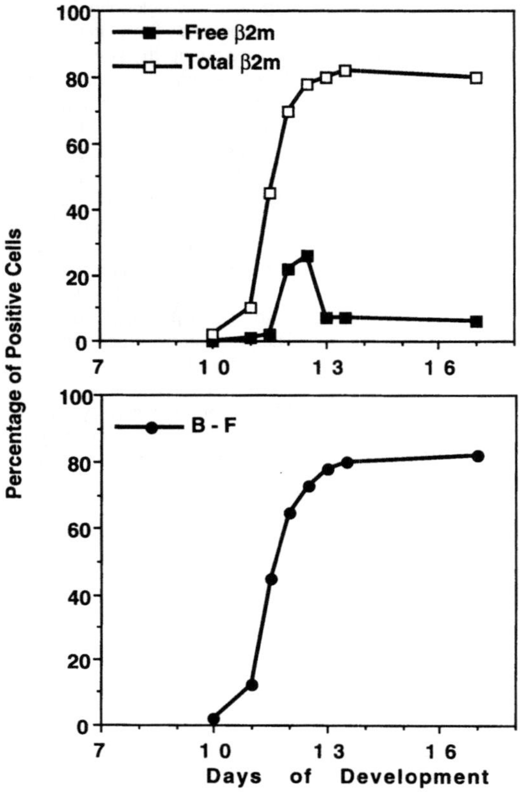

Fig. 2. Appearance of β2-microglobulin (β2m) and MHC class I antigen (B–F) during thymus development. Percentages of positive cells were determined by flow cytometry. Antibody F21-21 was used to detect surface β_2m positive cells, β_2m being either free or associated with the MHC class I heavy chain. Antibody F21-2 was used to detect MHC class I$^+$ (B–F) cells (DUNON et al. 1990a). Free β_2m positive cells were quantified on thymus sections by fluorescent staining of anti-body A4G2. This technique allowed detection of the intracellular pool of free β_2m

(CHEN et al. 1988) and they become the predominant thymocyte subpopulation by E17–18. Finally, the development of the Vβ2$^+$ thymocytes begins later, around E18, and is very similar to that of the Vβ1$^+$ thymocytes (CHEN et al. 1990).

Thus, from the time of the initial influx of thymocyte precursors during the sixth day of chick embryonic life, approximately 2 weeks are required to generate peak numbers of thymocyte progeny. Experimental analysis of thymocyte development in chick-quail chimeras indicates that the γδ, Vβ1 and Vβ2 T cell populations are produced sequentially from the first wave of thymocyte precursors, in agreement with TCR ontogeny (COLTEY et al. 1989). Thymocyte attrition and emigration of maturing T cells then contribute to a sharp decline of this intrathymic population (JOTEREAU et al. 1980; JOTEREAU and LE DOUARIN 1982). When it is measured in quail-chick chimeras, the duration of the first wave is approximately 21 days (COLTEY et al. 1987, 1989), whereas bone marrow transfer experiments in congenic chickens suggest a shorter duration of around 16 days for this wave and each of the subsequent waves (unpublished results).

Thymocyte transfer experiments in chick congenic strains indicate that γδ and Vβ1 thymocytes generated by the different waves of thymocyte precursors also colonize peripheral organs, such as the spleen and the intestine, in discrete

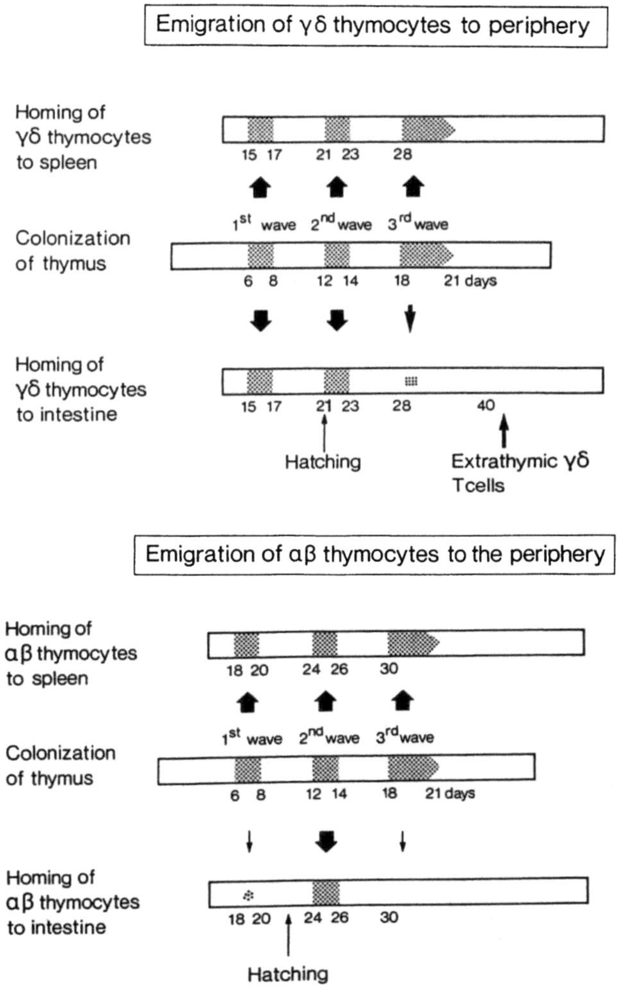

Fig. 3. Embryonic traffic of T lineage cells. The embryonic time periods of the colonization of the thymus by hematopoietic progenitor cells are shown. This phenomenon is followed by the homing of these cells after intrathymic differentiation to peripheral organs. The two schemes illustrate the colonization of the spleen and the intestine by γδ and αβ (Vβ1) thymocytes, respectively

waves (Fig. 3) (Dunon et al. 1993a,b; 1994). It is interesting to note that, according to the difference in time required for a precursor of the first wave of colonization to differentiate into a γδ or a Vβ1 T cell, the periods of emigration of γδ thymocyte and Vβ1 thymocytes differ and even alternate during development (Fig. 3 and unpublished results).

4 Intestinal Intraepithelial T Cells Are Derived from the Thymus During Development

A question that has received much attention in mammals is whether T cells can be generated in nonthymic locations. Analysis of this issue in chick-quail chimeras suggests that during embryonic life the generation of T cells is restricted to the thymic microenvironment (Coltey et al. 1987, 1989). The analysis of the peripheral T cell subpopulations in birds thymectomized at hatching reveals an interesting pattern. Although development of the αβ T-cell subpopulation in the circulation is only moderately compromised, a striking deficit of γδ cells is consistently observed in the thymectomized birds (Chen et al. 1989). This suggests that sustained thymic seeding of γδ cells is required for the development of the circulating pool of γδ cells. In these birds, thymectomized at hatching, the intestinal γδ population is not as severely compromised. This can be attributed to the high self-renewal potential of the intestinal γδ T cells (Dunon et al. 1993a) (see also below). In fact, conclusions about the origin of lymphocytes could not be obtained in animals thymectomized on hatching. The complete delineation of the thymus-dependent system of T cells and their functions required whole body irradiation in addition to thymectomy of the newly hatched chicken, since at hatching T cell migration from the thymus and the population of peripheral tissues are well underway (Cooper et al. 1965).

The developmental origin of chicken intestinal intraepithelial lymphocytes (iIELs) is an important issue, since these cells may play crucial immunological roles such as control of oral tolerance, control of bacterial colonization and elimination of damaged epithelial cells. In the mouse, experiments conducted in immunocompromised adult animals led few scientists to conclude that all murine γδ iIELs are generated extrathymically (Poussier et al. 1992; Poussier and Julius 1994). A more reasonable interpretation of these results is that in the mouse significant proportions of both γδ and αβ iIELs are of extrathymic origin (Lefrancois and Puddington 1995). In the chicken, adoptive cell transfer experiments showed that γδ thymocytes migrated to the intestinal epithelium (Fig. 4), whereas sorted $\gamma\delta^-$ thymocytes enriched for T cell precursors failed to give rise to intestinal γδ T cells (Dunon et al. 1993b). Injection of bone marrow cells and splenocytes from E13 donor embryos into thymectomized newly hatched recipients also indicated that avian γδ and Vβ1 iIELs include few, if any, T lymphocytes that are derived from extrathymic sources. In these experiments, these very few γδ and Vβ1 extrathymic T cells appeared only 1 or 2 months after hatching and in only half of the recipients (Dunon et al. 1993b, 1994). Finally, the fact that none of the Vβ1 and Jβ family members were used exclusively in the spleen or in the intestine argues also against an extrathymic origin of Vβ1 T cells. It can be concluded that chicken γδ and Vβ1 lymphocytes from spleen and intestine, including iIELs, are thymus-derived during chicken development. The extrathymic pathway exists, but its efficiency is practically null, at least during the first months of life, although we cannot exclude that it plays a more important role in the adult chicken. These

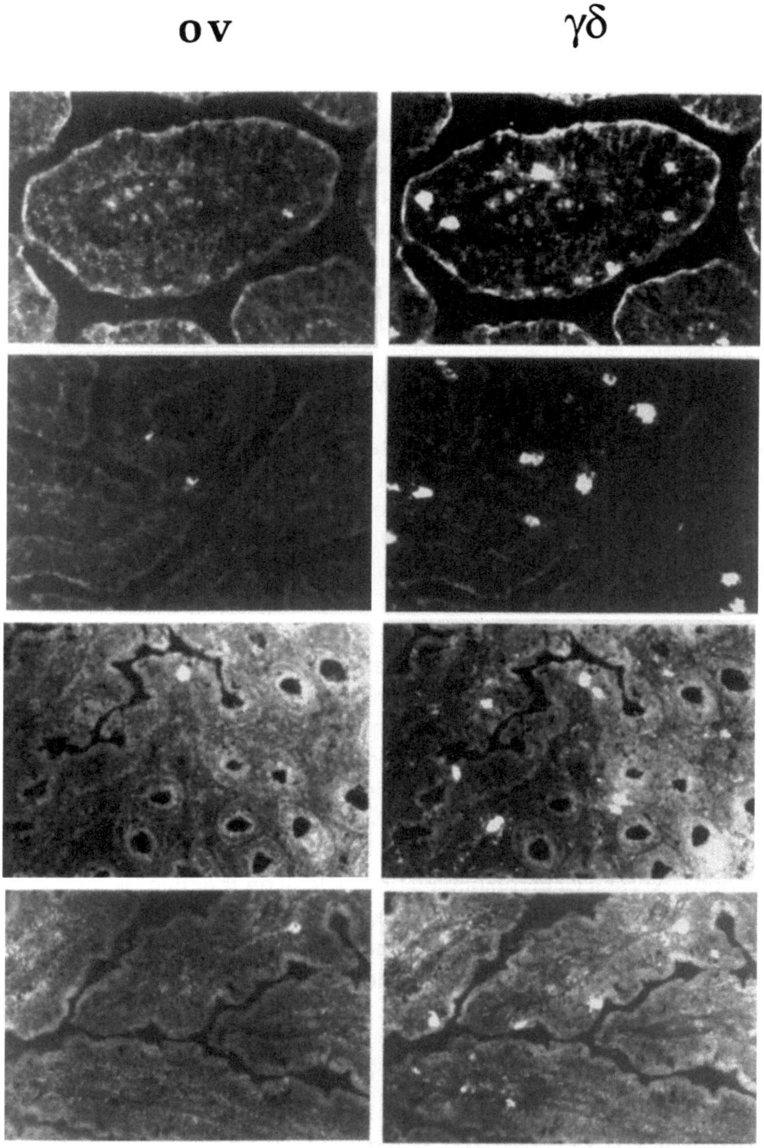

Fig. 4. Embryonic γδ thymocytes colonize the intestinal epithelium. Double immunofluorescence staining of frozen intestine sections; small intestine from 4-day-old H.B19ov - recipient chicks injected at day 16 of embryogenesis with 14 day old ov^+ thymocytes. *Left column*, ov antigen staining; *right column*, γδ T cell receptor staining

conclusions are in agreement with that of Lefrancois and Puddington who propose that the extrathymic function of the iIEL differentiation pathway is dependent on whether a thymus is present or not (LEFRANCOIS and PUDDINGTON 1995).

5 Homing Preferences of γδ Thymocytes

The γδ thymocytes derived from the first wave of precursors can be found in the spleen as early as E15 and in the intestine on the next day. Each developmental wave of γδ thymocytes homes to the spleen and the intestine with the same efficiency (DUNON et al. 1993a). However, γδ thymocytes persist longer in the intestine (more than two and a half months after emigration) than in the spleen. This greater survival time of intestinal γδ T cells is the result of the higher self-renewal potential of these cells. Indeed, the proliferation rate of intestinal γδ T cells reaches 1% per hour following 5-bromodeoxyuridine (BrdU) incorporation, whereas the proliferation rate of γδ splenocytes is only 0.3% per hour. These data suggest that once the population of self-renewing intestinal γδ T cells is established, it may require minimal replenishment.

In the spleen the γδ lymphocytes are located predominantly in the red pulp and are rarely found in germinal centers (BUCY et al. 1988). In the intestine of adult chicken, T cells are preferentially enriched in the epithelium, although 20%–30% of them can be found in the lamina propria. Histological analysis of embryonic intestine revealed that γδ T cells appear in the lamina propria by E15–16 (BUCY et al. 1988; DUNON et al. 1993a, b). Very few γδ T cells are present in the intestinal epithelium from E18 and E19 onwards, although many are still seen within the lamina propria near the basal membrane of the epithelium. γδ T cells enter the epithelium in large numbers at all levels of the villi in the immediate posthatching period (Fig. 4) (DUNON et al. 1993b). Analysis of intestine of thymectomized chicks suggests that γδ intraepithelial lymphocytes comigrate to some extent with the enterocytes, which move continuously towards the villus tip, from which they may be shed into the intestinal lumen.

An additional feature of the intestinal γδ T cells is that, although the γδ thymocytes seldom express CD8 molecules before leaving the thymus, approximately 40% of embryonic intestinal γδ T cells express CD8 on their surface (CHEN et al. 1988; DUNON et al. 1993a). It is not yet known whether this results from a specific homing of CD8γδ T cells to this organ and/or the acquisition of this marker during blood transportation.

6 Ontogeny of T Cell Receptor β Repertoires and Homing Preferences of αβ Thymocytes

The model of Vβ1 T cell differentiation and emigration during development (Fig.3) allowed exploration of the ontogeny of TCR Vβ1 repertoires generated by each wave of thymocyte precursors in thymus, spleen and intestine (DUNON et al. 1994). In the donor congenic chicken strain H.B19ov^{+}, 17 principal Vβ1 gene segments were expressed. Of these, four (Vβ1.1, Vβ1.2, Vβ1.8 and Vβ1.15) exist in two forms, each with one substituted amino acid. Vβ1.17 gene segment contains two different nucleotide sequences but they code for the same amino acids. Only eight Vβ1 members were detected in a single animal (unpublished results), which indicates that the studied strain is not homozygous chicken strains contain six members of the Vβ1 family and no pseudogenes were detected (COOPER et al. 1991). As in other strains, one Dβ and four Jβ segments were encountered (COOPER et al. 1991; DUNON et al. 1994; MCCORMACK et al. 1991; TJOELKER et al. 1990). During ontogeny TCR Vβ1 promoters are turned on by E10, 2 days before detection of TCR-β gene recombination (DUNON et al. 1995). The TCR-β rearrangement can start either by the Vβ1Dβ or the Dβ-Jβ step and is restricted to thymus at least during embryogenesis. Furthermore, alternatively spliced transcripts are found in the thymus and might lead to synthesis of an invariant truncated TCR-β chain containing the NH_2-terminal portion of the Vβ1 region followed by the Cβ region (DUNON et al. 1995). These invariant forms could play a role in thymocyte development.

Analysis of TCR Vβ1 repertoires revealed that all Vβ1 segments are simultaneously expressed as early as day 17 of embryogenesis and that the thymic Vβ1-Dβ-Jβ repertoires generated by each of the three waves of thymocyte progenitors have the same type of diversity with regard to the usage of Vβ1 and Jβ elements and to the length of the Vβ-Dβ-Jβ junction. However, comparison of Dβ-Jβ junctions during the three waves indicated an increase in the number of N nucleotides which was compensated by increased deletion of nucleotides at the 5'end of Jβ in the thymocytes and splenocytes from the third wave. According to the different criteria defined above, colonizations of the spleen and intestine by thymocytes are not characterized by preselection of the TCR Vβ1 repertoire.

A striking TCR β selection does seem to occur during intestinal homing, since it is often impossible to find Vβ2 T cells in intestinal tissue samples, whereas Vβ1 T cells are frequent (CHAR et al. 1990). In addition, the very low frequency of nonproductive Vβ1 rearrangements in the intestine suggests that negative selection can occur in this organ (DUNON et al. 1994). Vβ1 T cells are mainly located within the lamina propria, although a subpopulation of these (30%), which are mainly $CD8^+$, can be found in the epithelial layer. By contrast, the homing patterns of Vβ1 and Vβ2 thymocytes are similar in the spleen. Both populations home to the periarteriolar lymphatic sheaths where they tend to form dense aggregates, whereas fewer αβ lymphocytes are present in the red pulp (COOPER et al. 1991).

Adoptive cell transfers into congenic chickens show that the Vβ1 positive progeny of the first wave of thymocyte precursors home preferentially to the spleen, while the progeny of the second wave have the same affinity for both organs (Fig. 3). The T cell progeny of the third wave, and probably subsequent waves, showed pereferential homing to the spleen. This is consistent with the observation that the increase of N nucleotides at the Dβ-Jβ junction was observed in the progeny of the third wave of thymocyte precursors in the thymus and the spleen but not in the intestine (DUNON et al. 1994). In contrast to γδ IELs, Vβ1 IELs have a lower self-renewal potential than Vβ1 and γδ splenocytes (i.e., a proliferation rate of 0.3% per hour, unpublished result).

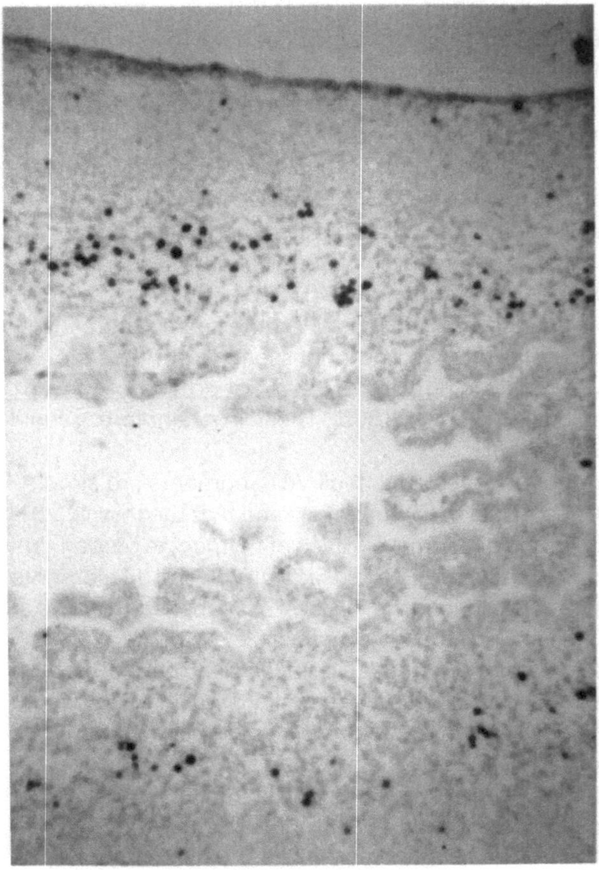

Fig. 5. In vitro binding of 14 day old thymocyte onto intestine sections of 16 day old embryos. The darker cells correspond to the bound thymocytes. They are specifically located on the lamina propria, which is the site of entry of thymocytes into the intestine

7 Perspectives

The emergence of the pattern underlying the migration of thymocyte precursors to the thymus, and the discovery that thymocyte progeny emigrate to peripheral organs in three waves during development, open different fields of investigation. For instance, the mechanisms which lead to the three discrete waves of thymus colonization await further analysis. An intriguing question is that of the physiological role of these discrete waves of thymus colonization, the consequence of which is an alternate emigration of $\gamma\delta$ and Vβ1 T cells to the periphery. The recent cloning of γ and δ TCR genes will soon lead to analysis of TCR repertoires of $\gamma\delta$ T cells during ontogeny, which will allow more detailed studies on thymus emigrants.

Another interesting field is the analysis at the molecular level of the homing process of lymphoid cells to different organs during embryogenesis. Most work has been done on this subject in mice, in which some adhesion molecules such as α6β1 integrin, CD44 and possibly L selectin have been suggested to play a role in thymus colonization. However, the details of this process are far from being known but, some aspects of the analysis could be performed in chicken. Adhesion molecules may also be involved in the homing of thymocytes to other organs. For example, a molecule involved in homing to the intestine could be the chicken A19 antigen expressed on IELs, which might be a homolog of the αEβ7 integrin (Haury et al. 1993). Recently, an in vitro binding assay of chicken thymocytes onto embryonic intestine sections has been set up (Fig.5). It can be used to complement the in vivo homing assay using congenic chicken strains and should thus allow further investigation of adhesion events supporting colonization of the intestine. Notably, it would be particularly interesting to determine the molecular difference between Vβ1 and Vβ2 thymocytes which results in the homing of the Vβ1 cells to the intestine, whereas Vβ2 cells do not. More importantly, it remains to be shown whether the chicken offers an experimental system which will lead to a discovery of a further general mechanism which functions also in mammals.

Acknowledgments. Part of this work was supported by the *Association pour la Recherche contre le Cancer* (ARC). The Basel Institute for Immunology was founded by and is supported by F. Hoffmann-La Roche and Co, Ltd, Switzerland. The authors thank Barbara Ecabert and J.-P. Dangy for technical assistance; Hanspeter Stahlberger for artwork; Hans Spalinger for photography.

References

Auerbach R (1961) Morphogenetic interactions in the development of the mouse thymus. Dev Biol 2: 271–284

Ben Silmane S, Houillier F, Tucker G, Thiery JP (1983) In vitro migration of avian hemoietic cells to the thymus. Cell Differ 13: 1–24

Bucy RP, Chen CH, Cihak J, Lösch U, Cooper MD (1988) Avian T cells expressing γδ receptors localize in the splenic sinusoids and the intestinal epithelium. J Immunol 141: 2200–2205

Bucy RP, Chen CL, Cooper MD (1990) Ontogeny of T cell receptors in the chicken thymus. J Immunol 144: 1161–1168

Champion S, Imhof BA, Savagner P, Thiery JP (1986) The embryonic thymus produces chemotactic peptides involved in the homing of hemopoietic precursors. Cell 44: 781–790

Char D, Sanchez P, Chen C, Bucy R, Cooper M (1990) A third sublineage of avian T cells can be identified with a T cell receptor-3-specific antibody: J Immunol 145: 3547–3555.

Chen C, Sowder J, Lahti J, Cihak J, Lösch U, Cooper M (1989) TCR3: a third T cell receptor in the chicken. Proc Natl Acad Sci USA 86: 2351–2355

Chen CH, Cihak J, Lösch U, Cooper MD (1988) Differential expression of two T cell receptors, TcR1 and TcR2, on chicken lymphocytes. Eur J Immunol 18: 539–543

Cihak J, Ziegler-Heitbrock HWS, Trainer H, Schranner I, Merkenschlager M, Lösch U (1988) Characterization and functional properties of a novel monoclonal antibody which identifies a T cell receptor in chickens. Eur J Immunol 18: 533–537

Coltey M, Jotereau FV, Le Douarin NM, (1987) Evidence for a cyclic renewal of lymphocyte precursor cells in the embryonic chick thymus. Cell Differ 22: 71–82

Coltey M, Bucy RP, Chen CH, Cihak J, Lösch U, Char D, Le Douarin NM, Cooper MD (1989) Analysis of the first two waves of thymus homing stem cells and their T cell progeny in chick-quail chimeras. J Exp Med 170: 543–557

Cooper MD, Peterson DA, Good RA (1965) Delineation of the thymic and bursal lymphoid system in the chicken. Nature 205: 143–146

Cooper MD, Chen-Lo HC, Bucy RP, Thompson CB (1991) Avian T cell ontogeny. Adv Immunol 50: 87–117

Dargemont C, Dunon D, Deugnier MA, Denoyelle M, Girault JM, Lederer F, Le KHD, Godeau F, Thiery JP, Imhof BA (1989) Thymotaxin, a chemotactic protein, is identical to β2-microglobulin. Science. 246: 803–806

Deugnier MA, Imhof BA, Bauvois B, Dunon D, Denoyelle M, Thiery JP (1989) Characterization of rat T cell precursors sorted by chemotactic migration toward thymotaxin. Cell. 56: 1073–1083

Dunon D, Imhof BA (1993) Mechanisms of thymus homing. Blood 81: 1–8

Dunon D, Kaufman J, Salomonsen J, Skjoedt K, Vainio O, Thiery JP, Imhof BA (1990a) T cell precursor migration towards β2-microglobulin is involved in thymus colonization of chicken embryos. EMBO J 9: 3315–3322

Dunon D, Salomonsen J, Skjødt K, Kaufman J, Imhof BA (1990b) Ontogenic appearance of MHC class I (B–F) antigens during embryogenesis. Dev Immunol 1: 127–135

Dunon D, Cooper MD, Imhof BA (1993a) Thymic origin of embryonic intestinal γ/δ cells. J Exp Med 177: 257–263

Dunon D, Cooper MD, Imhof BA (1993b) Migration patterns of thymus-derived gamma delta T cells during chicken development. Eur J Immunol 23: 2545–2550

Dunon D, Schwager J, Dangy JP, Cooper MD, Imhof BA (1994) T cell migration during development: homing is not related to TCR Vβ1 repertoire selection. EMBO J 13: 808–815

Dunon D, Schwager J, Dangy JP, Imhof BA (1995) Ontogeny of TCR Vβ1 exoression revealed novel invariant alternative transcripts. J Immunol 154: 1256–1264

Göbel TW, Chen CL, J Lahti J, Kubota T, Kuo CL, Aebersold R, Hood L, Cooper MD (1994) Identification of T-cell receptor alpha-chain genes in the chicken. Proc Natl Acad Sci USA 91: 1094–1098

Haury M, Kashara Y, Bucy R, Cooper M (1993) Intestinal lymphocytes in the chicken express an integrin-like antigen. Eur J Immunol 23: 313–319

Houssaint E, Mansikka A, Vainio O (1991) Early separation of B and T lymphocyte precursors in chick embryo. J Exp Med 174: 397–406

Imhof BA, Deugnier MA, Girault JM, Champion S, Damais C, Itoh T, Thiery JP (1988) Thymotaxin: a thymic epithelial peptide chemotactic for T-cell precursors. Proc Natl Acad Sci USA 85: 7699–7703

Imhof BA, Ruiz P, Hesse B, Palacios R, Dunon D (1991) EA-1, a novel adhesion molecule involved in the homing of progenitor T lymphocytes to the thymus. J Cell Biol 114: 1069–1078

Jotereau FV, Le Douarin NM (1982) Demonstration of a cyclic renewal of the lymphocyte precursor cells in the quail thymus during embryonic and perinatal life. J Immunol 129: 1869–1877

Jotereau FV, Houssaint E, Le Douarin NM (1980) Lymphoid stem cell homing to the early thymic primordium of the avian embryo. Eur J Immunol 10: 620–627

Klein J (1986) Natural history of the major histocompatibility complex. Wiley, New York, p 775

Lahti J, Chen C, Tjoelker L, Pickel J, Shat K, Calnek B, Thompson C, Cooper M (1991) Two distinct αβ T-cell lineages can be distinguished by the differential usage to T-cell receptor Vβ gene segments. Proc Natl Acad Sci USA 88: 10956–10960

Le Douarin NM (1978). Ontogeny of hematopoietic organs studied in avian embryo interspecific chimeras. In: *Differentiation of normal and neoplastic hematopoietic cells, vol 5* Clakson B, Marks PA, Till TE, (eds) Cold Spring Harbor Laboratory Press, Cold Spring Harbor, New York, pp 5–31

Le Douarin NM, Jotereau FV, Houssaint E, Thiery JP (1984). Primary lymphoid organ ontogeny in birds. In: Le Douarin NM, McLaren A (eds) Chimeras in developmental biology. Academic, London, pp 179–216

Lefrancois L, Puddington L (1995) Extrathymic intestinal T-cell development: virtual reality? Immunol Today 16: 16–21

Martin C, Beaupain D, Dieterlen-Lievre (1978) Developmental relationships between vitellin and intra-embryonic haemopoiesis studied in avian yolk sac chimeras. Cell Differ. 7: 115–130

McCormack W, Tjoelker L, Stella G, Postema C, Thompson C (1991) Chicken T-cell receptor β-chain diversity: an evolutionarily conserved Db-encoded glycine turn within the hypervariable CDR3 domain. Proc Natl Acad Sci USA 88: 7699–7703

Moore MAS, Owen JJT (1967) Experimental studies on the development of the thymus. J Exp Med 126: 715–722

Poussier P, Julius M (1994) Intestinal intraepithelial lymphocytes: the plot thickens. J Exp Med 180:1185–1189

Poussier P, Edouard P, Lee C, Binnie M, Julius M (1992) Thymus-independent development and negative selection of T cells expressing T cell receptor α/β in the intestinal epithelium: evidence for distinct circulation patterns of gut and thymus-derived T lymphocytes. J Exp Med 176: 187–199

Savagner P, Imhof BA, Yamada KM, Thiery JP (1986) Homing of hemopoietic precursor cells to the embryonic thymus: characterization of an invasive mechanism induced by chemotactic peptides. J Cell Biol 103: 2715–2727

Sowder JT, Chen CH, Ager LL, Chan MM, Cooper MD (1988) A large subpopulation of avian T cells express a homologue of the mammalian T gd receptor. J Exp Med 167: 315–321

Tjoelker L, Carlson L, Lee K, Lahti J, McCormack W, Leiden J, Chen C, Cooper M, Thompson C (1990) Evolutionary conservation of antigen recognition: the chicken T-cell receptor β chain. Proc Natl Acad Sci USA 87: 7856–7860

Vainio O, Veromaa TV, Eerola E, Toivanen P (1987). Characterization of two monoclonal antibodies against chicken T lymphocytes surface antigen. In: Immunology. Weber WT, Ewert DL, (eds) Avian Immunology. Liss, New York, pp 99–108

Zigmond SH (1978) Chemotaxis by polymorphonuclear leukocytes. J Cell Biol 77: 269–287

Zijlstra M, Bix M, Simister NE, Loring JM, Raulet DH, Jaenisch R (1990) β2- microglobulin deficient mice lack CD4–CD8+ cytolytic cells. Nature 344: 742–746

The Avian Model in the Study of Tolerance to Self

C. Corbel

1 Introduction

The problem of how tolerance to self is established is central in immunology. T lymphocytes responsible for cell-mediated immunity are generated in the thymus. During their intrathymic phase of differentiation, they acquire receptors (T cell receptors, TCRs) which enable them to recognize self-peptides in the context of self-MHC antigens (see Germain 1994 for review). The generally accepted view is that T cells potentially harmful to the individual, because they carry a TCR with high affinity to self-antigens, are eliminated soon after they emerge in the thymus (Burnet 1957; Lederberg 1959; see also von Boehmer 1990 for review). However, deletion of autoreactive clones within the thymus, although demonstrated for T cells with TCR-recognizing superantigens (Kappler et al. 1987; MacDonald et al. 1988) and in transgenic mice (Kisielow et al. 1988), is probably not complete. T lymphocytes able to recognize self-antigens actually exist in the post thymic, peripheral T cell compartment and can be activated in various autoimmune conditions. Other mechanisms capable of inactivating such peripheral T cells must exist to maintain the integrity of self-components in healthy individuals.

Institut d'Embryologie cellulaire et moléculaire du CNRS et du Collège de France, 49bis, Avenue de la Belle-Gabrielle, 94736 Nogent-sur-Marne Cédex, France

Taking advantage of the possibilities offered by the avian embryo model, we have devised an experimental paradigm through which we show that the epithelial component of the thymic rudiment is essential in tissue tolerance induction. This tolerance is not due to clonal elimination, but rather to a mechanism through which anti-self T cells are inactivated at the periphery. We also demonstrate that additional inhibitions exerted on T cells outside the thymus, in the peripheral tissues, concur with the intrathymic selection to protect self-components against potentially harmful T cells.

Our experimental system thus addresses the roles of both central and peripheral mechanisms of tolerance induction occurring in addition to the classical clonal elimination process.

2 Central Tolerance

Central tolerance refers to the discrimination between self and non-self which is mediated by a mechanism that occurs inside the thymus. Clonal deletion or death of self-reactive cells was demonstrated by means of monoclonal antibodies (Mabs) directed against a given variable region of the TCRβ subunit which recognizes superantigens and by using transgenic mice (KAPPLER et al. 1987; KISIELOW et al. 1988).

Clonal deletion is considered to be the major mechanism for inducing tolerance to antigens expressed in the thymus, although clonal anergy or functional inactivation was also shown to occur (RAMSDELL et al. 1989; ROBERTS et al. 1990). Anergy implies that autoreactive T cells expressing αβTCRs survive but are not able to respond to the antigen. It is accepted that these processes occur inside the thymus during development of immature T lymphocytes, but the type of stromal cells through which they are mediated is not clearly known. The microenvironment is which T cells differentiate is made up of two thymic components: (1) an epithelial component derived from the endoderm of pharyngeal pouches and (2) hematopoietic cells which colonize the thymic epithelium during the course of embryogenesis. In the 1980s, several groups addressed the question of the respective roles of epithelial and hematopoietic stromal cells in T cell differentiation and selection by trying to deplete the thymus of hematopoietic cells. The consensus view was that T cells with low affinity to self would be positively selected on the epithelium whereas cells with high affinity for self would be negatively selected by the dendritic cells of the medulla.

We took advantage of the avian embryonic model to further investigate the site of central tolerance induction. Indeed, it is possible to introduce definite embryonic areas into the avian embryo by using precise microsurgical techniques. Our aim was to graft thymic rudiments at the stage when they still consisted of endodermal buds, derived from the third and the fourth pharyngeal pouches, and prior to their colonization by hematopoietic precursor cells (HPCs). This was feasible, since the time at which the first HPCs enter the thymic epithelial rudiment has been previously demonstrated (LE DOUARIN and JOTEREAU 1973, 1975; COLTEY et al. 1987). Thymus development has been investigated in quail and chick embryos, and the timing of the three waves of colonization by HPCs was precisely established. At embryonic day 4.5 (E4.5) quail thymic primordia have not yet been colonized by HPCs. The first invasion of the thymic epithelium begins at E5 in the quail and at E6.5 in the chick and lasts for 24 and 36 h, respectively, in quail and chick embryos. These HPCs not only yield lymphocytes but also the MHC class II-positive dendritic cells and macrophages of the thymic medulla (GUILLEMOT et al. 1984).

2.1 Xenogeneic Wing Chimeras

The capacity of thymic epithelial cells to induce tolerance when introduced into an individual during embryogenesis was analyzed in a xenogeneic situation. An E3.5 quail limb bud was isotypically grafted into a E4 chick host. It developed normally in the embryo and remained healthy during the first few days after hatching. However, within the first 2 weeks after birth, acute rejection led to total destruction and auto amputation of the wing during the following month. Rejection always started before day 15 posthatching (OHKI et al. 1987).

To investigate the possible role of the thymic epithelial component in tolerance induction towards quail limb tissue, quail thymic epithelial rudiments were in situ implanted into a totally or partially thymectomized E4.5 chick host, which also received a quail limb bud from the same donor. In such chimeras a definitive state of tolerance or a significant delay in the onset of wing rejection was observed. Full tolerance to the grafted wing was obtained for most chimeras which were kept alive for 2–16 months. In the others, the first signs of rejection appeared between 18 and 83 days posthatching, i.e., longer than in the chimeras without thymic grafts. In some cases, reversible signs of rejection were observed (OKHI et al. 1987). Chimerism analysis of the thymus was performed using Feulgen staining and species-specific (quail and chick) anti-class II Mabs. In the chimeric thymus, the antibody directed against quail class II (B-L in birds) MHC antigens reacted with the epithelial cells in the cortex, whereas the medullary dendritic cells of hematopoietic origin expressed chicken class II antigens. A positive correlation could be established between tolerance induction and the presence of quail epithelial stroma in the host's thymus. The amount of thymic tissue in which the epithelial stroma was of quail origin in tolerant birds was found to be more than one third of the total volume of the thymus, as determined by the number of chimeric chick lobes found at dissection (OHKI et al. 1988).

2.2 Xenogeneic Bursal Chimeras

Tissue tolerance induced by grafting thymic epithelium has been obtained for somatic tissue other than the wing. This was shown for the epithelium of the bursa of Fabricius (Belo et al. 1989).

Xenogeneic quail-chick bursal chimeras were constructed by selective removal of the bursal primordium followed by in situ implantation of a bursal rudiment from a quail at the same developmental age, E5. At this stage, the bursal anlage is an endodermal bud surrounded by mesenchymal cells and not yet colonized by HPCs. It was previously shown that the bursa of Fabricius becomes colonized via a single influx of HPCs that takes place between E7 and E11 in the quail and between E8 and E14 in the chick (Houssaint et al. 1976). Such isotopic and isochronic grafts of the quail bursa of Fabricius rudiments into chick embryos resulted in the development of a chimeric bursa whose chick host B lymphocytes and accessory cells differentiated in a foreign quail epithelial environment. However, by the second week posthatching, the quail bursal stroma of these chimeric animals is subjected to acute immune rejection and is completely eliminated within the first month after birth (Belo et al. 1985; Corbel et al. 1987). Isotopic embryonic grafts of the thymic epitheliomesenchymal anlagen from the quail donor of the bursal rudiment were carried out at E4.5, following partial or complete thymectomy. As for the thymus-wing chimeras, the quail thymic epithelial stroma was accepted and invaded by chick HPCs, which differentiated into lymphocytes and dendritic cells. Tolerance of the foreign bursa was induced in such thymo-bursal chimeras (Belo et al. 1989). The thymuses were analyzed for chimerism. It was found that, except in two cases, the chick rudiment had not been completely removed. The thymic lobes developing in the chimeras were therefore made up of either quail or chick epithelial cells but rarely of a mixture of both (Table 1). Tolerance occurred provided that lobes with quail epithelial cells formed at least one third of the total volume of the thymic tissue. This indicates that two populations of peripheral T cells were generated, one which differentiated

Table 1. Correlation between the presence of quail thymic epithelial cells and the tolerance of the grafted quail bursa

Chimera number	Thymic lobes with epithelial cells of:			Tolerance of the BF
	Quail type only	Chick type only	Quail and chick only	
.37	10	0	0	+
121	14	0	0	+
138	10	3	0	+
154	0	6	3	+
407	16	0	2	+
409	11	1	1	+
410	5	1	3	+
123	0	14	0	–

BF, bursa of Fabricius.

in contact with the quail and the other with the chick thymic epithelium. Nevertheless tolerance of quail tissue resulted.

3 Peripheral Tolerance

Since many antigens are not represented in the thymus, such as tissue-specific antigens, the establishment of tolerance must occur in the post thymic environment. The existence of peripheral tolerance induction has been demonstrated by means of transgenic mice constructed in order to direct the expression of a given antigen in extrathymic tissues. Mice expressing class I (MORAHAN et al. 1989; MILLER et al. 1989; HAMMERLING et al. 1991; SCHÖNRICH et al. 1991). or class II (LO et al. 1989; SARVETNICK et al. 1988; BÖHME et al. 1989; MURPHY et al. 1989; MILLER et al. 1990) MHC antigens or viral proteins (ADAMS et al. 1987; ROMAN et al. 1990; OHASHI et al. 1991; LO et al. 1992) under the control of tissue-specific promoters are generally tolerant to the respective target tissue or organ, provided that expression starts early in development. In most cases, the mechanism responsible for tolerance induction is anergy. However, in the periphery, as inside the thymus, clonal deletion can occur. Clonal deletion of mature T cells has been shown for antigens encoded by endogenous mouse mammary tumor virus (WEBB et al. 1990).

Thus, different mechanisms seem to exist for peripheral tolerance induction. These may be dependent on the site where the antigen is expressed, its amount and the cell type bearing it.

Besides the transgenic model, we demonstrated that foreign antigens introduced into the avian embryo and expressed locally (and not in the thymus) can induce tolerance to adult skin grafts of a similar haplotype during the entire life span of the animal. Such experiments performed in birds support the existence of peripheral mechanisms for inducing tissue tolerance (CORBEL et al. 1990; MARTIN et al. 1991).

3.1 Allogeneic Wing Chimeras

Allogeneic wing chimeras, constructed by in situ transplantation of a limb bud from a chick embryo of the B4 or B12 MHC haplotype into a (B15 x B21) F1 chick recipient, were studied from birth for periods of 5–11 months posthatching (Fig. 1). A prolonged state of tolerance was observed in most animals since the grafted wing was tolerated for more than 2 months. In general, when signs of rejection could be visualized they were slight and characterized only by a moderate edema, followed by scab formation and cicatrization. Therefore, early

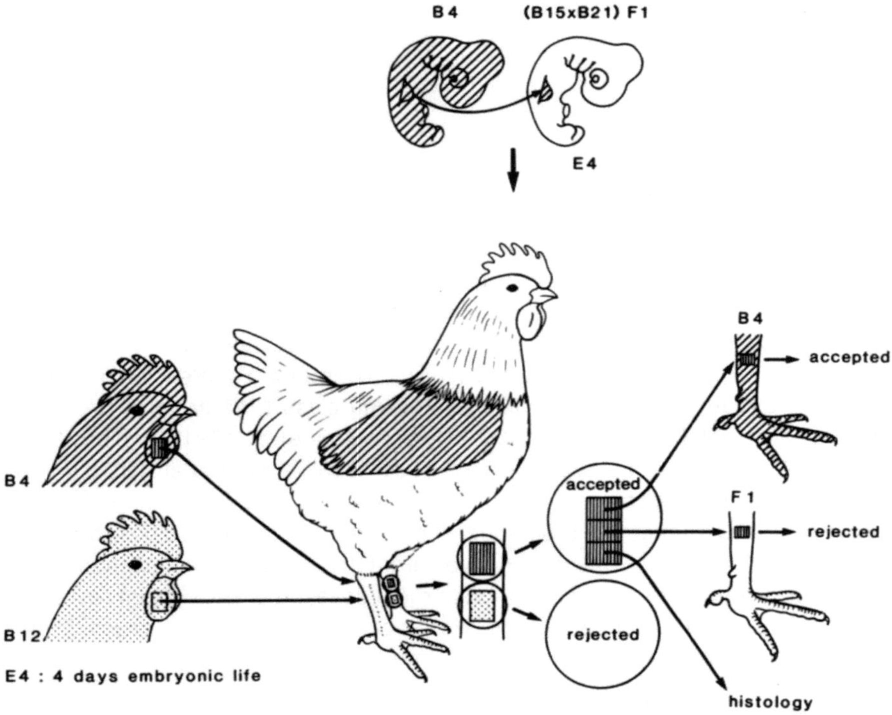

Fig. 1. Allogeneic wing chimera. The right limb bud of a B4 chick embryo was substituted to its counterpart in a (B15 x B21)F1 chick embryo, at E4. Two wattle grafts were performed 2–3 months later, one from a B4 and the other from a B12 haplotype. The tolerated B4 skin graft was cut in three fragments. Two of them were regrafted, one onto a B4 chick, which was accepted, the other onto a F1 chick, which was rejected. The third fragment was processed for histology. (From CORBEL et al. 1990).

grafting of embryonic limb buds in allogeneic combinations resulted in long-term tolerance with slight signs of chronic rejection. Moreover, it induced an in vivo tolerance state for adult skin grafts from a donor with the same MHC haplotype as the wing. The tolerated allogeneic skin grafts of the donor wing type were not only perfectly healthy but retained their ability to be accepted or rejected in an MHC-specific fashion if grafted in a naive recipient (Fig. 1). This confirms that the adult skin grafts were tolerated up to 8 months without having lost their antigenicity. In two cases, a second donor type skin graft was performed 1 month after the first one and was also perfectly tolerated.

Comparable results have been obtained in amphibians by FLAJNIK et al. (1985). *Xenopus* grafted with an allogeneic eye anlage at the tailbud stage became tolerant to the grafted eye and to the donor type skin graft.

In our bird model, we showed that tolerance did not extend to the mixed leukocyte reaction (MLR) between host and donor T cells, since the proliferative responses of blood T cells from all chimeras tested against wing donor MHC haplotype stimulators were comparable to those elicited by T cells of control ungrafted chicks (Corbel et al. 1990). Moreover, tissue tolerance based on unresponsiveness, which operates extrathymically can occur between different species provided they are closely related in taxonomy. Like allogeneic grafts in chicken, a chick or guinea-fowl limb bud grafted into a quail at E4 is tolerated and induces adult skin graft tolerance without modifying the MLR response (Martin et al. 1991). In all of these experiments, "split tolerance" was observed. Split tolerance, i.e., the lack of correlation between assays of T lymphocytes in vitro (MLR or cell-mediated lympholysis) or and in vivo (such as tissue acceptance), has already been reported.

3.2 Allogeneic Bursal Chimeras

We then searched whether peripheral tolerance could be induced by different types of embryonic tissues, such as an endomesodermal organ, the bursa of Fabricius.

We grafted the epithelio-mesenchymal rudiment of the bursa of Fabricius between histoincompatible chickens at E5 and found that: (1) the transplant is normally colonized by HPCs from the host and (2) that a normal contingent of B cells is produced. The grafted bursa is tolerated after birth for a few weeks (4–8 or more) but is, in all cases, rejected by an immune mechanism which can in no case be compared to the physiological involution process which occurs at 4–5 months of age. In these allogeneic bursa chimeras, the tolerant state is relatively short-lived as compared to that observed in wing chimeras (Martin et al. 1994). Moreover, even in the first month after birth, when the bursa is still healthy, skin grafts of the same MHC haplotype are promptly rejected (Table 2).

Therefore, different organ grafts are not equally competent to induce host tolerance even if they are performed early in development, that is before the host's immune system has started to develop. The differences between the immunological status of the limb and that of the bursa may be due to the size of the graft and the nature of the tissue-specific antigens. The fact that the bursa is smaller than a chicken wing may account for the difference observed in the postnatal behavior of the two types of grafts.

3.3 Allogeneic Thymic Chimeras

The same protocol to test induction of tolerance by thymic epithelium grafts in the same allogeneic combinations was used as a control. Seven chimeras were obtained by grafting in situ E4.5 thymic epithelium into a histoincompatible, partially thymectomized, E5 chick embryo. All the allogeneic thymic chimeras remained healthy for several months (Table 3). MLRs were performed with peripheral blood lymphocytes (PBLs) when the chimeras were 2–3 months old, and all exhibited proliferative responses against donor type stimulator cells

Table 2. In vivo tolerance assayed by skin graft acceptance in allogeneic bursal chimeras (from MARTIN et al. 1994)

Chimera	Age[a] (days)	Skin graft rejection (days)		
MHC haplotype of the graft		MHC haplotype of the host	MHC haplotype of the bursa graft	MHC haplotype of third party
GBO 6091 (B12)	33	ND	13	7
GBO 6092 (B4)	32	ND	9	7
GBO 570	30	–[b]	8	9
(B4)	106	ND	9	9
GBO 391 (B4)	46	–[c]	10	10

A normal (B15 x B21) F1 control chick rejected B12 and B4 skin grafts at day 11.
A skin graft from host haplotype was accepted until sacrifice at 60 days.
ND not done.
[a] age of the chimera when the skin were perfomed.
[b] tolerated until sacrifice at 90 days.
[c] tolerated until sacrifice at 15 days.

Table 3. In vivo tolerance assayed by skin graft acceptance in thymic chimeras (from MARTIN et al. 1994)

Chimera	Age[a]	Tolerance Of Skin Graft (+/–)[b]	
		Donor Type	Third Party
B4 → (B15 x B21) F1			
TOP 26	123	+ (43 days)	–
3286	80	+ (13 months)	–
B12 → (B15 x B21) F1			
TOP 31	102	+ (135 days)	–
40	133	+(13 months)	–
47	135	+(13 months)	–
53	142	+(13 months)	–
55	135	+ (12 months)	–

[a] age of the chimera when skin grafts were performed (days).
[b] + indicates that skin grafts were perfectly tolerated (duration in parentheses).
– means rejection; third party skin grafts were rejected from 8 to 15 days postgrafting.

comparable to those obtained against a third-party graft (Fig. 2). However, skin grafts from the donor MHC haplotype were permanently accepted. Although the chimeras rejected a third-party skin graft at 8–15 days postgrafting, the donor type skin grafts were tolerated up to the time of sacrifice (Table 3).

4 Conclusions

The analysis of avian embryonic chimeras, in which the endodermal epithelial thymic rudiment of a chick embryo was partly or completely substituted by its

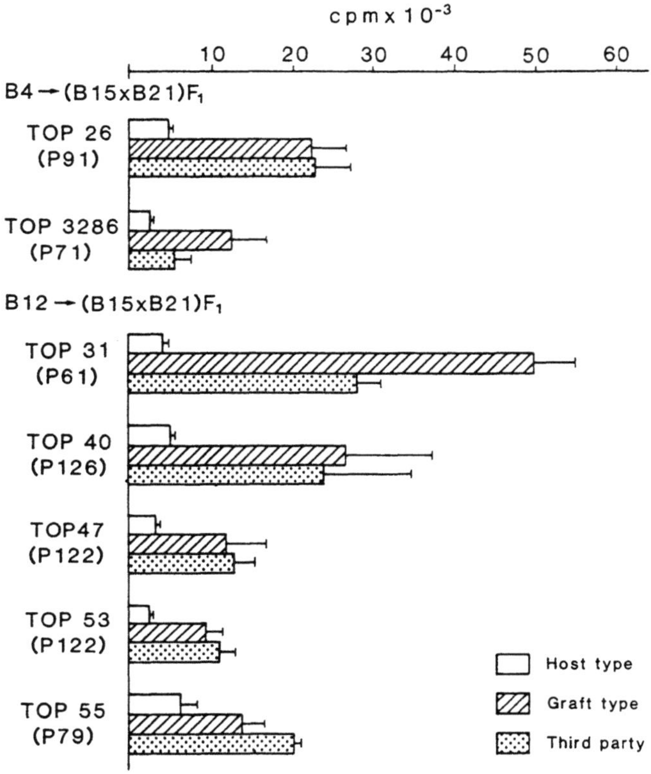

Fig. 2. Mixed leukocyte reaction of blood T lymphocytes from allogeneic thymic chimeras. The age of the chimeras (*TOP*) when the assay was performed is given in *brackets* (*P*, posthatching). Results are expressed as the mean cpm ± SD for three replicate wells

quail counterpart, led to the notion that central tolerance to tissue grafts can be induced by the epithelial component of the thymus. In these chimeras, the chick host permanently tolerates all types of tissues including those of the wing (ectomesodermal) or of the bursa of Fabricius (endomesodermal; OHKI et al. 1987, 1988; BELO et al. 1989; CORBEL et al. 1989). Moreover, the persistence of two thirds of the normal thymus of the chick host did not prevent tolerance to quail tissue grafts to be induced by one third of the total thymic lobes in which the epithelium was of quail origin (OHKI et al. 1988; BELO et al. 1989). This finding suggests the presence of regulatory suppressive T cells. Tissular tolerance has also been obtained in the chick in allogeneic MHC combinations between host and donor thymic epithelium (HOUSSAINT et al. 1986; MARTIN et al. 1994).

The capacity of the thymic epithelium to induce tissue-specific tolerance was confirmed in mice in a model using nude mice engrafted at birth with an allogeneic thymic epithelium taken before HPC colonization (LE DOUARIN et al. 1989; SALAÜN et al. 1990, 1992; COUTINHO et al. 1993). The possibility of transfering tolerance by CD4+ T cells from a thymic epithelium grafted nude mouse to a naive nude recipient (Modigliani et al. 1995) strongly suggests that the state of tolerance is due to differentiation in the thymus of cells able to prevent T cell activation at the periphery.

In this series of experiments, we showed that a certain level of tolerance can be induced into a host by grafting peripheral tissues early in embryonic life. Such is the case for the wing rudiment. Tolerance, however, occurs only if host and donor belong to the same species and is not complete, since transient signs of rejection often appear in this type of combination. Not only is the grafted wing tolerated in these chimeras but the latter also accept skin grafts of the same haplotype. In vitro tolerance tested in MLRs show that activation of peripheral lymphocytes of the host by donor stimulator cells remains intact in this situation. Moreover, all types of organs are not equally able to induce tolerance in an allogeneic situation. Thus, the bursa of Fabricius appears as a poor tolerogen, since grafting the bursal rudiment early in embryonic life protracts only slighly its viability in posthatching life and fails to induce tolerance to tissues of a different embryonic origin such as the skin.

Acknowledgments. I thank Pr. N. Le Douarin for critical reading of the manuscript. This work was supported by the CNRS and grants from INSERM and DRET.

References

Adams T, Alpert S, Hanahan D (1987) Non-tolerance and autoantibodies to a transgenic self antigen expressed in pancreatic β cells. Nature 325: 223–228

Belo M, Martin C, Corbel C, Le Douarin N (1985) A novel method to bursectomize avian embryos and obtain quail-chick bursal chimeras. I. Immunochemical analysis of such chimeras by using species-specific monoclonal antibodies. J Immunol 135: 785–794

Belo M, Corbel C, Martin C, Le Douarin NM (1989) Thymic epithelium tolerizes chickens to embryonic grafts of quail bursa of Fabricius. Int Immunol 1: 105–112

Böhme J, Haskins K, Stecha P, van Ewijk W, Lemeur M, Gerlinger P, Benoist C, Mathis D (1989) Transgenic mice with I-A on islet cells are normoglycemic but immunologically intolerant. Science 244: 1179–1183

Burnet MF (1957) A modification of Jerne's theory of antibody production using the concept of clonal selection. Aust J Sci 20: 67–69

Coltey M, Jotereau F, Le Douarin N (1987) Evidence for a cyclic renewal of lymphocyte precursor cells in the embryonic chick thymus. Cell Differ 22: 71–82

Corbel C, Belo M, Martin C, Le Douarin NM (1987) A novel method to bursectomize avian embryos and obtain quail-chick bursal chimeras. II Immune response of bursectomized chicks and chimeras and post-natal rejection of the grafted quail bursas. J Immunol 138: 2813–2821

Corbel C, Belo M, Martin C, Le Douarin NM (1989) Grafts of the bursal primordium in Quail-Chick chimeras are tolerated after implantation of thymic epithelium. Prog Clin Biol Res 307: 31–43

Corbel C, Martin C, Ohki H, Coltey M, Hlozanek I, Le Douarin MN (1990) Evidence for peripheral mechanisms inducing tissue tolerance during ontogeny. Int Immunol 2: 33–40

Coutinho A, Salaün J, Corbel C, Bandeira A, Le Douarin N (1993) The role of thymic epithelium in the establishment of transplantation tolerance. Immunol Rev 133: 225–240

Flajnik MF, du Pasquier L, Cohen N (1985) Immune responses of thymus/lymphocyte embryonic chimeras: studies on tolerance and major histocompatibility complex restriction in Xenopus. Eur J Immunol 15: 540–547

Germain RN (1994) MHC-dependent antigen processing and peptide presentation: providing ligands for T lymphocyte activation. Cell 76: 287–299

Guillemot FP, Oliver PD, Peault BM, Le Douarin NM (1984) Cells expressing Ia antigens in the avian thymus. J Exp Med 160: 1803–1819

Hammerling GJ, Schonrich G, Momburg F, Auphan N, Malissen M, Malissen B, Schmitt-Verhulst A-M, Arnold B (1991) Non-deletional mechanisms of peripheral and central tolerance: studies with transgenic mice with tissue-specific expression of a foreign MHC class I antigen. Immunol Rev 122: 47–67

Houssaint E, Belo M, Le Douarin N (1976) Investigations on cell lineage and tissue interactions in the developing. bursa of Fabricius. Dev Biol 53: 250–264

Houssaint E, Torano A, Ivanyi J (1986) Split tolerance induced by chick embryo thymic epithelium allografted to embryonic recipients. J Immunol 136: 3155–3158

Kappler JW, Wade R, White J, Kushnir E, Blackman M, Bill J, Roehm N, Marrack P (1987) A T cell receptor V-beta segment that imparts reactivity to a class II major histocompatibility complex product. Cell 49: 263–271

Kisielow P, Bluthmann H, Staerz UD, Steinmetz M, von Boehmer H (1988) Tolerance in T-cell receptor transgenic mice involves deletion of non mature CD4+8+ thymocytes. Nature 333: 742–745

Le Douarin NM, Jotereau F (1973) Origin and renewal of lymphocytes in avian embryos thymuses studied in interspecific combinations. Nature New Biol 246: 25–26

Le Douarin NM, Jotereau FV (1975) Tracing of cells of the avian thymus through embryonic life in interspecific chimeras. J Exp Med 142: 17–40

Le Douarin NM, Corbel C, Martin C, Coltey M, Salaün J (1989) Induction of tolerance by embryonic thymic epithelial grafts in birds and mammals. Cold Spring Harbor Symp Quant Biol 54: 777–787

Lederberg J (1959) Genes and antibodies. Science 129: 1649–1653

Lo D, Burkly AL, Flavell RA, Palmiter RD, Brinster RL (1989) Tolerance in transgenic mice expressing class II major histocompatibility complex on pancreatic acinar cells. J Exp Med 170: 87–104

Lo D, Freedman J, Hesse S, Palmiter RD, Brinster RL, Sherman LA (1992) Peripheral tolerance to an islet cell-specific hemagglutinin transgene affects both CD4+ and CD8+ T cells. Eur J Immunol 22: 1013–1022

MacDonald HR, Schneider R, Lees RK, Howe RC, Acha-Orbea H, Festenstein H, Zinkernagel RM, Hengartner H (1988) T cell receptor Vβ use predicts reactivity and tolerance to Mls^a-encoded antigens. Nature 332: 40–45

Martin C, Ohki-Hamazaki H, Corbel C, Coltey M, Le Douarin NM (1991) Successful xenogeneic transplantation in embryos: induction of tolerance by extrathymic chick tissue grafted into quail. Dev Immunol 1: 265–277

Martin C, Belo M, Le Douarin NM, Corbel C (1994) A study of peripheral tolerance through embryonic grafts of the bursal epithelial rudiment between MHC-distinct chick embryos. Int Immunol 6: 795–804

Miller JFAP, Morahan G, Allison J, Bhathal PS, Cox KO (1989) T-cell tolerance in transgenic mice expressing major histocompatibility class I molecules in defined tissues. Immunol Rev 107: 109–123

Miller J, Daitch L, Rath S, Selsing E (1990) Tissue specific expression of allogeneic class II MHC molecules induces neither tissue rejection nor clonal inactivation of alloreactive T cells. J Immunol 144: 334–341

Modigliani Y, Thomas-Vaslin V, Bandeira A, Coltey M, Le Douarin N, Coutinho A, Salaün J (1995) Lymphocytes selected in allogeneic thymic epithelium mediate dominant tolerance towards tissue graft of the thymic epithelium haplotype. Proc Natl Acad Sci USA 92: 7555–7559

Morahan G, Allison J, Miller JFAP (1989) Tolerance of class I histocompatibility antigens expressed extrathymically. Nature 339: 622–624

Murphy KM, Weaver CT, Elish M, Allen PM, Loh DY (1989) Peripheral tolerance to allogeneic class II histocompatibility antigens expressed in transgenic mice: evidence againts a clonal-deletion mechanism. Proc Natl Acad Sci USA 86: 10034–10038

Ohashi PS, Oehen S, Buerki K, Pircher H, Ohashi CT, Odermatt B, Malissen B, Zinkernagel RM, Hengartner H (1991) Ablation of "tolerance" and induction of diabetes by virus infection in viral antigen transgenic mice. Cell 65: 305–317

Ohki H, Martin C, Corbel C, Coltey M, Le Douarin NM (1987) Tolerance induced by thymic epithelial grafts in birds. Science 237: 1032–1035

Ohki H, Martin C, Coltey M, Le Douarin NM (1988) Implants of quail thymic epithelium generate permanent tolerance in embryonically constructed quail/chick chimeras. Development 104: 619–630

Ramsdell F, Lantz T, Fowlkes BJ (1989) A non deletional mechanism of thymic self tolerance. Science 246: 1038–1041

Roberts J, Sharrow S, Singer A (1990) Clonal deletion and clonal anergy in the thymus induced by cellular elements with different radiations sensitivities. J Exp Med 171: 935–940
Roman LM, Simons LF, Hammer RE, Sambrook JF, Gething M-JH (1990) The expression of influenza virus hemagglutinin in the pancreatic β cells of transgenic mice results in autoimmune diabetes. Cell 61: 383–396
Salaün J, Bandeira A, Khazaal I, Calman F, Coltey M, Coutinho A, Le Douarin N (1990) Thymic epithelium tolerizes for histocompatibility antigens. Science 247: 1471–1474
Salaün J, Bandeira A, Khazaal I, Burlen-Defranoux O, Thomas-Vaslin V, Coltey M, Le Douarin N, Coutinho A (1992) Transplantation tolerance is unrelated to superantigen-dependent deletion and anergy. Proc Natl Acad Sci USA 89: 10420–10424
Sarvetnick N, Liggit D, Pitts SL, Hansen SE, Stewart TA (1988) Insulin-dependent diabetes mellitus induced in transgenic mice by ectopic expression of class II MHC and interferon-gamma. Cell 52: 773–782
Schönrich G, Kalinke U, Momburg F, Malissen M, Schmitt-Verhulst A-M, Malissen B, Hammerling GJ, Arnold B (1991) Down-regulation of T cell receptors on self-reactive T cells as a novel mechanism for extrathymic tolerance induction. Cell 65: 293–304
von Boehmer H (1990) Developmental biology of T cells in T cell-receptor transgenic mice. Annu Rev Immunol 8: 531–556
Webb S, Morris C, Sprent J (1990) Extrathymic tolerance of mature T cells: clonal elimination as a consequence of immunity. Cell 63: 1249–1256

Avian Natural Killer Cells

T.W.F. Göbel[1], C.H. Chen[2], and M.D. Cooper[3]

1 Introduction

Natural Killer (NK) cells represent a distinct lymphoid lineage. These cells were initially defined functionally by their ability to kill certain virally infected cells and tumor cells (reviewed in Trinchieri 1989; Ritz et al. 1988). This unique function and its evolutionary conservation led to the classification of NK cells as part of the innate immune system. Recently, renewed interest in NK cells has lead to better understanding of NK cell phenotype, cytokine secretion and reactivity, developmental origin, functional capabilities, as well as the receptors mediating cytotoxicity (Moretta et al. 1994).

We have studied avian embryonic NK cells that were identified by their cytoplasmic CD3 (cy. CD3) expression and undetectable cell surface CD3/ T cell receptor (TCR) (Bucy et al. 1990). These cells have been designated TCR0 cells

[1] Basel Institute for Immunology, Grenzacherstr. 487, 4005 Basel, Switzerland
[2] Division of Developmental and Clinical Immunology, Department of Microbiology, University of Alabama at Birmingham, 378 Wallace Tumor Institute, Birmingham, AL 35294-3300, USA
[3] Division of Developmental and Clinical Immunology, Departments of Microbiology, Medicine, Pediatrics, University of Alabama at Birmingham, and Howard Hughes Medical Institute, 378 Wallace Tumor Institute, Birmingham, AL 35294-3300, USA

and this term will be used in this review. NK cell activity can be detected throughout vertebrate evolution (Evans and McKinney 1991), suggesting that these cells have evolved prior to T or B lymphocytes. A comparative phylogenetic approach to analyze NK cell development and function may therefore provide novel insights into the biology of lymphoid cells in higher vertebrates.

2 Phenotype and Genotype of Natural Killer Cells

2.1 Antigens Expressed Mainly by Natural Killer Cells

A phenotypic comparison of TCRO, T cells and B cells is shown in Table 1. Heterologous anti-asialo GM1 (AGM1) antisera were among the first reagents used to detect mouse and rat NK cells. Although AGM1 stains other cell types, NK cells are contained in the AGM1 bright population (Tiberghien et al. 1990). Lillehoj (1989) reported that chicken intraintestinal epithelial lymphocytes (IELs) are also reactive with AGM1 antibodies.

CD56 (N-CAM) is a widely used marker for human NK cells, which can be further subdivided according to high or low CD56 expression (Baume et al. 1992). Cells of different chicken tissues, i.e., peripheral blood lymphocytes (PBLs), thymus, spleem, IEL, TCR0 cells are unreactive with antisera raised against chicken N-CAM (Göbel, unpublished result). Similarly, mouse NK cells may not express CD56 molecules, but are identified by the NK 1.1 marker, a member of the NKR-P1 gene family (Yokoyama and Seaman 1993). Members of this family are expressed by NK cells and some T cells in mouse, rat and humans Brissette-Storkus et al. 1994).

Table 1. Phenotypic comparison of TCR0, T cells and B cells

Antigen	TCR0 cell	T cell	B cell
CT1	–	+	–
CD3	+cy	+	–
CD4	–	(+)	–
CD8α	(+)	(+)	–
CD8β	–	(+)	–
TCR	–	+	–
CD28	–	+	–
CD45	+	+	+
IL2R	+	(+)	–
αEβ7	–	(+)	–
CD6	(+)	(+)	–
CD56	–	–	–
IgL	–	–	+
MHCII	–	(+)	+

cy cytoplasmic expression only ; 0, only expressed by subsets.

An avian marker similar to the NKR-P1 family is not available yet. However, BERNOT et al. (1994) have recently cloned a gene which shares homologies with this C-type lectin supergene family. It encodes a type II membrane protein with a predicted molecular weight of 28 kDa. By Southern blot analysis, multiple copies of the gene could be detected. While these are characteristics of the NKR-P1 family, the gene is expressed in many tissues. The generation of monoclonal antibodies (mAbs) against the gene product should clarify if it is an avian homologue of the mouse NK1.1 antigen.

CD16, the low affinity FcγR, is expressed by mouse, rat and human NK cell (TRINCHIERI 1989). Two highly related genes encode either a transmembrane form on NK cells or a phosphatidyl-inositol linked form on neutrophils (LANIER et al. 1989a; RAVETCH and PERUSSIA 1989). TCR0 cells were found to bind heat aggregated chicken IgG, reflecting the presence of a similar low affinity receptor (GÖBEL et al. 1994a). The CD16α chain is associated with either a ζ–ζ homodimer or a ζ-FcεRγ heterodimer (LANIER et al. 1989b; VIVIER et al. 1991). These dimers are crucial for signal transduction. They are linked to $p56^{lck}$ and Zap 70 protein tyrosine kinases (SALCEDO et al. 1993; CONE et al. 1993; VIVER et al. 1993) and are tyrosine phosphorylated upon ligand binding (O'SHEA et al. 1991). It will be important to characterize the IgG receptor on chicken NK cells and its possible association with CD3 chains.

HNK1 (CD57), another marker on human NK cells (ABO and BALCH 1981), has been used in immunohistology of avian tissues. Cells with NK cell morphology could be detected with HNK1 mAb in the intraintestinal epithelium (VERVELDE and JEURISSEN 1993). In contrast, TCR0 cells were HNK1$^-$, a discrepancy which could be explained by the rapid loss of HNK1 molecules during in vitro culture (RITZ et al. 1988). HNK1 mAb also reacts with other types of chicken cells (PEAULT et al. 1987) as is the case in mammals.

Two avian NK cell reactive mAbs have been described (CHUNG and LILLEHOJ 1991). These stained different percentages of PBLs and partially inhibited spontaneous cytotoxicity. Molecular analysis will provide more information about these antigens.

2.2 Antigens Shared with T Cells

Depletion of T or B cells using lysis by heterologous antisera and complement was first used to demonstrate that NK cell cytotoxicity was unaffected by this treatment (SHARMA and OKAZAKI 1981). The recent development of avian T cell specific mAbs (reviewed in COOPER et al. 1991; CHEN et al. 1990) allowed a more detailed analysis of antigens shared by T cells and NK cells in the chicken.

By using the CT3 mAb (avian CD nomenclature reviewed in RATCLIFFE et al. 1993) for immunohistology, a subset of embryonic splenocytes was found to express cytoplasmic, but not surface, CD3 (BUCY et al. 1990). The cy. CD3 molecules in the TCR0 cells were biochemically similar to those in T cells. The CT3 mAb preciptated unreduced protein bands of 32 kDa, 20 kDa and 17kDa in both cell types, but an additional band of 19 kDa was visible only in T cells. Northern blot analysis using the cDNA probe encoding the 19 kDa CD3 protein (BERNOT and AUFFRAY 1991) revealed transcripts present in TCR0 cells (GÖBEL et al. 1994a). The lack of the 19 kDa protein, but presence of mRNA transcripts is reminiscent of human embryonic NK cell clones. They express cy. CD3δ and CD3ε molecules, but CD3γ is expressed at very low levels and comigrates with CD3ε due to partial glycosylation (PHILLIPS et al. 1992; LANIER et al. 1992a).

The 32 kDa protein band represents a homo- or heterodimer of two 18 kDa chain CD3 chains. In fact, CD3ε–ε homodimers have been described to form in NK cells (LANIER et al. 1992a) and T cells (BLUMBERG et al. 1990). Alternatively, the 32 kDa protein could represent a CD3 ζ–ζ homodimer, which seems unlikely since in mammals the ζ chains associate with CD3 chains only in the presence of TCR molecules and thus should not be precipitated with the CT3 mAb. The physiological significance of the cy. CD3 molecules in NK cell development or function and the issue of associated molecules need to be further analyzed. Interestingly, in mice with high copy numbers of a human CD3ε transgene, T cell and NK cell development was blocked (WANG et al. 1994).

CD3 associates with TCR molecules to form a complex receptor on T cells (CHEN et al. 1986). However, mAbs against TCRαβ and TCRγδ do not stain TCR0 cells and TCRα (GÖBEL et al. 1994b) or TCRβ (TJOELKER et al. 1990) specific cDNA probes do not react with TCR0 RNA in Northern blot analysis. Furthermore, the TCRβ genes are in germline configuration (GÖBEL et al. 1994a)

CD8 may be found on subsets of TCR0 cells depending on their localization (see Sect. 3). The increase in CD8 frequency observed during culture of embryonic splenocytes might reflect the activation status of these cells grown in the presence of cytokines. The CD8 α and β chains form either αα homodimers or αβ heterodimers on T cells (TREGASKES et al. 1995). In contrast, TCR0 cells, like mammalian NK cells, express the CD8 αα homodimer, exclusively (GÖBEL et al. 1994a).

Development and growth of NK cells is highly dependent on interleukin 2 (IL2). The IL2 receptors (IL2Rs) are found on mammalian NK cells either as an intermediate IL2R composed of the βγ chains or a high affinity IL2R consisting of all three chains (αβγ) (VOSS et al. 1992). The mAb INN-CH-16 (SCHAUENSTEIN et al. 1988), which may recognize the chicken IL2R α chain, stains virtually all TCR0 cells (GÖBEL et al. 1994a). Ontogenic studies using this mAb also detect $IL2R^+$ cells early in ontogeny in the spleen (FEDECKA-BRUNER et al. 1991a).

We have recently characterized a mAb, designated S3, which identifies an avian CD6 candidate (GÖBEL et al. 1994c). CD6 has been described as a pan-T cell marker in humans. Interestingly, the S3 mAb stains some TCR0 cells and a low percentage of embryonic spleen cells (Göbel, unpublished result). Since T cells begin to express this antigen at a late stage during thymic maturation, this mAb could be a useful tool to enrich embryonic NK cells.

CD28, a costimulatory molecule on T lymphocytes, has recently been characterized in chickens (YOUNG et al. 1994). In contrast to IL2 stimulated human $CD28^+$ NK cells, TCR0 cells are $CD28^-$. TCR0 cells are also negative for A19, a homing/

retention molecule that is selectively expressed on IELs and most likely represents an $\alpha E \beta 7$ integrin (HAURY et al. 1993; KASAHARA et al. 1993). TCR0 cells are also negative for CT1 (CHEN et al. 1984), a marker found on early thymocytes (GÖBEL, unpublished result).

2.3 Other Phenotypic Markers

TCR0 cells, which resemble activated NK cells, do not express MHC class II molecules (GÖBEL et al. 1994a). Mouse NK cells are also MHC class II$^-$, whereas activated human and all rat NK cells are MHC class II$^+$ (TRINCHIERI 1989). All TCR0 cells are CD45$^+$, but they lack surface immunoglobulin and K1 (KASPERS et al. 1993), a marker for chicken macrophages and thrombocytes.

3 Morphology and Tissue Distribution of Natural Killer Cells

Detailed analysis of the tissue distribution of avian NK cell is hindered by the lack of highly specific markers. Nevertheless, studies of NK cell distribution have been conducted on the basis of their unique morphology, function and combination of phenotypic markers.

Like the mammalian NK cells, TCR0 cells are large granular lymphocytes. They have a high cytoplasm to nucleus ratio, relative abundance of cell organelles and typical electron dense granules (GÖBEL et al. 1994a). Using morphological criteria, BACK (1972) described IELs as a heterogeneous cell population consisting of 77% lymphocytes and 22% "globule" leukocytes and 1% eosinophils. The globule leukocytes are unaffected by thymectomy or bursectomy (BACK 1970a,b) are larger than IELs, variable in size, highly refractile and contained intensively acidophilic staining granules (KITAGAWA et al. 1988). Using the human HNK1 marker, VERVELDE and JEURISSEN (1993) detected large granulated, dendritic cells in both the epithelium and lamina propria of the colon.

Spontaneous cytotoxicity, as the hallmark of NK cells, was also used to study their tissue distribution. Substantial levels of cytotoxicity were detectable in leukocyte preparations of IELs and spleen after 4 h assays and in thymic or bursal cell suspensions after 16 h incubations. Interestingly, cells of the chicken strains SC and FP exhibited different levels of overall cytotoxic capacity (LILLEHOJ and CHAI 1988). The most potent cytotoxic cells were always localized in the gut, where jejunal or ileal IELs exhibited higher NK cell levels than duodenal or cecal IELs (CHAI and LILLEHOJ 1988).

When BUCY et al. (1990) used the cy. CD3 and surface CD8 expression as markers for NK cell tisue distribution, cy. CD3$^+$ cells were found in the spleen of 8 day old chick embryos (E8), and at E11 and E12 in bursa and IELs, respectively.

The splenic population expanded rapidly to reach peak levels at E14, when they comprised 10% of all nucleated cells. CD8 was expressed by 50% to 70% of the splenic TCR0 cells, 10% of the bursal counterparts and the majority of IEL TCR0 cells. In addition, FEDECKA-BRUNER et al. (1991a) detected $IL2R^+$ cells in increasing frequencies during ontogeny, and these probably included embryonic TCR0 cells.

Once T cell seeding to the periphery has taken place, distinction of T cells and TCR0 cells is limited to indirect calculation of surface $CD3^-$, cy. $CD3^+$ cells which express the CD8αα homodimer. It is unknown whether all NK cells in adult chicken express cy. CD3 components and, in fact, cy. CD3ε is found only in activated human NK cells (LANIER et al. 1992a). However, TCR0 cells are readily detected in the IELs of adult chickens, where they comprise 20%–40% of all lymphocytes (BUCY et al. 1990). In contrast, PBLs and spleen contain relatively few TCR0 cells. Furthermore almost 100% of splenic lymphocytes and PBLs express either B cell, T cell or thrombocyte markers.

4 Developmental Relationship of Natural Killer Cells and T Cells

Like cytotoxic T lymphocytes, TCR0 cells lyse targets, express cy. CD3 proteins and a subset expresses surface CD8. In addition, both cell types respond to T cell derived growth factors and express the IL2R. However, TCR0 cells are surface $CD3^-$ and express CD8αα homodimers.

Using the chick-quail chimera system, BUCY et al. (1989) further investigated the relationship of T cells and TCR0 cells. In these experiments E3 quails were grafted with E6 chicken spleens or bursa fragments. This time point was chosen to prevent the possibility of grafting thymus derived cells, since the first wave of precursors homing to the thymus is not detected before E 6.5. Six days later TCR0 were found in quail spleen, bursa and thymus. The thymic TCR0 cells did not express the early thymocyte antigen CT1. In contrast, T cell maturation was not supported by the quail thymic microenvironment. In addition, whereas early thymectomy of chickens led to a dramatic reduction of all peripheral T cell subsets, TCR0 cells were unaffected (COOPER, unpublished result). Taken together, these experiments demonstrated the thymus independence of TCR0 cells. This is similar to mice, in which NK cells are present in T cell deficient nude mice, SCID mice and RAG2 knockout mice (LANIER et al. 1992b).

However, a common precursor of all lymphoid cells most likely exists. This was elegantly demonstrated by a knockout of the transcription factor Ikaros; B, T and NK cells were not detectable, but other leukocytes were unaffected (GEORGOPOULOS et al. 1994). In addition, the isolation of a $Fc\gamma RII/III^+$ thymocyte and fetal bone marrow population, which developed as T cells when injected intrathymically or as NK cells when cultured with IL2, illustrated the dependence on different microenvironments during lymphoid development (RODEWALD et al. 1992; MOINGEON et al. 1993).

5 Functional Characterization of Avian Natural Killer Cells

5.1 Effector Cells and Cytotoxicity

Crude cell preparations of splenocytes, IELs and other tissues have mainly been used as effector cells (SHARMA and SCHAT 1991). In order to enrich for NK cells, SCHAT et al. (1987) cultured adult splenocytes in the presence of soluble factors produced by mitogen activated spleen cells. These were characterized as large granular lymphocytes with capacity for spontaneous cytotoxicity (SCHAT et al. 1986, 1987). When TCR0 cells were cultured is a similar system, large inhibitory molecules had to be removed from the growth factor preparation to stimulate TCR0 growth (GÖBEL et al. 1994a). The differences could reflect the developmental status of embryonic vs adult cells.

The polyclonal nature and the limited life span of the cultured cells are major limitations of both systems. Using the avian reticuloendotheliosis virus (REV-T) in combination with the chicken syncitial virus (CSV), MARMOR et al. (1993) could transform $CD8^+$, $CD3/TCR^-$ cells, suggeting that TCR0 cells can be transformed by this lymphotropic virus. Recently, we have generated REV-T transformed, bone marrow derived cell lines which express cy. CD3 molecules but no B or T cell specific surface markers and thus resemble the TCR0 phenotype. Some of these cell lines were potent producers of macrophage activating factor and interferon like activity (GÖBEL and KASPERS 1995).

The effector cells mediating antibody dependent cellular cytotoxicity as an alternate function of NK cells have been characterized as Fc receptor-bearing non-T, non-B cells, which were adherent to nylon wool or plastic and resistant to 2000 rad irradiation (CHI and THORBECKE 1981; FLEISCHER 1980). These are some of the characteristics of TCR0 cells.

The mechanism leading to recognition of target cells and subsequent cytotoxicity are largely unknown for avian NK cells. Recent data for mammalian NK cells support a model in which cytotoxicity is regulated by inhibitory and stimulatory events. Binding of NK cell receptors to MHC class I molecules on target cells results in a dominant inhibition of NK cell cytotoxicity. When MHC class I molecules on target cells are lost as a consequence of viral infection or malignant transformation, the NK cell receptor cannot bind to its ligand and other activating signals (like FcγR binding) trigger the NK cell to kill the target (LANIER and PHILLIPS 1992; MORETTA et al. 1992, 1994). The cellular cytotoxicity is mediated by either perforin mediated lysis or FAS mediated triggering of apoptosis in the target (KÄGI et al. 1994).

5.2 Natural Killer Cells and Disease

Natural Killer cell activity has been implicated as a major limiting factor in avian viral (Mareks disease, rotavirus infections) and protozoan (coccidiosis) infections (Sharma and Schat 1991). Infection with the Mareks disease virus (MDV), a herpes virus, causes T cell lymphomas in the chicken (Calnek and Witter 1984). MDV resistant chicken strains may have elevated NK cell levels upon MDV infection, whereas decreased NK cell cytotoxicity was found in MDV susceptible chicken strains (Sharma 1981). In another model using subclinical, self-limiting infections with rotavirus, Myers and Schat (1990) found that IELs were able to lyse rotavirus infected kidney cells independent of prior infection and MHC haplotype. Finally, primary Eimeria infections of chickens decreased NK cell activity in IELs, but upon challenge infection, NK cell cytotoxicity and AGM1$^+$ cells increased (Lillehoj 1989).

Graft vs host reactions (GVHR) have traditionally been studied in chick embryos, because of the ease of injecting donor cells into the large veins of the yolk sac. The GVHR can easily be measured by the increased weight of the spleen 4 or 5 days after injection. Host cells contributing to the splenomegaly included CD8$^+$ TCR$^-$ cells (Fedecka-Bruner et al. 1991b) We found that $\gamma\delta$ T cells, TCR0 cells and other MHC class II$^+$ cells of host origin are recruited in the GVHR lesions (Tsuji et al. manuscript in preparation).

6 Conclusion

Once regarded as a part of the innate immune system with nonspecific function, NK cells are now characterized as a highly specialized, independent lymphoid lineage. The generation of lymphocyte specific mAb has enabled a detailed phenotypical description of chicken NK cells. However, unique markers of NK cells are still not available. In fact, it is still an open question as to whether these antigens exist. Relative to the wealth of knowledge about T and B cell development and function, there is a paucity of information concerning these aspects of NK cell physiology. Provided that T cells evolved as "mirror images of NK cells" (Versteeg 1992), most obviously demonstrated by the shared expression of CD3, their similar phenotype, function and cytokine responsiveness, a detailed phylogenetic analysis of NK cell biology in different species is essential for the comprehensive understanding of T and NK cell development.

Acknowledgments. We would like to thank Drs. J. Kaufman and O. Vainio for critically reading the manuscript. The Basel Institute for Immunology was founded and is supported by F. Hoffmann-La Roche & Co. Ltd., CH 4005 Basel, Switzerland. This work was supported by grants AI 30879 and CA 13148 awarded by USPHS National Institute of Health. MDC is a Howard Hughes Medical Institute Investigator.

References

Abo T, Balch CM (1981) A differentiation antigen of human NK and K cells identifed by a monoclonal antibody (HNK-1). J Immunol 127: 1024–1029

Back O (1970a) Studies on the lymphocytes in the intestinal epithelium of the chicken. III. Effect of thymectomy. Int Arch Allergy Appl Immunol 39: 192–200

Back O (1970b) Studies on the lymphocytes in the intestinal epithelium of the chicken. IV. Effect of bursectomy. Int Arch Allergy Appl Immunol 39: 342–351

Back O (1972) Studies on the lymphocytes in the intestinal epithelium of the chicken, I: ontogeny. Acta Pathol Microbiol Scand [A] 80: 84–90

Baume DM, Robertson MJ, Levine H, Manley TJ, Schow PW, Ritz J (1992) Differential response to interleukin 2 define functionally distinct subsets of human natural killer cells. Eur J Immunol 22: 1–6

Bernot A, Auffray C (1991) Primary structure and ontogeny of an avian CD3 transcript. Proc Natl Acad Sci USA 88: 2550–2554

Bernot A, Zoorob R, Auffray C (1994) Linkage of a new member of the lectin supergene family to chicken Mhc genes. Immunogenet 39: 221–229

Blumberg RS, Ley S, Sancho J, Lonberg N, Lacy E, McDermott F, Schad V, Greenstein JL, Terhorst C (1990) Structue of the T-cell antigen receptor. Evidence for two CD3 ε subunits in the T-cell receptor-CD3 complex. Proc Natl Acad Sci USA 87: 7220–7224

Brissette-Storkus C, Kaufman CL, Pasewicz L, Worsey HM, Lakomy R, Ildstad ST, Chambers WH (1994) Characterization and function of the NKR-P1 dim/T cell receptor-aβ^+ subset of rat T cells. J Immunol 152: 388–396

Bucy RP, Coltey M, Chen C-LH, Char D, Le Douarin NM, Cooper MD (1989) Cytoplasmic CD3$^+$ surface CD8$^+$ lymphocytes develop as a thymus-independent lineage in chick-quail chimeras. Eur J Immunol 19: 1449–1455

Bucy RP, Chen C-H, Cooper MD (1990) Development of cytoplasmic CD3$^+$/T cell receptor-negative cells in the peripheral lymphoid tissues of chickens. Eur J Immunol 20: 1345–1350

Calnek BW, Witter RL (1984) Marek's disease. In: Hofstad MS, Barnes JH, Calnek BW, Reid WM, Yoder HE (eds) Diseases of poultry. Iowa State University Press, Ames, pp 326–340

Chai J-Y and Lillehoj HS (1988) Isolation and functional characterization of chicken intestinal intra-epithelial lymphocytes showing natural killer cell activity against tumour target cells. Immunology 63: 111–117

Chen CH, Chanh TC, Cooper/MD (1984) Chicken thymocyte specific antigen identified by monoclonal antibodies: ontogeny, tissue distribution and biochemical characterization. Eur J Immunol 14: 385–391

Chen CH, Ager LL, Gartland GL, Cooper MD (1986) Identification of a T3/T cell receptor complex in chickens. J Exp Med 164: 375–380

Chen C-LH, Bucy RP, Cooper MD (1990) T cell differentiation in birds. Semin Immununol 2: 79–86

Chi DS and Thorbecke GJ (1981) Cytotoxicity to allogeneic cells in the chicken. III. Antibody-dependent cell-mediated cytotoxicity in normal and agammaglobulinemic chickens. Cells Immunol 64: 258–266

Chung K-S, Lillehoj HS (1991) Development and functional characterization of monoclonal antibodies recognizing chicken lymphocytes with natural killer cell activity. Vet Immunol Immunopath 28: 351–363

Cone JC, Lu Y, Trevillyan JM, Bjorndahl JM, Phillips CA (1993) Association of the p56^{lck} protein tyrosine kinase with the FcγRIIIA/CD16 complex in human natural killer cells. Eur J Immunol 23: 2488–2497

Cooper MD, Chen C-LH, Bucy RP, Thompson CB (1991) Avian T cell ontogeny. Adv Immunol 50: 87–117

Evans DL, McKinney EC (1991) Phylogeny of cytotoxic cells. In: Warr GW (ed) Phylogenesis of immune functions. CRC Press, Boca Raton, Florida, pp 215–239

Fedecka-Bruner B, Penninger J, Vaigot P, Lehmann A, Martinez-A. C, Kroemer G (1991a) Development expression of IL-2-receptor light chain (CD25) in the chicken embryo. Dev Immunol 1: 237–242

Fedecka-Bruner B, Vaigot P, Dèsveaux-Chabgrol J, Gendreau M, Kroemer G, Dieterlen-Lièvre F (1991b) T-lymphocyte subsets in the embryonic spleen undergoing a graft-versus-host reaction. Dev Immunol 1: 163–168

Fleischer B (1980) Effector cells in avian spontaneous and antibody-dependent cell-mediated cytotoxicity. J Immunol 125: 1161–1166

Georgopoulos K, Bigby M, Wang J-H, Molnar A, Wu P, Winandy S, Sharpe A (1994) The Ikaros gene is required for the development of all lymphoid lineages. Cell 79: 143–156

Göbel TWF, Kaspers B (1995) In vitro generation of chicken cell lines with NK cell phenotype (abstract). 9th international congress of immunology, San Francisco, USA

Göbel TWF, Chen C-LH, Shrimpf J, Grossi CE, Bernot A, Bucy RP, Auffray C, Cooper MD (1994a) Characterization of avian natural killer cells and their intracellular CD3 protein complex. Eur J Immunol 24: 1685–1691

Göbel TWF, Chen CH, Lahti J, Kubota T, Kuo C, Aebersold R, Hood L, Cooper MD (1994b) Identification of T-cell receptor α-chain genes in the chicken. Proc Natl Acad Sci USA 91: 1094–1098

Göbel TWF, Chen CH, Cooper MD (1994c) Characterization of an avian CD6 candidate (abstract). 12th European immunology meeting, Barcelona, Spain

Haury M, Kasahara Y, Schaal S, Bucy RP, Cooper MD (1993) Intestinal T lymphocytes in the chicken express an integrin-like antigen. Eur J Immunol 23: 313–319

Kägi D, Vignaux F, Ledermann B, Bürki K, Depraetere V, Nagata S, Hengartner H, Golstein P (1994) Fas and perforin pathways as major mechanisms of T cell-mediated cytotoxicity. Science 265: 528–530

Kasahara Y , Chen CH, Göbel TWF, Bucy RP, Cooper MD (1993) Intraepithelial lymphocytes in birds. In: Kiyono H, McGhee JR (eds) Mucusal immunology: intraepithelial lymphocytes. Raven, New York, pp 163–174

Kaspers B, Lillehoj HS, Lillehoj EP (1993) Chicken macrophages and thrombocytes share a common cell surface antigen defined by a monoclonal antibody. Vet Immunol Immunopathol 36: 333–346

Kitagawa H, Hashimoto Y, Kon Y, Kudo N (1988) Light and electron microscopic studies on chicken intestinal globule leukocytes. Jpn J Vet Res 36: 83–117

Lanier LL, Phillips JH (1992) Natural killer cells. Curr Op in Immunol 4: 38–42

Lanier LL, Phillips JH, Testi R (1989a) Membrane anchoring and spontaneous release of CD16 (FcγR III) by natural killer cells and granulocytes. Eur J Immunol 19: 775–778

Lanier LL, Yu G, Phillips JH (1989b) Co-association of CD3ζ with a receptor (CD16) for IgG Fc on human natural killer cells. Nature 342: 803–806

Lanier LL, Chang C, Spits H, Phillips JH (1992a) Expression of cytoplasmic CD3ε proteins in activated human adult natural killer (NK) cells and CD3γ, δ, ε complexes in fetal NK cells. J Immunol 149: 1876–1880

Lanier LL, Spits H, Phillips JH (1992b) The developmental relationship between NK cells and T cells. Immunol Today 13: 392–395

Lillehoj HS (1989) Intestinal intraepithelial and splenic natural killer cell responses to Eimerian infections in inbred chickens. Infect Immun 57: 1879–1884

Lillehoj HS, Chai J-Y (1988) Comparative natural killer cell activites of thymic, bursal, splenic and intestinal intraepithelial lymphocytes of chickens. Dev Comp Immunol 12: 629–643

Marmor MD, Benatar T, Ratcliffe MJH (1993) Retroviral transformation in vitro of chicken T cells expressing either α/β or γδ T cell receptors by reticuloendotheliosis virus strain T. J Exp med 177: 647–656

Moingeon P, Rodewald H-R, McConkey D, Mildonian A, Awad K, Reinherz EL (1993) Generation of natural killer cells from both FcγRII/III^{+} and FcγRII/III^{-} murine fetal liver progenitors. Blood 82: 1453–1462

Moretta L, Ciccone E, Moretta A, Höglund P, Öhlèn C, Kärre K (1992) Allorecognition by NK cells: nonself or no self? Immunol Today 13: 300–306

Moretta L, Ciccone E, Mingari MC, Biassoni R, Moretta A (1994) Human natural killer cells: origin, clonality, specificity, and receptors. Adv Immunol 55: 341–380

Myers TJ and Schat KA (1990) Natural killer cell activity of chicken intraepithelial leukocytes against rotavirus-infected target cells. Vet Immunol Immunopathol 26: 157–170

O'Shea JJ, Wiessman AM, Kennedy ICS, Ortaldo JR (1991) Engagement of the natural killer cell IgG Fc receptor results in tyrosine phosphorylation of the ζ chain. Proc Natl Acad Sci USA 88: 350–354

Peault B, Chen CH, Cooper MD, Barbu M, Lipinski M, LeDouarin N (1987) Phylogenetically conserved antigen on nerve cells and lymphocytes resembles myelin-associated gycoprotein. Proc Natl Acad Sci USA 84: 814–818

Phillips JH, Hori T, Nagler A, Bhat N, Spits H, Lanier LL (1992) Ontogeny of human natural killer (NK) cells: Fetal NK cells mediate cytolytic function and express cytoplasmic CD3ε, δ proteins. J Exp Med 175: 1055–1066

Ratcliffe MJH, Boyd R, Chen C, Vainio O (1993) Avian CD nomenclature workshops, Montreal, June 1992, Budapest, August, 1992 and Tours, September 1992, Vet Immunol Immunopathol 38: 375–386

Revetch FV, Perussia B (1989) Alternative membrane forms of FcγRIII(CD16) on human natural killer cells and neutrophils. J Exp Med 170: 481–497

Ritz J, Schmidt RE, Michion J, Hercend T, Schlossman SF (1988) Characterization of functional surface structures on human natural killer cells. Adv Immunol 42: 181–211

Rodewald H-R, Moingeon P, Lucich JL, Dosiou C, Lopez C, Lopez P, Reinherz EL (1992) A population of early fetal thymocytes expressing FcγRII/III contains precursors of T lymphocytes and natural killer cells. Cell 69: 139–150

Salcedo TW, Kurosaki T, Kanakaraj P, Ravetch JV, Perussia B (1993) Physical and functional association of $p56^{lck}$ with FcγRIIIA (CD16) in natural killer cells. J Exp Med 177: 1475–1480

Schat KA, Calnek BW, Weinstock D (1986) Cultivation and characterisation of avian lymphocytes with natural killer cell activity. Avian Pathol 15: 539–556

Schat KA, Calnek BW, Weinstock D (1987) Cultured avian lymphocytes with natural killer cell activity. Proc Clin Biol Res 238: 157–169

Schauenstein K, Krömer G, Hàla K, Böck G, Wick G (1988) Chicken-activated-T-lymphocyte-antigen (CATLA) recognized by monoclonal antibody INN-CH 16 represents the Il-2 Receptor. Dev Comp Immunol 12: 823–831

Sharma JM (1981) Natural killer cell activity in chickens exposed to Marek's disease virus: inhibition of activity in susceptible chickens and enhancement of activity in resistant and vaccinated chickens. Avian Dis 25: 882–890

Sharma JM, Okazaki W (1981) Natural killer cell activity in chickens: target cell analysis and effect of antithymocyte serum on effector cells. Infect Immun 31: 1078–1085

Sharma JM, Schat KA (1991) Natural immune functions. In: Sharma JM (ed) Avian cellular immunology. CRC Press, Boca Raton, pp 51–70

Tiberghien P, Longo DL, Wine JW, Alvord WG, Reynolds CW (1990) Anti-asialo GM1 antiseum treatment of lethally irradiated recipients before bone marrow transplantation: evidence that recipient natural killer cell depletion enhances survival, engraftment, and hematopoietic recovery. Blood 76: 1419–1430

Tjoelker LW, Carlson: M, Lee K, Lahti J, McCormack WT, Leiden JM, Chen CH, Cooper MD, Thompson CB (1990) Evolutionary conservation of antigen recognition: the chicken T-cell receptor β-chain, Proc Natl Acad Sci USA 87: 7856–7860

Tregaskes CA, Kong FK, Paramithiotis E, Chen CH, Ratcliffe MJH, Davison TF, Young JR (1995) Identification and analysis of the expression of CD8αβ and CD8αα isoforms in chickens reveals a major TCRγδ CD8αβ subset of intestinal intraepithelial lymphocytes. J Immunol 154: 4485–4494

Trinchieri G (1989) Biology of natural killer cells. Adv Immunol 47: 187–376

Versteeg R (1992) NK cells and T cells: mirror images? Immunol Today 13: 244–247

Vervelde L, Jeurissen SHM (1993) Postnatal development of intra-epithelial leukocytes in the chicken digestive tract: Phenotypical characterization in situ. Cell Tissue Res 274: 295–301

Vivier E, Rochet N, Kochan JP, Presky DH, Schlossman SF, Anderson P (1991) Structural similarity between Fc receptors and T cell receptors. Expression of the γ-subunit of FcεRI in human T cells, natural killer cells and thymocytes. J Immunol 147: 4263–4270

Vivier E, Da Silva AJ, Ackerly M, Levine H, Rudd CE, Anderson P (1993) Association of a 70-kDa tyrosine phosphoprotein with the CD16: ζ: γ complex expressed in human natural killer cells. Eur J Immunol 23: 1872–1876

Voss SD, Sondel PM, Robb RJ (1992) Characterization of the interleukin 2 receptor (IL-2R) expressed on human natural killer cells activated in vivo by IL-2: association of th p64 IL-2Rγ chain with the IL-2R β chain in functional intermediate affinity IL-2R. J Exp Med 176: 531–541

Wang B, Biron C, She J, Higgins K, Sunshine M-J, Lacy E, Lonberg N, Terhorst C (1994) A block in both early T lymphocyte and natural killer cell development in transgenic mice with high-copy numbers of the human CD3ε gene. Proc Natl Acad Sci USA 91: 9402–9406

Yokoyamma WM, Seaman WE (1993) The Ly-49 and NKR-P1 gene families encoding lectin-like receptors on natural killer cells: The NK gene complex. Annu Rev Immunol 11: 613–635

Young JR, Davison TF, Tregaskes CA, Rennie MC, Vainio O (1994) Monomeric homologue of mammalian CD28s expressed on chicken T cells. J Immunol 152: 3848–3851

Ontogeny of Hematopoiesis in the Avian Embryo: A General Paradigm

F. Dieterlen-Lievre, I. Godin, and L. Pardanaud

1 Introduction

Experimental studies carried out on birds have yielded major discoveries of general interest. In the present report, we review the features of developing hematopoiesis that have been established in the avian model and then discuss their general significance for higher vertebrates, in light of recent data concerning mammalian embryos.

The ontogeny of the hematopoietic system is characterized by the cooperation of cells from different origins, stromal cells that create distinct microenvironments and hematopoietic stem cells (HScs). HSCs constitute a self-renewable pool. In contrast to what occurs in several other systems, for instance the skin or the intestinal epithelium, where a permanent turnover also proceeds from self-renewable stem cells, HSCs do not have an easily identified anatomical location and are capable, once mobilized, of migrating extensively. Furthermore, HSCs represent a minority among the crowds of differentiating cells, they are difficult to detect, and their stem cell quality must be established through functional tests. Consequently, their emergence during embryonic life is still far from completely understood. According to Moore and Owen's hypothesis (1967), HSCs are thought to originate in the yolk sac, one of the embryo's appendages, which is the earliest blood-forming organ during development. From experimental evidence in

Institut d' Embryologie cellulaire et moléculaire du CNRS et du Collège de France , 49bis, av. de la Belle Gabrielle, 94736 Nogent-sur-Marne Cedex, France

the chicken and mouse embryo, indicating that cells traffic between hematopoietic organ rudiments (MOORE and OWEN 1965; OWEN and RITTER 1969), these authors concluded that stromal and hematopoietic cells had distinct origins and attributed a central role of the yolk sac as a unique producer of HSCs. This model has been reevaluated, on the basis of experimental results in bird, amphibian and, very recently, mouse and even human embryos.

The avian embryo has been particularly helpful in unraveling the ontogeny of the hematopoietic system because of the quail/chick marker system (LE DOUARIN 1969). This system allows species-specific identification of cells, tissues, rudiments or embryonic territories in ovo. Due to the absence of the immune system at the stages concerned in the experiments and to the taxonomic proximity of the two species, development follows close or identical to normal patterns. The quail cells may be identified at any stage by the presence of a heavy heterochromatin mass associated with the nucleolus (or several masses depending on the cell type), and in the case of hematopoietic and endothelial cells (ECs), with quail-specific monoclonal antibodies MB1 and QH1 (PÉAULT et al. 1983; PARDANAUD et al. 1987). The quail/chick system has been extensively used to study the developing hematopoietic system. Results from transplantations of organ rudiments confirmed and extended MOORE and OWEN's conclusions about developmental relationships between blood cell progenitors and stromal cells. All hematopoietic progenitors were found to have an extrinsic origin and to colonize each rudiment beginning at a precise stage of development. In the case of the thymus, stem cells enter the anlage according to a cyclic rhythm, the precision of which is astonishing (LE DOUARIN and JOTEREAU 1973; JOTEREAU et al. 1980).

2 Yolk Sac and Embryo Make Distinct Contributions to Hematopoiesis in Birds

The yolk sac, comprised of an endodermal and a mesodermal layer, is an extension of the embryo's intestine, to which it is appended. During the first 2 days, the embryo lies flat on the yolk with its three germ layers superimposed. The mesoderm, which gives rise to the blood and vascular system, lies in the middle, at first as a single layer. The yolk sac does not exist as such yet: the extraembryonic area is continuous at that time with the central area where the embryo is developing. As morphogenesis becomes initiated, the three germ layers rise to delimit the embryo and the median mesodermal layer splits into two layers. The superficial layer or somatopleural mesoderm is associated with ectoderm, and these two layers give rise to the body wall and the limb. The deep layer, or splanchnopleural mesoderm, is associated with endoderm; they give rise to the internal organs (Fig. 1). The splitting process is responsible for the formation of the coelom and the yolk sac, whose wall comprises the lower mesodermal layer and endoderm.

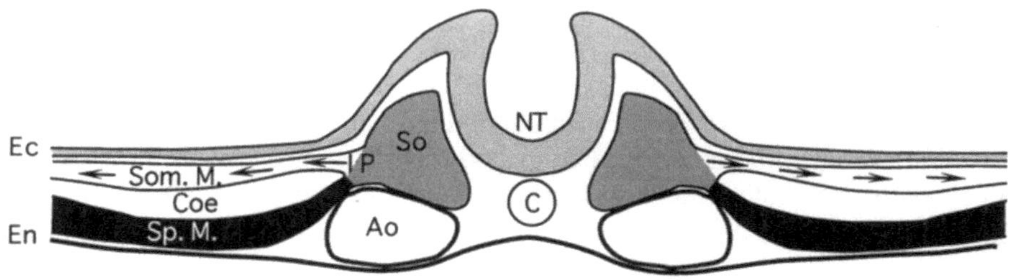

Fig. 1. The germ layers in the avian embryo truncal region at the neurulation period (E2). The mesoderm is subdivided into compartments. Cells from the somite (*So*) are destined to give rise to dermis, striated muscle, bone and cartilage. The intermediate plate (*IP*) is the forerunner of the kidney. The lateral plate is split into a somatopleural layer (*Som.M.*), associated with ectoderm (*Ec*) and a splanchnopeural layer (*Sp.M.*) associated with endoderm (*En*). The *dark shading* symbolizes the endothelium of the aortae (*Ao*) and the splanchnopeural mesoderm due to produce hematopoietic stem cells and endothelial cells. Somatopleural mesoderm does not produce these lineages and becomes colonized by endothelial precursors from the somite, whose migration is represented by *arrows*. *C*, notocord; *Coe*, coelome; *NT*, neural tube

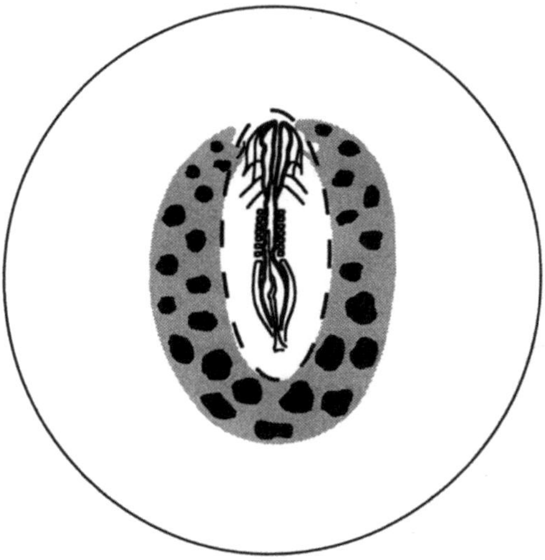

Fig. 2. "Yolk sac chimera." The blastodisc which develops on the surface of the yolk displays a central or pellucid area where the embryo, represented by its neural tube, somites and heart anlage, forms. The vascular area, where blood islands are symbolized, surrounds the pellucid area. The *stippled line* between these two areas indicates the place where a surgical suture is performed between explanted embryonic area and recipient blastodisc

An experimental scheme in which the embryonic area from a quail is grafted onto the extraembryonic area of a chick blastodisc has been devised by MARTIN (1972) (Fig. 2). The developing quail embryo is thus associated with a chick yolk sac. When we began investigating hematopoiesis in these chimeras, the yolk sac was held to have a central role. Thus, it was completely unexpected that the hematopoietic organs of these chimeras turned out to be purely quail in both stromal and blood forming cells (DIETERLEN-LIÈVRE 1975; MARTIN et al. (1978). Circulating blood was analyzed in these chimeras by means of polyclonal antibodies directed against red blood cells of chick or quail. Chick erythrocytes made up the blood until E5, then became progressively diluted by quail: this meant that stem cells from the chick yolk sac were replaced by stem cells formed in the embryo (Beaupain et al. 1979). These results were confirmed in chick-chick chimeras, built according to the same model, which could be raised past hatching until adulthood. It was clear in these chimeras that yolk sac blood progenitors became extinct without contributing to a permanent self-renewable pool (Lassila et al. 1979). Interestingly a very similar evolution has been uncovered in amphibians chimeras (Chen and Turpen 1995).

3 The Paraaortic Region

The early avian embryo displays diffuse hematopoietic processes, namely, blood cell formation that occurs outside of discrete organs. These processes take place successively in two distinct sites, both related to the aorta. At E3-4, clusters of cells appear to bud from the endothelium of the aorta into the lumen. These clusters are always restricted to the ventrolateral aspects of the blood vessel. At E6–8, large foci develop in the dorsal mesentery, ventral to the aorta. A specific role of the paraaortic region could be demonstrated by heterospecific grafts (Dieterlen-Lièvre 1984) and clonal cultures (Cormier et al. 1986; Cormier and Dieterlen-Lièvre 1988). In the latter system, many cells from the paraaortic region were hematopoietic progenitors, while cells obtained from the rest of the embryo did not display any progenitors (Cormier et al. 1986; Cormier and Dieterlen-Lièvre 1988).

4 A Homologous Region from Mouse and Human Embryos Has a Priviledged Role in the Developing Hematopoietic System

Several hematopoietic organs sequentially carry out blood cell forming functions during embryonic and fetal development. With the exception of the yolk sac, all these organs have to be colonized by extrinsic stem cells. We decided to focus our efforts on the period immediately preceding liver colonization, a process that begins between the stages of 28–32 pairs of somites (Fig. 3). Taking precedence from the sites identified in birds, we tested tissues localized in the neighborhood of the aortae, i.e., the paired vessels that are going to fuse to make the dorsal aorta, at the stages of 10–25 somites. The dissected area comprises the gut endoderm and surrounding mesoderm within which the aortae and omphalomesenteric artery are embedded. We call these structures the "paraaortic splanchnopleura" (Pa-Sp). The potentialities of these tissues were tested either in vivo or in vitro. In the first instance, involving grafts under the kidney capsule of SCID adult hosts, some lymphoid subsets of the immunodeficient hosts were restored

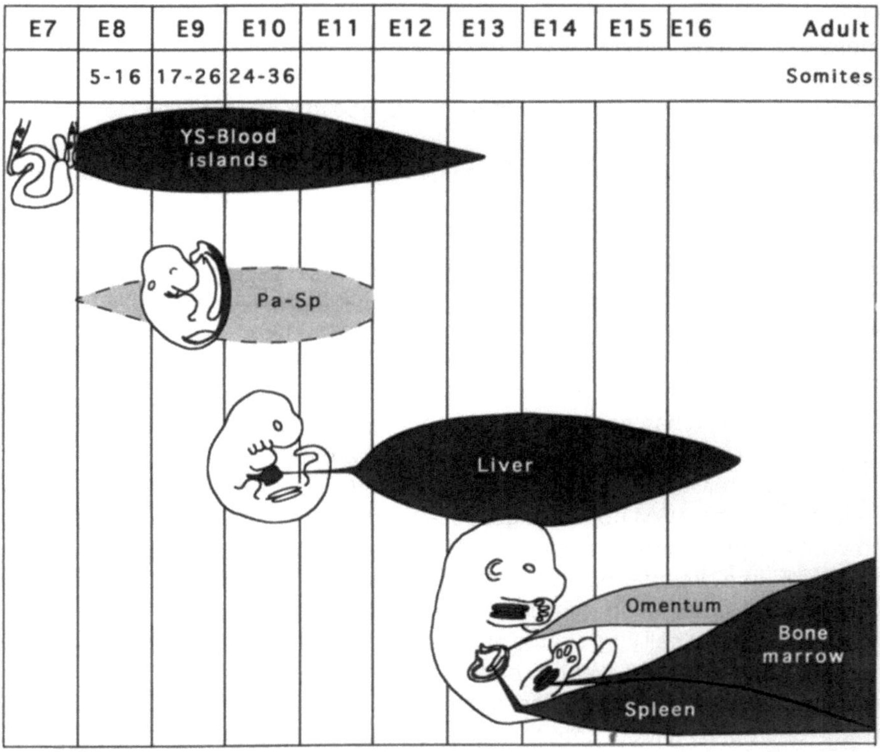

Fig. 3. Successive hematopoietic sites in the mouse embryo. *Shaded areas* symbolize the periods when the sites are active. *Pa-Sp* (paraaortic splanchnopleura) comprises the digestive endoderm and associated mesoderm. The omentum, a connective tissue lamina which supports the stomach and spleen, displays hematopoiesis and has been attributed a special role in hematopoietic sytem ontogeny. Thymus has not been indicated in this scheme

(Godin et al. 1993). Using the in vitro approach, the emergence of cells in different compartments could be precisely followed. The culture was carried out in two phases. During the first phase, cells obtained from the dissociated Pa-Sp (about 1000 cells at the ten pairs of somites stage) were cultivated for 10 days on stromal cell line S17 in the presence of cytokines interleukin-3 (IL3), IL7 and stem cell factor (SCF) (Cumano et al. 1992); each progenitor became amplified, giving rise to clones that were comprised of up to several thousand cells. The number of clones gave the measure of the number of progenitors present in the cell suspension at the start of the culture. At the stage of ten somites, one occasional progenitor was observed in the Pa-Sp. From the stage of 15 somites, this structure regularly contained progenitors and ten of these cells were present per embryo at the 25 somite stage. Interestingly, cells from the yolk sac, dissociated and seeded under these conditions, also contained progenitors whose numbers followed similar kinetics. In contrast, the remainder of the embryo after yolk sac and Pa-Sp have been trimmed away contained no clonogenic cells. This evolution is parallel with that in birds, and the differential distribution clearly results from the parallel emergence of progenitors in the Pa-Sp and the yolk sac. If colonization was involved, progenitors would be present in the blood, hence in the remainder of the embryo. The culture technique involved a second step, in which the clones were divided into three lots, submitted to conditions favoring myeloid differentiation and B or lymphoid differentiation. Cells from one clone were perfectly capable of differentiating according to any of the imposed differentiation pathways, thus leading to the conclusion that the originally amplified progenitors are totipotent (Godin et al. 1995). The functional tests further permit the allocation of progenitors to the postumbilical area of the embryos, where a cytological analysis detected aggregated cells within several arteries (aorta, omphalomesenteric artery, vitelline artery) and intramesenteric blood island-like groups of cells (Garcia-Porrero et al. 1995).

Elaine Dzierzak's group in Great Britain is carrying out investigations on the same problem and has detected progenitors capable of restoring irradiated mice in the medium and long term. These progenitors are found in the same region but at a slightly later date (Medvinsky et al. 1993; Müller et al. 1994). This region, designated as the AGM (for aorta-gonads-mesonephros), derives from the Pa-Sp and contains progenitors at stages of 31–40 pairs of somites.

Finally, in a recent piece of work, Manuella Tavian and Bruno Péault, at our institute, have obtained impressive pictures of the human embryo aorta, whose endothelium bears cell aggregates at 35 days of gestation. These aggregates are tightly associated to the ventral endothelium and look strikingly similar to the avian intraaortic aggregates. The aggregated cells bear the CD34 antigen, a marker shared by blood progenitors and ECs but are devoid of endothelial markers. These cells also bear the CD45 antigen (common leukocyte antigen). In the yolk sac and in fetal liver, $CD34^+$ cells are much sparser than in the aorta.

5 Relationship Between the Endothelial Network and Hematopoiesis

The developmental relationships betwen the blood and endothelial lineages are unclear. A presumptive common progenitor, the hemangioblast, was hypothesized many years ago; however, there is no experimental proof in favor of it. The existence of a hemangioblast is suggested by cytological aspects in very immature yolk sac blood islands and intraaortic aggregates. Associations and similarities between cells of the blood and endothelial lineages are confined to the early period of development.

At the period of organogenesis, the two lineages are independent, as demonstrated in the quail/chick model, in which the origin of the cells can be identified with monoclonal antibody QH1. Judah Folkman, in 1974, defined "angiogenesis" as the process of vascular invasion from preexisting ECs that is necessary for tumor development. At the beginning of development, some mesodermal cells become committed into the endothelial pathway; this neoformation is distinct from angiogenesis and has been called vasculogenesis (RISAU and LEMMON 1988; PARDANAUD et al. 1989). We have been able to show, by means of quail-chick transplantations, that these two processes operate during organogenesis and that the vascular network develops by means of one or the other (vasculogenesis in the internal organ rudiments, angiogenesis in limb buds and body wall). In contrast, HSCs colonize both types of rudiments, for instance, bone marrow or spleen, as already mentioned above.

If transplantation involves the primordial germ layers, rather than organ rudiments, it is found that the splanchnopleure (endoderm + mesoderm) is capable of producing both ECs and HSCs. The capacity to produce HSCs becomes lost in the splanchnic mesoderm when it is incorporated in organ rudiments; at that period HSC production becomes restricted to the paraaortic mesoderm. Somatopleure (ectoderm + mesoderm), by contrast, behaves like the limb bud and is colonized by precursors from both lineages.

Where do the precursors that colonize the somatopleure come from? Somites have the capacity of producing ECs (NODEN 1989; WILTING et al. 1995; PARDANAUD and DIETERLEN-LIÈVRE 1995). When somites are grafted orthotopically, these ECs vascularize the body wall and the kidney but never invade splanchnopleural derivatives. In the head, which is devoid of somites, the paraaxial mesoderm essentially carries out the production of the endothelial network that irrigates the brain and face (COULY et al. 1995).

6 Conclusions

Descriptive and experimental data indicate that, in very young embryos, ontogenesis of the blood system entails a period during which the mesoderm gives rise to hematopoietic and endothelial precursors. The emergence of cells belonging to these two lineages occurs relatively ubiquitously in the mesodermal layer in contact with the endoderm. In a first stage, HSCs arise essentially in peripheral areas, that is, in the yolk sac. In a second stage, the central area, that of the embryo proper, also produces HSCs, and these colonize the definitive hematopoietic organs. From the period of organogenesis onwards, ECs and HSCs appear to be independent lineages. Figure 4 schematizes the sequential emergence of two generations of HSCs in birds. These two generations have no developmental relationship. On the other hand it is impossible to assert whether the generation indicated as CSH2 is responsible for seeding the definitive hematopoietic system. Diffuse hematopoiesis, which occurs in the paraaortic region until E8, may correspond to the continuous emergence of HSCs from mesodermal precursors. In this hypothesis, it is likely that commitment changes as the homing environment evolves. For instance, these cells may undergo erythropoiesis at E3 and colonize the thymus at E6.5–8.

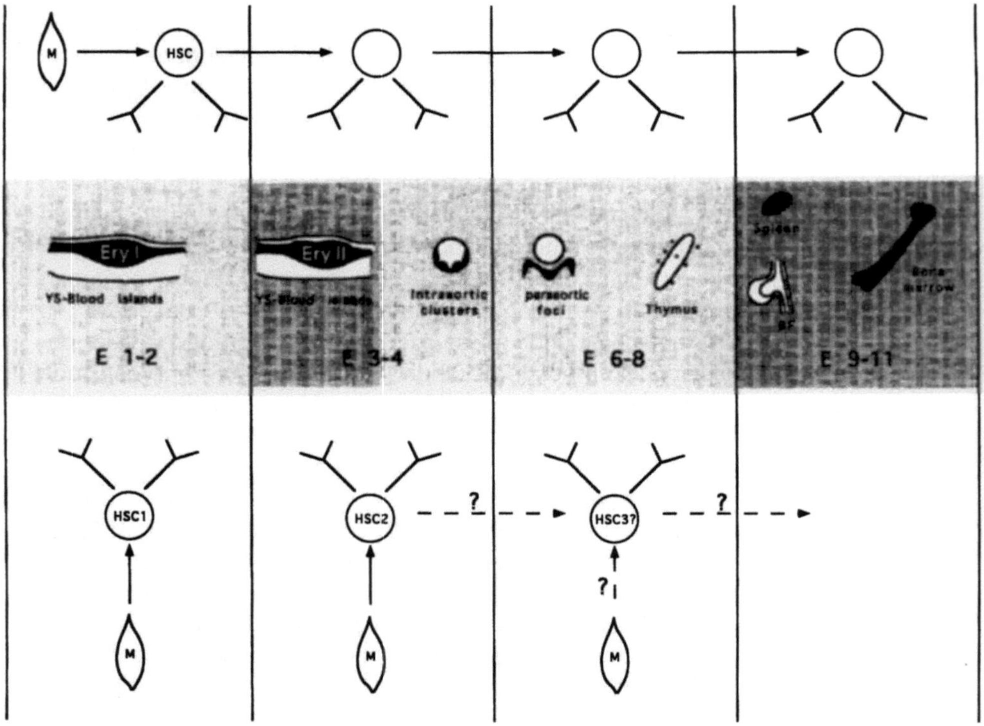

Fig. 4. Interpretations regarding early hematopoietic processes in the avian embryo. The *central shaded band* depicts successive steps. Primitive erythropoisesis (*Ery I*) is carried out in the extraembryonic blood islands. At E3–4 secondary erythropiesis (*Ery II*) is initiated in parallel in yolk sac blood islands and in the aortic lumen. At E6–8 hematopoiesis occurs in the dorsal mesentery, ventral to the aorta, which is schematized by a *circle*; at the same time hematopoietic stem cells (*HSC*) colonize the thymus rudiment. Colonization of bone marrow, where the HSC pool resides in the adult, begins only later, at E10.5. The *upper band* schematizes MOORE and OWEN's hypothesis, namely, the early unique emergence of self renewable HSC from the extraembryonic mesoderm. The *lower band* schematizes the two independant HSC generations that have been demonstrated in yolk sac chimeras. HSC2 emerge simultaneously in the embryo and yolk sac, but yolk sac HSC2 engage in erythropoiesis and their progeny disappears: they do not participate in definitive organ colonization. The period during which HSC may be produced from mesodermal progenitors has not been defined experimentally. It is possible that HSC produced at E3–4 into the aortic lumen engage in erythropoiesis without contributing to a permanent self-renewable pool. If such is the case, the progeny of intraaortic cluster HSC and of paraaortic HSC would be distinct generations

In the case of mouse or human embryo, it is not possible to conclude as forcefully about the origin of definitive HSCs. The present state of the art does demonstrate that a generation of independent HSCs follows the early yolk sac generation. This new generation appears both in the yolk sac and the embryo, so that the ontogeny of the mammalian hematopoietic system looks very similar to that of birds.

References

Beaupain D, Martin C, Dieterlen-Lièvre F (1979) Are devlopmental hemoglobin changes related to the origin of stem cells and site of erythropoiesis? Blood 53: 212–225

Chen XD, Turpen JB (1995) Intraembryonic origin of hepatic hematopoiesis in Xenopus laevis. J Immunol 154: 2557–2567

Cormier, F, Dieterlen-Lièvre F (1988) The wall of the chick embryo aorta harbors M-CFC, G-CFC and BFU-E. Development 102: 279–285

Cormier F, de Paz P, Dieterlen-Lièvre F (1986) In vitro detection of cells with monocytic potentiality in the wall of the chick embryo aorta. Dev Biol 118: 167–175

Couly G, Coltey P, Eichmann A, Le Douarin NM (1995) The angiogenic potentials of the cephalic mesoderm and the origin of brain and head blood vessels. Mech Dev 53: 97–117

Cumano A, Paige CJ, Iscove NN, Brady G (1992) Bipotential precursors of B cells and macrophages in murine fetal liver. Nature 356: 612–615

Dieterlen-Lièvre F (1975) On the origin of haemopoietic stem cells in the avian embryo: an experimental approach. J Embryol Exp Morphol 33: 607–619

Dieterlen-Lièvre F (1984) Emergence of intraembryonic blood stem cells studied in avian chimeras by means of monoclonal antibodies. Dev Comp Immunol Suppl 3: 75–80

Folkman J (1974) Tumour angiogenesis. Adv Cancer Res 19: 331–358

Garcia-Porrero J, Godin I, Dieterlen-Lièvre F (1995) Potential intraembryonic hemogenic sites at preliver stages in the mouse. Anat Embryol (Berl) 192: 425–435

Godin I, Garcia-Porrero J, Coutinho A, Dieterlen-Lièvre F, Marcos MAR (1993) Paraaortic splanchnopleura from early mouse embryo contains B1a-cell progenitors. Nature 364: 67–70

Godin I, Dieterlen-Lièvre F, Cumano A (1995) Emergence of multipotent hematopoietic cells in the yolk sac and paraaortic splanchnopleura of 8.5 dpc mouse embryos. Proc Natl Acad Sci (USA) 92: 773–777

Jotereau, FV, Houssaint E, Le Douarin NM (1980) Lymphoid stem cell homing to the early thymic primordium of the avian embryo.Eur J Immunol 10: 620–627

Lassila O, Martin C, Dieterlen-Lièvre F, Nurmi TE, Eskola J, Toivanen P. (1979) Is the yolk sac the primary origin of lymphoid stem cells? Transplant Proc 11: 1085–1088

Le Douarin NM (1969) Particularités du noyau interphasique chez la Caille japonaise (Coturnix coturnix japonica). Utilisation de ces particularités comme 'marquage biologique' dans les recherches sur les interactions tissulaires et les migrations cellulaires au cours de l'ontogenèse. Bull Biol Fr Belg 103: 435–452

Le Douarin NM, Jotereau FV (1973) Origin and renewal of lymphocytes in avian embryo thymuses studied in interspecific combinations. Nature New Biol 246: 25–27

Martin C (1972) Technique d'explantation in ovo de blastodermes d'embryons d'oiseaux. C R Soc Biol 116: 283–285.

Martin C, Beaupain D, Dieterlen–Lièvre F (1978) Developmental relationships between vitelline and intra-embryonic haemopoiesis studied in avian 'yolk sac chimaeras'. Cell Differ 7: 115–130

Medvinsky AL, Samoylina NL, Müller AM, Dzierzak EA (1993) An early pre-liver intraembryonic source of CFU-S in the developing mouse. Nature 364: 64–66

Moore MAS, Owen JJT (1965) Chromosome marker studies on the development of the haemopoietic system. Nature 208: 958–989

Moore MAS, Owen JJT (1967) Stem cell migration in developing myeloid and lymphoid systems. Lancet 2: 658–659

Müller AM, Medvinsky A, Strouboulis J, Grosveld F, Dzierzak E (1994) Development of hematopoietic stem cell activity in the mouse embryo. Immunity 1: 291–301

Noden DM (1989) Embryonic origins and assembly of embryonic blood vessels. A. Rev Pulmon Dis 140: 1097–1103

Owen JJT, Ritter MA (1969) Tissue interaction in the development of thymus lymphocytes. J Exp Med 129: 431–442

Pardanaud L, Dieterlen-Lièvre F (1995) Does the paraxial mesoderm of the avian embryo have hemangioblastic capacities? Anat Embryol (Berl) 192: 301–308

Pardanaud L, Altmann C, Kitos P, Dieterlen-Lièvre F, Buck C (1987) Vasculogenesis in the early quail blastodisc as studied with a monoclonal antibody recognizing endothelial cells. Development 100: 339–349

Pardanaud L, Yassine F, Dieterlen-Lièvre F (1989) Relationship between vasculogenesis, angiogenesis and haemopoiesis during avian ontogeny. Development 105: 473–485
Péault B, Thiery JP, Le Douarin NM (1983) Surface marker for hemopoietic and endothelial cell lineages in quail that is defined by a monoclonal antibody. Proc Natl Acad Sci USA 80: 2976–2980
Risau W, Lemmon V (1988) Changes in the vascular extracellular matrix during embryonic vasculogenesis and angiogenesis. Dev Biol 125: 441–450
Wilting J, Brand-Saberi B, Huang R, Zhi Q, Köntges G, Ordahl CP, Christ B (1995) Angiogenic potential of the avian somite. Dev Dyn 202: 165–171

Chicken MHC Molecules, Disease Resistance and the Evolutionary Origin of Birds

J. KAUFMAN[1] and H.J. WALLNY[1,2]

1 Introduction

The major histocompatibility complex (MHC) of the typical mammal is a very large genetic region that encodes many molecules of different structures and functions. However, the MHC was discovered and is presently defined by the presence of two multigene families with some highly polymorphic (that is, multiallelic) members. These encode the so-called classical class I and class II cell surface glycoproteins that bind antigenic peptides and present them for specific recognition by T lymphocytes of the immune system (reviewed in KLEIN 1986; LAWLOR et al. 1990).

The large number of common alleles (that is, the high polymorphism) of these genes is unprecedented among vertebrate loci and is thought to be, in large part, the result of attempts to resist newly arising pathogen variants, which themselves appear in an attempt to evade the immune system. However, there are theoretical, experimental and observational grounds to suggest that in mammals only a very small part of this polymorphism is directly due to resistance to infectious pathogens (reviewed in KLEIN 1987; LAWLOR et al. 1990; POTTS and WAKELAND 1993). For example, all strong disease associations with the human

[1]Basel Institute for Immunology, Grenzacherstrasse 487, 4005 Basel, Switzerland

[2]Present address: Biotest Pharma GmbH, Landsteiner Str. 5, 63303 Dreieich, Germany

MHC involve autoimmune diseases, with the best example of an association with an infectious disease only being detected with very large population sizes and careful statistics (HILL et al. 1991).

In this report, we summarize some data and arguments to suggest that the chicken MHC (and, presumably, the MHC of the typical bird) differs from the MHC of mammals in two important ways. First, it is simpler and smaller, with many MHC haplotypes having only one class I and one class II molecule dominantly expressed on hematopoietic cells. Second, as a consequence of having a simple MHC, resistance and susceptibility to particular infectious pathogens are determined by the peptide-binding specificity of the MHC molecules of particular MHC haplotypes. Finally, we consider some possibilities for the origin and maintenance of the simple organization of the chicken MHC, with the proposal that it is part of a phenomenon that is widespread in the avian genome.

2 The Chicken MHC Is Compact and Simple Compared to the Typical Mammalian MHC

The MHC of the typical mammal, based on very detailed analyses of the mouse and human MHCs, is a vast genetic region divided by frequent recombination into one region that encodes primarily but not exclusively class I genes, another region that encodes class II genes as well as some genes involved in antigen processing, and a region in between that encodes many structurally unrelated genes that collectively have been called class III region genes (KLEIN 1986; CAMPBELL and TROWSDALE 1993; TROWSDALE 1993, 1995).

Our current view of the chicken MHC chromosome (Fig. 1) is loosely based on various genetic maps (involving backcross families, strains with fixed recombinational events, and strains with aneuploidy, BUMSTEAD and PALYGA 1992; KOCH et al. 1983; HALA et al. 1976; BLOOM and BACON 1985) as well as the physical analysis of five cosmid clusters (GUILLEMOT et al. 1988; KAUFMAN et al. 1991b).The MHC (defined by serology and graft rejection as the B locus) and the nucleolar organizer region (*NOR*) are located together on a microchromosome (number 16 or 17), based on chicken strains with aneuploid karyotypes (BLOOM and BACON 1985; BLOOM et al. 1987). Recombinant strains divide this chromosome into a *B-F/B-L* region and a *B-G* region. In these strains, the *B-F/B-L* region determines rapid allograft rejection, encodes at least one class I, one class II and one *B-G* antigen, and includes the cosmid cluster I. The *B-G* region encodes the erythrocyte *B-G* antigens generally used for serological typing (*EaB*) and includes the *B-G* cosmid clusters V and VI (PINK et al. 1977; GOTO et al. 1988; KAUFMAN et al. 1989, 1991b). In the backcross families, the serologically determined MHC (*EaB*) cosegregates with particular class I α and class II β restriction fragment length polymorphisms (RFLPs) (which represent cluster I) and not with others (which at least in part determine the *Rfp-Y* locus that includes cosmid clusters II/IV and III), whereas the

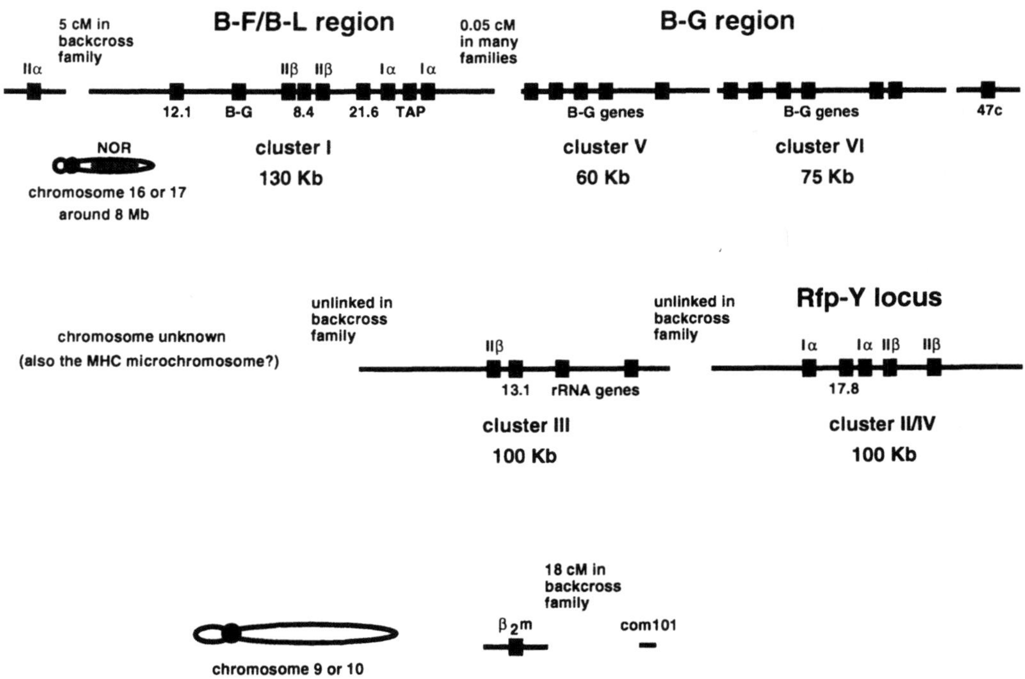

Fig. 1. Our current view of the chicken MHC and genomic regions containing related genes. Gene assignments on cluster I, III and II/IV from GUILLEMOT et al. (1988) with modification from KAUFMAN, (unpublished). Gene assignments on cluster V and VI modified from KAUFMAN et al. (1991b). Genetic distances between the *B-F/B-L* and *B-G* regions from CRONE and SIMONSEN (1987); between the class II α gene and *B-F/B-L* region from KAUFMAN et al. (1995a); between the β_2-microglobulin (β_2m) gene and other genes on the linkage map from RIEGERT et al. (1995); and between clusters III, IV and the rest of the linkage map from MILLER et al. (1994b) and KAUFMAN. (unpublished). Chromosomal assignments from BLOOM and BACON (1985) and RIEGERT et al. (1995)

class II α gene is located roughly 5 centimorgans from the *EaB* (MILLER et al. 1994b; KAUFMAN et al. 1995a).

As yet, these data have not been integrated to give a single map. It is not yet published whether the cosmid clusters II/IV and III are present on the MHC microchromosome, although the presence of rRNA genes on cluster III makes it likely. In addition, the precise parent strains for the recombinant strains are either not available anymore or are uniformative for the *Rfp-Y* locus and the class II α gene (KAUFMAN, unpublished). No RFLPs for the *B-G* cosmids were found in the backcross families, and the *Rfp-Y* locus was not linked to any other linkage group in the study (MILLER et al. 1994b). A semi-educated guess is that the class II α gene is on one side of cluster I, which is closely linked to clusters V and VI, whereas clusters II/IV and III are embedded in or on the far side of the *NOR* (a highly repetitive region which should be highly recombinogenic) (Fig. 1, KAUFMAN et al. 1995b).

Does this mean that the chicken MHC is spread over a large area, or, at the least, split into several parts? This depends on the definition of an MHC, a concept that is argued over and often revised by those considering themselves authorities on the subject. The traditional definition, inherent in the name, is that the MHC contains those polymorphic class I and class II genes that are responsible for rapid allograft rejection, although there is no consensus as to whether this should be only those particular genes, or the smallest region containing those genes, or an extended region of genes (mostly involved in antigen presentation but not necessarily polymorphic) that travel together in evolution. In one sense, the chicken MHC allows us to ask the question which genes do actually remain together in evolution. (Parenthetically, we have recently suggested that the MHC might be defined in another way, as a region of the genome in which polymorphism is tolerated or even encouraged, with the MHC molecules being merely the most obvious and earliest discovered polymorphic genes of the region, KAUFMAN et al. 1995b).

For the chicken, it is quite clear that the regions tightly linked with the erythrocyte B-G antigens (the *B* locus) are responsible for rapid allograft rejection. It is of course possible that the regions associated with the *Rfp-Y* locus do in fact encode polymorphic MHC antigens that determine rapid graft rejection—that is to say, the *Rfp-Y* locus is part of the MHC (although separated from the B locus) or represents another MHC. However, there are no reports of highly polymorphic loci leading to rapid allograft rejection unlinked to the *B* complex. Moreover, the RFLPs reported for the *Rfp-Y* locus (MILLER et al. 1994a) almost certainly describe polymorphism outside of the coding regions of the genes, and those cluster II/IV and cluster III class II β gene alleles that have been sequenced show very low polymorphism (ZOOROB et al. 1993). Finally, the class I α and class II β molecules that have been detected as highly expressed in blood and spleen cells are not encoded in the *Rfp-Y* locus (KAUFMAN, unpublished). So we have taken the position, until moved by further evidence, that the MHC, in the sense of a highly polymorphic region that encodes the molecules responsible for rapid allograft rejection, is represented by a portion of cosmid cluster I within the *B-F/B-L* part of the *B* locus (KAUFMAN et al. 1995b).

If we are correct in our view of the real MHC being confined to a portion of cluster I, then it is very small compared to the MHC of the typical mammal—in size, in number of genes and in kinds of genes. At the moment, there are only two class II β genes, two class I α genes and a TAP gene defined in a 50 kilobasepair region. There are several other genes packed in this region, and it will be interesting to determine whether they correspond to genes known to be within the MHC of mammals, particularly genes from the class III region (such as C4, C2, factor B, HSP70, TNF and MIC genes) and from the antigen presentation region (LMP, nonclassical class II, and RING3 genes). It may be that the chicken MHC contains only those genes which are most favored to stay together in evolution and thus may approach a "minimal essential MHC" (KAUFMAN et al. 1995b)

In any case, recombinations that separate the genes encoding the serologically defined class I and class II antigens have never been observed in breeding families (although putative recombinants are found by comparison of different strains) (CRONE and SIMONSEN 1987). So it is possible that, in contrast to mammals, the MHC of different strains of chickens really exist as stable haplotypes.

3 The Chicken MHC Class I Molecules Are Polymorphic Both in Sequence and Level of Expression

We have begun to examine peptides associated with chicken MHC class I molecules (KAUFMAN et al. 1995b). In brief, we isolated the class I molecules from erythrocytes and from spleen cells using monoclonal antibodies to chicken β_2-microglobulin (β_2m), separated the peptides by reverse phase chromatography and analyzed the sequences of the total pool of peptides and of individual dominant self-peptides. We have found that: (1) most of the peptides are octamers with a few nonamers; (2) the peptides are derived from chicken proteins; (3) the sequences of the dominant peptides conform to the consensus sequences from the peptide pools; (4) for most chicken MHC haplotypes there is only one dominant motif: and (5) class I molecules from different chicken MHC haplotypes have different motifs.

Most of the motifs for chicken class I molecules are similar to those already known from mammals, but one is quite different (Table 1). The major class I molecules from our *B15* and *B19* strains have peptide motifs much like that of HLA-B27, the human allele closely associated with ankylosing spondylitis, with a large positive charged arginine at position 2 and with some kind of hydrophobic residue at the end of the peptide. These two chicken alleles are known to be serologically similar (SIMONSEN et al. 1982) and, in fact, have similar sequences in the peptide-binding regions. Together they are present in a large proportion of chickens found in commercial egg-laying flocks (SIMONSEN et al. 1982); whether

Table 1. Anchor residues of peptides found bound to chicken class I molecules

MHC type	Peptide residue									
	1	2	3	4	5	6	7	8	9	(10)	(11)
H-2 K^d		YF							IL		
K^k		E						I			
L^d		P							LMF		
HLA-A2		LM							VL		
A68		VT							K	(K)	(K)
B7		P							L		
B27		R							LYFMRL		
B35		PY							FMILY		
B-F15		R						Y	(Y)		
B-F19		R						(YPL)	(YPL)		
H-2 D^b					N				MIL		
K^b					YF			LMIF			
B-F12					VI			V			
H-2 D^d		G	PLM						LFMI		
HLA-B8			K		KR				L		
B37		DE			VMI			LFIM	(LFIM)		
B-F4		DE			DE			E			

Dominant residues (anchor residue motifs) found in peptides eluted from human and mouse MHC class I molecules (FALK and ROETSCHKE 1993) and chicken MHC class I molecules (KAUFMAN et al. 1995b).

this is due to founder effect, selection due to some disease or to overall efficiency of this motify is not clear. The major class I molecule from our *B12* strain (CB) has a peptide motif much like the mouse alleles H-2K^b and H-2D^b, with hydrophobic residues at position 5 and at the end of the peptide. The major class I molecule from *B4* (CC strain) has a motif unlike any yet described in mammals, with three negatively charged anchor residues at position 2, position 5 and the end of the peptide. In particular, all COOH-terminal anchor residues in mammals have been hydrophobic or positively charged.

We had difficulty in determining a motif for the class I molecules from the *B21* haplotype isolated with monoclonal antibodies to β_2m, in part because there seems to be more than one class I molecule expressed and in part because there is a lower level of expression. The difference in expression level between the *B21* strains and the *B4, B12, B15* and *B19* strains is profound—erythrocytes have seven to ten fold less, thrombocytes have five to seven fold less and leukocytes two to five fold less class I on the surface (Fig. 2). In the MHC recombinant strains, this difference is determined by the *B-F/B-L* region of the chicken MHC (which includes cluster I). Some other haplotypes (like *B2, B6* and *B14*) have intermediate levels of class I molecules on the cell surface. These results have been confirmed for many strains kept at three different institutes, including *B21* haplotypes from other genetic backgrounds, such as different egg-laying strains, meat-producing strains, and wild birds from Borneo.

The basis of the different levels of class I expression in these haplotypes is not yet clear. Analysis of Concanavalin A(Con A) treated blasts from *B21* compared to *B4, B12, B15* and *B19* animals shows no large changes in the amount of cell surface class I after incubation with specific peptides or incubation at lower temperatures. So it would appear not to be a difference in the amounts of appropriate peptides, as would be found in a TAP defect. Nor did we find remarkable differences in the amount of class I molecules, association with β_2m, or maturation of N-linked glycans on class I molecules in pulse-chase experiments using Con A blasts; however, fewer class I molecules appeared on the surface of *B21* Con A blasts than on the other four haplotypes. Perhaps the class I molecules from *B21* are secreted, but more likely we are dealing with a novel mechanism.

4 Particular Chicken MHC Haplotypes Can Determine Resistance and Susceptibility to Particular Infectious Pathogens

There are no good examples of a mammalian MHC haplotype conferring overwhelming susceptibility to a particular infectious pathogen (Klein 1987). We believe that the complexity of the typical mammalian MHC, with multigene families encoding classical class I and class II molecules, results in every MHC haplotype conferring a greater-or-lesser degree of resistance to most every

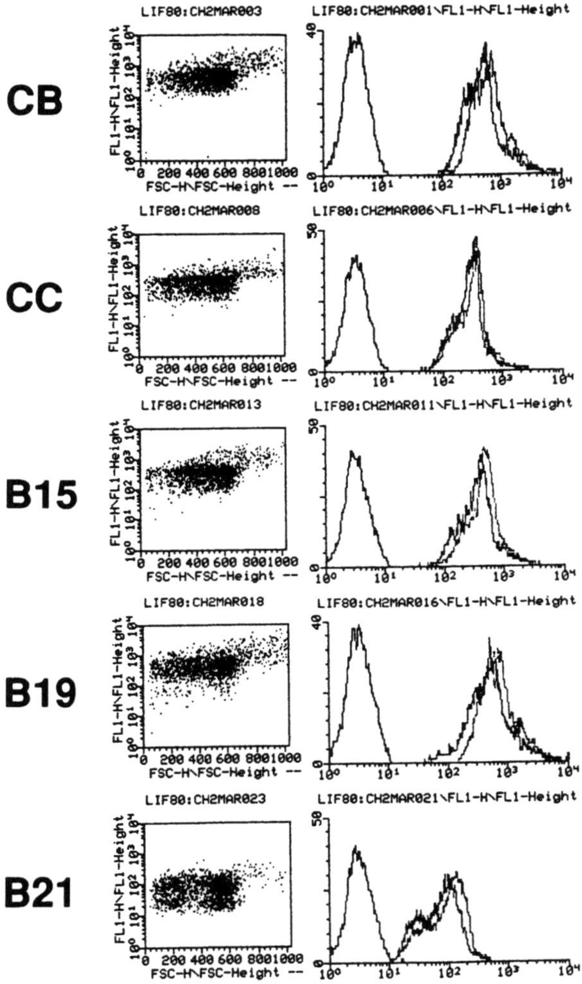

Fig. 2. Flow cytometry of erythrocytes from five strains of chicken stained with mouse monoclonal antibodies to chicken class I (F21-2), to chicken β_2m (F21-21) or to antigens not present in chickens (negative control) followed by fluorescein-conjugated goat antibodies to mouse immunoglobulin. *Left*, dot-plot with fluorescence of anti-chicken class I on the *abscissa* and forward scatter on the *ordinate*. *Right*, histogram of number of cells on the *abscissa* with fluorescence of negative control antibody compared to anti-class I and anti-β_2m antibodies. Relative fluorescence was measured on a log scale using a FACscan (Becton-Dickenson)

pathogen. This means that pathogens exert a relatively weak selection for polymorphism in mammals; in fact, there are other mechanisms to explain the high polymorphism of mammals (Kaufman et al. 1995b).

In contrast to the situation in mammals, there are many examples of particular chicken MHC haplotypes that confer striking resistance or susceptibility to particular infectious pathogens (Schat 1987; Dietert et al. 1990; Plachy et al. 1992). The pathogens are mostly small viruses (like retroviruses), and in such cases, the MHC-dependent resistance and susceptibility may be simply explained by the fact that there is only a single dominantly expressed class I molecule in many chicken MHC haplotypes, so that selection by the pathogen depends on the peptide-binding specificity of a single MHC molecule, rather than on many MHC molecules as in a typical mammal.

Rous sarcoma virus (RSV) is a good example of a small virus for which there are resistant and susceptible chicken MHC haplotypes. It is a classic transforming retrovirus with four genes (*gag, env, pol* and *src*) (SCHWARTZ et al. 1983). Chickens develop tumors which progress in some strains leading to death of the chickens and which regress in other strains with nearly complete survival of the chickens. The resistance to RSV depends on a functioning immune system, is genetically dominant and is determined by the *B-F/B-L* region (that includes cluster I) of the MHC (SCHAT 1987; DIETERT et al. 1990; PLACHY et al. 1992).

The MHC congenic inbred chicken strains CB and CC are a well-studied system for such disease resistance. The CB chicken with the MHC haplotype *B12* is a tumor regressor that is resistant to RSV, whereas the MHC congenic CC chicken with the MHC haplotype *B4* is a tumor progressor that is susceptible to RSV (PLACHY et al. 1992). A scan for octamer and nonamer peptides that could be produced from the four proteins encoded by the relevant RSV strain shows that there are 17 peptides that have the motif of the major class I molecule of *B12* compared to two peptides with the motif of the major class I molecule of *B4*. Of course it is still necessary to prove that *B12* class I gene will confer resistance to RSV in a CC chicken (for instance, by making the appropriate transgenic chickens), but it seems likely to us that this very simple model could explain the MHC-determined resistance.

The simple explanation that the peptide-binding specificity of a dominantly expressed class I molecule determines resistance to an infectious pathogen is unlikely to hold for large pathogens, since among the many proteins of a large pathogen an appropriate peptide will be found for almost any specificity. But in fact, the most famous example of MHC association with an infectious pathogen is the association of *B21* with Marek's disease virus, which is a large herpes virus (SCHAT 1987; DIETERT et al. 1990; PLACHY et al. 1992).

The Marek's disease virus first cytolyticly infects B cells and macrophages resulting in viremia, but eventually latently infects T cells fairly rapidly leading to lethal T cell tumors in many strains. This disease is very important in economic terms, and even though most chickens in poultry farms are vaccinated, epidemics of Marek's disease occur regularly enough to be a concern to the poultry industry. Chicken strains with the MHC haplotype *B21* have an amazing degree of resistance to Marek's disease (as much as 95% survival), whereas other haplotypes are associated with lesser or no resistance to the disease. Interestingly, both Marek's disease and serological relatives of the *B21* haplotype are found in the Red Jungle fowl of Borneo, which is a presumed ancestor of the domestic chicken. So this disease has a strong and immediate selection on survival in contemporary populations as well as a long history of persistence in chicken populations (PAZDERKA et al. 1975; BRILES et al. 1977, 1983; SIMONSEN et al. 1982; SCHAT 1987; DIETERT et al. 1990; PLACHY et al. 1992)

Thus far, we have found a near-perfect inverse correlation between the level of class I expression and resistance to Marek's disease virus—*B21* haplotypes have low expression and high resistance, *B2, B6* and *B14* have intermediate levels of expression and intermediate levels of resistance, and the rest have high levels of expression and low resistance to Marek's disease. There are a couple of exceptions which may be due to the effects of other genetic loci. In particular, lines 6_2 and 7_2 are both MHC haplotype *B2* and, as expected, both express moderate levels of class I molecules. However, 6_1 is moderately resistant and 7_2 is quite sensitive to Marek's disease, and it has been reported that this is due to a locus unlinked with the MHC, which might encode a susceptibility gene (Crittenden et al. 1972; Gallatin and Longenecker 1979).

If the level of class I expression truly is a major factor in determining resistance and susceptibility to Marek's disease, then the mechanism of resistance is naturally of interest. One possibility is class I is directly involved in infection, so that lower cell surface concentrations on cells presenting antigen lead to fewer infected T cells. Another possibility is that class I is involved in elimination of infected cells, particularly by natural killer (NK) cells, since at least some NK cells recognize the reduction in level of specific class I molecules on the surface of cells (Moretta et al. 1992). In fact, the few data in the literature indicate that *B21* chickens have a higher level of NK cell activity than other chicken strains (Sharma and Okazaki 1981; Göbel et al., this volume). In this light, *B21* would be regarded as an MHC that encoded class I molecules designed to elicit maximum cytotoxic NK activity rather than maximum cytotoxic T cell activity. The MHC haplotypes with moderate expression of class I molecules would elicit moderate levels of both cytotoxic NK and T cell activity, much like the levels that might be present in heterozygotes of *B21* with an MHC haplotype with high expression of class I molecules. If the balance of NK and T cell cytotoxic activity is determined in this manner, then the level of expression of particular class I molecules could represent a polymorphism that is selected by pathogens during evolution, perhaps as important as the actual binding specificity.

5 The Compact Nature of the Chicken MHC May Be a General Feature of the Avian Genome That Appeared with Microchromosomes in the Lineage Leading to Birds

The two notions propounded above, that the chicken MHC is more compact and simple than the MHC of the typical mammal and that this has profound effects on the ability of an individual animal to respond to infectious pathogens, lead directly to the twin questions of origin and maintenance of these features in birds. First, did birds (rather than mammals) diverge from the original ancestral condition, and if so, how and why? Second, if the avian condition is less favorable for survival than the mammalian condition, then why haven't birds simply expanded their MHC?

The available evidence indicates that the MHCs of mammals and amphibians share certain features that are different from those of birds, as though birds diverged markedly from a common ancestor. These features are not restricted to the MHC, but are characteristic of much of the avian genome. For instance, like the class I α and class II β genes of the MHC, the chicken β_2m gene has small introns and some regions with a remarkably high content of G and C nucleotides (and is also located on a microchromosome) (Kaufman et al. 1992; Riegert et al. 1995). Also, the chicken immunoglobulin heavy and light chain loci as well as the T cell receptor α and β chain loci are quite compact, with fewer genes, small introns, small intergenic distances (and some G+C bias), like the MHC genes (Reynaud et al. 1987, 1989; Tjoelker et al. 1990; Göbel et al. 1994; Chen et al., this volume).

In fact, the avian genome is roughly a third of the size of typical mammalian and amphibian genomes, and is mostly composed of microchromosomes, which are generally G+C rich. Unlike mammals and amphibians, most birds have a very similar karyotype, with roughly the same number of microchromosomes and with the few avian macrochromosomes (roughly the size of typical mammalian or amphibian chromosomes) having similar banding patterns (Shields 1982; Tegelström et al. 1983; Auer et al. 1987; Holmquist 1989; Christidis 1990).

We have proposed (Kaufman et al. 1991a; Kaufman and Salomonsen 1993; Riegert et al. 1995) that the lineage leading to birds was afflicted with an unknown agent that led to the random deletion of DNA over evolutionary time. This hypothetical agent might have directly deleted DNA from the genome (for instance, a transposon or a mutation in some enzyme involved in DNA synthesis, repair or recombination). Alternatively, an outside agent might have selected for a host response that resulted in DNA deletion. In either case, this random process would remove extra DNA in introns and intergenic regions, remove "nonessential" genes including extra members of multigene families, and remove some (thus far unidentified) sequence elements responsible for the maintenance of isochores, particularly those A+T rich, gene poor regions that are represented by Giemsa-staining chromosomal bands. The result would be a smaller genome composed mainly of microchromosomes, most of which would be G+C rich. Random deletion of essential genes would leave few survivors, and presumably the process stopped with the founder of the birds, resulting in their homogeneous karyotypes.

While such a scenario could explain many of the features of the avian genome, much of the data is still fragmentary or at low resolution. For instance, it is still possible that microchromosomes have both G+C rich isochores with small genes and A+T rich isochores with big genes, but that the bands are too small to detect by staining chromosomes. So we need to know how tight the correlation between microchromosomal location and compact G+C rich genes and loci really is. It is also still possible that the details of gene organization are not the same in different birds, even though the banding patterns of the macrochromosomes and the number of microchromosomes are similar. So we need to know the linkage maps of several distantly related birds. And, even if our understanding of the data is correct, there are other explanations to consider. For instance, it might be that the avian genome is in a dynamic equilibrium (as proposed by Holmquist 1989), with the process of diminution balanced by augmentation, so that it is constantly changing, but at a level of resolution that has not yet been examined.

Whatever the model for the origin of the small avian genome, the maintenance of a simple MHC is an important issue. If there is a big advantage to having a reasonably sized multigene family of MHC molecules (since the individual would be less at risk), then why don't chickens expand the MHC genes as mammals have done; in fact, why don't they just up-regulate the expression of the MHC genes they already have? In other words, what compensating pressure might there be for a simple MHC?

One possibility that has been suggested to us is that the greater the number of MHC molecules, the greater the propensity for autoimmune disease and the birds have simply opted for a different mixture of reactivity and autoreactivity than mammals. Another possibility is that Marek's virus is such an important disease that a simple MHC is required to ensure easy regulation of the level of expression. A final possibility that we have considered is that the number of MHC genes is limited by the number of T cell receptor genes—since the avian T cell receptor loci are so small and simple compared to mammals, too many highly expressed MHC molecules means too many T cells are deleted during selection in the thymus. If this last idea is correct, then we must find out why the T cell receptor genes do not simply expand. Clearly there is much work yet to be done to understand, in this case, whether the chicken or the egg came first.

6 Summary

Birds, like mammals, have highly a polymorphic MHC that determines strong allograft rejection. However, chickens have a much smaller, more compact and simpler MHC than mammals, as though the MHC has been stripped down to the essentials during evolution. The selection pressure on a single MHC gene should be much stronger than on a large multigene family, and, in contrast to mammals, there are a number of viral diseases for which resistance and susceptibility are determined by particular chicken MHC haplotypes. We have determined the peptide motifs for the dominant class I molecules from a number of chicken MHC haplotypes, which may explain some disease associations quite simply. Other disease associations, like the famous examples with Marek's disease, may be due to polymorphism in the level of expression of MHC class I molecules. We believe that the compact and simple nature of the MHC is due to the presence of microchromosomes in birds and suggest that the evolutionary origin of birds has been strongly influenced by the emergence of microchromosomes.

Acknowledgments. We would like to thank Thomas Göbel and Ollie Vainio for critical reading of the manuscript, and Hans Spalinger for photography. The Basel Institute for Immunology was founded and is supported by Hoffman-La Roche, Basel, Switzerland.

References

Auer H, Mayr B, Lambrou M, Schleger W (1987) An extended chicken karyotype, including the NOR chromosome. Cytogenet Cell Genet 45: 218–221

Bloom S, Bacon L (1985) Linkage of the major histocompatibility (B) complex and the nucleolar organizer region in the chicken. J Heredity 76: 146–154

Bloom SE, Briles WE, Briles RW et al. (1987) Chromosomal localization of the major histocompatibility (B) complex (MHC) and its expression in chickens aneuploid for the major histocompatibility complex/ribosomal deoxyribonucleic acid microchromosome. Poult Sci 66: 782–789

Briles WE, Stone H, Cole R (1977) Marek's disease: effect of B histocompatibility alloalleles in resistant and susceptible chicken lines. Science 195: 193–195

Briles WE, Briles RW, Stone HA (1983) Resistance to a malignant lymphoma in chicken is mapped to a subregion of the major histocompatibility (B) complex. Science 219: 977–979

Bumstead N, Palyga J (1992) A preliminary linkage map of the chicken genome Genomics 13: 690–697

Campbell RD, Trowsdale J (1993) The human MHC. Immunol Today 14: 349

Christidis L (1990) Aves, In: John B (ed) Animal cytogenetics, vol 413B. Borntraeger, Berlin, pp 1–116

Crittenden L, Muhm R, Burmester B (1972) Genetic control of susceptibility to avian leukosis complex. 2. Marek's disease. Poult Sci 51: 261–267

Crone M, Simonsen M (1987) Avian major histocompatibility complex: a birds eye view. In: Toivanen A. Toivanen P (eds) Avian immunology: basis and practice, vol 2. CRC Press, Boca Raton FL, pp 25–42

Dietert R, Taylor R, Dietert M (1990) The chicken Major Histocompatibility complex: structure and impact on immune function, disease resistance and productivity. In: Basta O (ed) MHC, differentiation antigens and cytokines in animals and birds. Monographs in animal immunology, vol 1. Barlab, Backsburg VA, pp 7–26

Falk K, Rotzschke O (1993) Consensus motifs and peptide ligands of MHC class I molecules. Semin Immunol 5: 81

Gallatin W, Longenecker B (1979) Expression of genetic resistance of an oncogenic herpesvirus at the target level. Nature 280: 587–589

Göbel T, Chen C–L, Lahti J, Kubota T, Kuo C-L, Aebersold R, Hood L, Cooper M (1994) Identification of T-cell receptor α-chain genes in the chicken. Proc Natl Acad Sci USA 91: 1094–1098

Goto R, Miyada C, Young S, Wallace R, Abplanalp H, Bloom S, Briles W, Miller M (1988) Isolation of a cDNA clone from the B-G subregion of the chicken histocompatibility (B) complex. Immunogenetics 27: 102–109

Guillemot F, Billault A, Pourquie O, Behar G, Chausse A-M, Zoorob R, Kreiblich G, Auffray C (1988) A molecular map of the chicken major histocompatibility complex: the class II β genes are closely-linked to the class I genes and the nucleolar organizer. EMBO J 7: 2775–2785

Hala K, Vilhelmova M, Hartmanova J (1976) Probable crossing-over in the B blood group of chickens. Immunogenetics 3: 97–103

Hill AVS, Allsopp C, Kwiatowski D, Antsey N, Twumasi P, Rowe A, Bennet S, Brewster D, McMichael A, Greenwood B (1991) Common West African HLA antigens are associated with protection from severe malaria. Nature 352: 595–600

Holmquist G (1989) Evolution of chromosome bands: molecular ecology of noncoding DNA. J Mol Evol 28: 469–486

Kaufman J, Salomonsen J (1993) What in the dickens is with those chickens? Res Immunol 144: 495–519

Kaufman J, Salomonsen J, Skjoedt K (1989) B-G cDNA clones have multiple small repeats and hybridize to both chicken MHC regions. Immunogenetics 30: 440–451

Kaufman J, Salomonsen J, Riegert P, Skjødt K (1991a) Using chicken class I sequences to understand how xenoantibodies crossreact with MHC-like molecules in nonmammalian vertebrates. Am Zool 31: 570–579

Kaufman J, Skjødt K, Salomonsen J (1991b) The B-G multigene family of the chicken MHC. Crit Rev Immunol 11: 113–143

Kaufman J, Andersen R, Avila D, Engberg J, Lambris J, Salomonsen J, Welinder K, Skjødt K (1992) Different features of the MHC class I heterodimer have evolved at different rates: chicken B-F and β_2 microglobulin sequences reveal invariant surface residues. J Immunol 148: 1532–1546

Kaufman J, Bumstead N, Miller M, Riegert P. Salomonsen J (1995a) The chicken class II α gene is located outside of the B complex. In: Davison TF, Bumstead N, Kaiser P (eds) Advances in avian immunology. Carfax, Abingdon, UK, pp 119–127

Kaufman J, Völk H, Wallny H-J (1995b) A "minimal essential MHC" and an "unrecognized MHC": two extremes in selection for polymorphism. Immunol Rev 143: 63–88
Klein J (1986) Natural history of the major histocompatibility complex. Wiley, New York
Klein J (1987) Origin of major histocompatibility complex polymorphism: the trans-species hypothesis. Hum Immunol 19: 155–162
Koch C, Skjoedt K, Toivanen A, Toivanen P (1983) New recombinants within the MHC (B-complex) of the chicken. Tissue Antigens 21: 129–137
Lawlor D, Zemmour J, Ennis P, Parnam P (1990) Evolution of class I MHC genes and proteins: from natural selection to thymic selection. Annu Rev Immunol 8: 23–64
Miller MM, Goto R, Zoorob R, Auffray C, Briles WE (1994a) Regions of homology shared by Rfp-Y and major histocompatibility B complex genes. Immunogenetics 39: 71–73
Miller MM, Goto R, Bernot A, Zoorob R, Auffray C, Bumstead N, Briles WE (1994b) Two MHC class I and two MHC class II genes map to the chicken Rfp-Y system outside the B complex. Proc Natl Acad Sci USA 91: 4397–4401
Moretta L, Ciccone E, Moretta A, Höglund P, Öhlen C, Kärre K (1992) Allorecognition by NK cells: nonself or no self? Immunol Today 13: 300–306
Pazderka F, Longenecker B, Law G, Stone H, Ruth R (1975) Histocompatibility of chicken populations selected for resistance to Marek's disease. Immunogenetics 2: 93–100
Pink JRL, Croege W, Hala K, Miggiano VC, Ziegler A (1977) A threelocus model for the chicken major histocompatibility complex. Immunogenetics 5: 203–216
Plachy J, Pink JRL, Hala K (1992) Biology of the chicken MHC (B complex). Crit Rev Immunol 12: 47–79
Potts W, Wakeland E (1993) Evolution of MHC genetic diversity: a tale of incest, pestilence and sexual preference. Trends Genet 9: 408
Reynaud C-A, Anquez V, Grimal H, Weill J-C (1987) A hyperconversion mechanism generates the chicken light chain preimmune repertoire. Cell 48: 379–388
Reynaud C-A, Dahan A, Anquez V, Weill J-C (1989) Somatic hyperconversion diversifies the single V_H gene of the chicken with a high incidence in the D region. Cell 59: 171–183
Riegert P, Andersen R, Bumstead N, Döhring C, Dominguez-Steglich M, Engberg J, Salomonsen J, Schmidt M, Skjødt K, Schwager J, Kaufman J (1996) The chicken β_2-microglobulin gene is located on a non-MHC microchromosome: a small G + C rich gene with X and Y boxes in the promoter. Proc Natl Acad Sci USA (in press)
Schat K (1987) Immunity in Marek's disease and other tumors. In: Toivanen A, Toivanen P (eds) Avian immunolgy: basis and practice, vol 2. CRC Press, Boca Raton FL, pp 101–128
Schwartz D, Tizard R, Gilbert W (1983) Nucleotide sequence of Rous Sarcoma virus. Cell 32: 853
Sharma JM, Okazaki W (1981) Natural killer cell activity in chickens: target cell analysis and effect of antithymocyte serum on effector cells. Infect Immun 31: 1078
Shields G (1982) Comparative avian cytogenetics: a review. Condor 84: 45–58
Simonsen M, Crone M, Koch C, Hala K (1982) The MHC haplotypes of the chicken. Immunogenetics 16: 513–532
Tegelström H, Ebenhard T, Ryttman H (1983) Rate of karyotype evolution and speciation in birds. Hereditas 98: 235–239
Tjoelker L, Carlson L, Lee K, Lahti J, McCormack W, Leiden J, Chen CL, Cooper M, Thopson C (1990) Evolutionary conservation of antigen recognition: the chicken T-cell receptor β chain. Proc Natl Acad Sci USA 87: 7856–7860
Trowsdale J (1993) Genomic structure, function in the MHC. Trends Genet 9: 117–122
Trowsdale J (1995) "Both man and bird and beast": comparative organization of MHC genes. Immunogenetics 41: 1–17
Zoorob R, Bernot G, Renoir DM, Choukri F, Auffray C (1993) Chicken major histocompatibility complex class II B genes: analysis of interallelic and inter-locus sequence variance. Eur J Immunol 23: 1139–1145

Acute Avian Leukemia Viruses as Tools to Study Hematopoietic Cell Differentiation

K.M. McNagny and T. Graf

1 Introduction

Hematopoiesis is a multistep process during which an uncommitted, self-renewing progenitor divides and differentiates along one of eight distinct lineages, each with a unique function and each expressing a unique complement of proteins enabling them to carry out these functions. In adult vertebrates, multipotent hematopoietic precursors reside within the microenvironment of the bone marrow where they undergo a progressive maturation and commitment to one of several lymphoid and nonlymphoid cell lineages. Regulation of normal hematopoietic development requires an intricate set of cues from stromal cells involving signaling via soluble growth factors, cell-cell and cell-extracellular matrix interactions (Quesenberry 1992). Analysis of the cellular intermediates in these differentiation pathways has been greatly facilitated by the development of in vitro colony forming assays for multipotent and monopotent hematopoietic precursors and the identification of cytokines which regulate their growth and differentiation (Metcalf 1984, 1988). Nevertheless, analysis of the molecular mechanisms governing differentiation and commitment to a given lineage has been technically difficult as bone marrow precursors are present in relatively low frequency and reagents allowing their isolation have only recently been developed. In addition, obtaining clonal precursor cells in sufficient numbers for

Differentiation Programme, European Molecular Biology Laboratory, Meyerhofstrasse 1, 69117 Heidelberg, Germany biochemical analysis has been problematic since these precursors have a limited life span in vitro.

In vitro transformation of hematopoietic cells by oncogene-containing retroviruses represents an alternative approach to probe hematopoietic differentiation processes. Because of the unusually high susceptibility of avian species to the proliferation-inducing effects of oncogenes, a large variety of acutely transforming avian retroviruses have been isolated from diseased animals. These viruses exhibit a remarkable specificity in that they selectively transform specific subsets of hematopoietic cells and thereby allow their expansion and analysis as homogeneous populations (GRAF and BEUG 1978; MOSCOVICI and GAZZOLO 1982; BEUG and GRAF 1989). Their transforming capacity is encoded in disregulated forms or "gain of function mutants" of cellular proto-oncogenes which affect both growth and differentiation. The analysis of the mechanism by which these viral oncogenes transform specific hematopoietic cells thus provides insights not only into mechanisms of leukemogenesis but also into normal growth control and differentiation and the role of hematopoietic specific proto-oncogenes. In particular, the development of temperature sensitive (ts) mutants of viral oncogenes offers a genetic approach to the study of hematopoiesis, consisting of the expansion of oncogene-transformed cells at the permissive temperature and subsequent inactivation of the oncoprotein by shifting these cells to the nonpermissive temperature.

Here we will provide a brief overview of the acutely transforming avian leukemia viruses and their utility as models to study hematopoietic cell differentiation and oncogenic transformation. Particular attention will be devoted to the E26 virus which has been the primary focus of the authors' laboratory in recent years.

2 Naturally Occurring Acute Leukemia Viruses

Representative examples of a variety of acute avian leukemia viruses, their oncogenes, and the phenotype of the cells they transform are presented in Table 1. They can be grouped into viruses which predominantly transform erythroid cells (AEV-type viruses), macrophages (MC29), lymphoid cells (REV-T), monoblasts (AMV), and myeloblasts and multipotent progenitors (E26). The initial isolation and characterization of these viruses have been reviewed in detail elsewhere (GRAF and BEUG 1978). All were isolated from birds which spontaneously developed leukemias or sarcomas and predominantly induce the same type of hematopoietic neoplasms as the original tumor when inoculated into recipient animals. Each virus contains sequences of cellular genes (proto-oncogenes) which exhibit truncations and/or point mutations. These genes were probably acquired as the result of integration of an ancestral provirus in proximity to a proto-oncogene and subsequent splicing into the viral genome during production of transcripts for virion packaging (see LUCIW and LEUNG 1992 for a detailed review on

Table 1. Acute avian leukemia viruses

Virus strain	Oncogene	Function of protein encoded by cellular counterpart	Transformed hematopoietic cell type	Reference for the original oncogene identification
AEV-ES4	v-*erbA*	Thyroid hormone receptor, plus	Erythroblasts	VENNSTRÖM and BISHOP 1982
	v-*erbB*	EGF/TGFα tyrosine kinase receptor		
AEV-H	v-*erbB*	EGF/TGFα tyrosine kinase receptor	Erythroblasts	YAMAMOTO et al. 1983
S13	v-*sea*	Receptor-type tyrosine kinase	Erythroblasts	BEUG et al. 1985a
MC29 OK10 CMII	v-*myc*	HLH-type transcription factor	Macrophages	ROUSSEL et al. 1979 BISTER et al. 1979
MH2	v-*myc*	HLH-type transcription factor, plus	Macrophages	COLL et al. 1983
	v-*mil*(v-*raf*)	MAP kinase kinase		JANSEN et al. 1983 KAN et al. 1983
REV_T	v-*rel*	NF-κB-related transcription factor	B cells and T cells	STEPHENS et al. 1983
E26	v-*myb*	HTH-type transcription factor as an in frame fusion with	Myeloblasts, and multilineage precursors (T-MEP)	ROUSSEL et al. 1979
	v-*ets*	HTH-type transcription factor		LEPRINCE et al. 1983 NUNN et al. 1983
AMV	v-*myb*	HTH-type transcription factor	Monoblasts	ROUSSEL et al. 1979

viral replication). For most strains it was observed that sequential passage from animal to animal resulted in viruses which induced leukemia with a progressively higher frequency (from 15% incidence to 100% after seven in vivo passages with AEV-ES4) and with much shorter latency periods (from months to weeks, reviewed in GRAF and BEUG 1978). This "in vivo selection" or "evolution" toward a virus with a more aggressive transforming potential probably reflects an accumulation of further point mutations in the transduced oncogene due to the error-prone process by which retroviruses propagate (HU and TEMIN 1990; PATHAK and TEMIN 1990a, b; ZHANG and TEMIN 1993). In addition, some viruses (AEV-ES4, MH2, and E26) were selected which had acquired a second oncogene, presumably by a second recombination event. For these virus strains it has been demonstrated that both oncogenes are required for their full leukemogenic potential (reviewed in BEUG and GRAF 1989). This "in vivo selection" has recently been exploited to identify some of the critical events needed for oncogene activation by infecting animals with retroviruses carrying nonleukemogenic versions of oncogenes and subsequent analysis of the activating mutations in these oncogenes in the rare animals which develop leukemia (METZ and GRAF 1991b; HRDLICKOVA et al. 1994).

2.1 ErbA, ErbB Viruses

Three well studied viruses transform predominantly avian erythroid cells: AEV-ES4, this strain is probably identical to AEV-R), AEV-H, and S13. The prototype virus, AEV-ES4, contains both v-*erbA* and v-*erbB* oncogenes (reviewed in METZ 1994). When injected into young animals this virus induces a rapid erythroleukemia and transforms cells in vitro with an immature erythroid (CFU-E) phenotype (SAMARUT and GAZZOLO 1982). The transformed cells express low levels of erythroid markers including α and β globins, the transcription factors GATA-1 and NFE-2, δ-aminolevulinic acid synthase (ALA-S), band 4.1, histone 5 (H5), carbonic anhydrase as well as cell surface antigens characteristic of erythroid precursors (BEUG et al. 1979; ZENKE et al. 1988; reviewed in BEUG and GRAF 1989).

The relative contributions of v-*erbA* and v-*erbB* to erythroid transformation have been analyzed in great detail. Experiments to address the function of v-*erbA* have been performed by comparing AEV-ES4 with the AEV-H isolate which lacks v-*erbA* or with an AEV-ES4 virus in which v-*erbA* had been deleted, or with AEV-ES4 mutants carrying a temperature sensitive lesion in v-*erbB* (BEUG et al. 1982; 1985b; FRYKBERG et al. 1983; YAMAMOTO et al. 1983; CHOI et al. 1986; Table 2). These experiments showed that v-*erbB* is sufficient for the transformatiion of erythroblasts and for the abrogation of the erythropoietin requirement. For example, cells transformed by v-*erbB*^{ts} and shifted to the nonpermissive temperature become dependent on erythropoietin for their survival and growth and, in addition, differentiate into erythrocyte-like cells. The primary effect of v-*erbA* also appears to be the transformation of erythroid cells as constructs that lack v-*erbB* can

Table 2. Conditional alleles of avian leukemia virus oncogenes

Virus strain (mutant)	Oncogene	Phenotype of transformed cells		References
		Before differentiation	After differentiation	
AEV-ES4 (ts34)	*erb*A, *erb*B^{ts}	Erythroblasts	Erythrocytes	GRAF et al. 1978 BEUG et al. 1982
S13	*sea*^{ts}	Erythroblasts	Erythrocytes	KNIGHT et al. 1988
REV_T	*rel*-ER	B cells	Dendritic cells, granulocyte-like cells	BOEHMELT et al. 1992 BOEHMELT et al. 1995
	rel-ER	Spleen cells of unknown phenotype	Cells of unknown phenotype	CAPOBIANCO and GILMORE 1993
	rel^{ts}	Spleen cells of unknown phenotype	Cells of unknown phenotype	WHITE et al. 1995
E26 (ts1.1)	*myb*-ets^{ts}	Multi-lineage progenitor (E-MEP)	Erythrocytes, eosinophils myeloblasts	GOLAY et al. 1988 KRAUT et al. 1994
		Promyelocytes	Promyelocytes	GOLAY et al. 1988
E26 (ts21)	*myb*^{ts}-ets	Thromboblast--multi-lineage progenitor (T-MEP)	Thrombocytes	GRAF et al. 1992 FRAMPTON et al. 1995
		Myeloblasts	Macrophages	BEUG et al. 1984

induce the proliferation of erythroid cells in virus infected bone marrow under certain conditions (GANDRILLON et al. 1989; CASINI and GRAF, submitted). In contrast to v-*erbB* transformants, however, these cells are blocked at an early stage of differentiation and are dependent on anemic serum (as a source of erythropoietin) for growth (GANDRILLON et al. 1989).

v-*erbA* encodes a mutated version of the thyroid hormone receptor-α (SAP et al. 1986; WEINBERGER et al. 1986; reviewed in BEUG et al. 1994; METZ 1994). The normal function of this protein in response to hormone is to transactivate genes containing thyroid hormone response elements. In the absence of hormone (triiodothyronine, T_3) the protein is inactive and may even repress transcription (BARETTINO et al. 1993; HERMANN et al. 1993). v-*erbA* was first shown to selectively repress a subset of genes important for erythroid maturation including ALA-S, the anion transporter band 3, and carbonic anhydrase (ZENKE et al. 1988, 1990; SCHROEDER et al. 1990; PAIN et al. 1990). More recent studies suggest that v-*erbA* constitutively binds hormone response elements but has lost the ability to bind thyroid hormone and therefore acts to repress rather than enhance transcription of target genes (DAMM et al. 1989; SAP et al. 1989; DISELA et al. 1991).

v-*erbB* represents a truncated and mutated version of the epidermal growth factor/transforming growth factor-α (EGF/ TGFα) receptor tyrosine kinase (DOWNWARD et al. 1984; LAX et al. 1988). The role of the cellular gene is to transduce mitogenic transmembrane signals in response to ligand binding. The viral form of this protein is highly mutated and also lacks the ligand binding domain of the cellular receptor. It appears to be constitutively active and thus provides a constant proliferative signal. The observation that "v-*erbA*-only" viruses fail to induce leukemia (GADRILLON et al. 1989) is consistent with the idea that the proliferative effect of v-*erbB* is the primary force driving transformation by AEV-ES4. More recently, however, using a virus with a higher replication efficiency, it could be shown that v-*erbA* alone is sufficient to induce lethal leukemia, suggesting that the interplay between v-*erbB* and v-*erbA* is more complex (CASINI and GRAF, submitted).

It has been shown that c-*erbB* and *c-erbA* are normally expressed during erythroid differentiation (PAIN et al. 1991; SCHROEDER et al. 1992; reviewed in BEUG et al. 1994, and see also MÜLLNER and BEUG this issue) and therefore transformation by AEV probably directly interferes or mimics the function of these proto-oncogenes. Indeed, it has recently been shown that a combination of TGFα (a natural ligand for c-*erbB* the encoded TGFα receptor) and estradiol (an activator of the estrogen receptor which functionally resembles the c-*erbA* encoded thryoid receptor) are able to support the long-term in vitro proliferation of a subset of erythroid progenitors (SCHROEDER et al. 1993; HAYMAN et al. 1993; see also Müllner and Beug, this issue). However, it must be noted that these cells only represent a fraction of the erythroid progenitors in normal bone marrow and their contribution to erythropoiesis is not clear (STEINLEIN et al. 1995). In summary, analysis of AEV has led to a fairly detailed picture about the transforming mechanisms of nuclear- and plasma membrane-type receptor oncogenes and their cooperation in disregulating erythroid cell proliferation and differentiation.

erbB is not the only receptor tyrosine kinase with the ability to transform erythoid cells. For example the v-*sea* oncogene of the S13 erythroleukemia virus (Beug et al. 1985a; and see Table 1) encodes a protein related to but distinct from the hepatocyte growth factor receptor, another receptor tyrosine kinase (Huff et al. 1993). Temperature sensitive mutations in the v-*sea* oncogene behave quite similarly to those *erbB* (Knight et al. 1988) suggesting that these two proteins transform erythroid cells by subverting the same signal transduction pathway.

2.2 Myc Viruses

A number of v-*myc* containing viruses (MC29, CMII, OK10, and MH2) induce "myelocytomatosis" or "endotheliomas" in-vivo and transform macrophage-like cells in vitro (reviewed in Graf and Beug 1978; Graf and Stéhelin 1982; Payne 1992; Table 1). In addition, these viruses induce hepatocarcinomas, adenocarcinomas, and sarcomas. The MH2 virus is unique among this group in that it has transduced v-*myc* and a second oncogene named v-*mil* (Coll et al. 1983; Jansen et al. 1983; Kan et al. 1983). c-Mil is the avian equivalent of the mammalian c-Raf serine/threonine kinase and acts as a MAP kinase (Moelling et al. 1984; Sutrave et al. 1984; Denhez et al. 1988). The hematopoietic cells transformed in vitro by v-*myc* containing viruses closely resemble normal avian macrophages based on morphology, adherence, phagocytic activity, expression of Fc receptors, the Colony-Stimulating Factor (CSF)-1 receptor, MHC class II antigens, and other myeloid specific surface antigens (Beug et al. 1979; Gazzolo et al. 1979; Graf et al. 1981, 1986).

As with the two oncogenes carried by the AEV virus, MH2 derivatives carrying v-*myc* or v-*mil* alone have been constructed to decipher the contributions of the two oncogenes to transformation (Graf et al. 1986). The conclusions from these studies are that neither the MH2 form of v-*myc* nor v-*mil* on their own are able to efficiently cause leukemia (although v-*myc* alone will transform macrophages in vitro). In addition, "v-*mil*-only" constructs are unable to transform myelomonocytic cells in vitro, although they weakly transform erythroid cells (Kahn et al. 1986). v-*mil* has been shown to contribute to the leukomogenic potential of MH2 virus by abrogating the dependence of myeloid cells on exogenous growth factors for proliferation. Thus macrophages transformed by "v-*myc*-only" viruses were shown to require chicken myelomonocytic growth factor (cMGF) for proliferation (cMGF is an avian growth factor with distant homology to mammalian granulocyte-CSF and interleukin-6; Leutz et al. 1984, 1989) while macrophages that coexpress v-*mil* do not require cMGF (Graf et al. 1986). The function of v-*mil* appears to be induction of cMGF production as culture supernatants from MH2 transformants supported the growth of "*myc*-only" transformants and this effect could be inhibited by the addition of neutralizing antibodies to cMGF (Graf et al. 1986). It is therefore likely that v-*mil* mimics a normal signal involved in cytokine production by macrophages and results in factor independence due to autocrine growth stimulation.

2.3 Rel Virus Strain T

The oncoprotein encoded by the reticuloendotheliosis virus strain T (REV-T), v-Rel, is a member of the NF-κB/dorsal family of transcription factors (reviewed in GILMORE 1991; BOSE 1992; KABRUN and ENRIETTO 1994). All members of this family contain an $NH_{2\alpha}$–terminal "Rel homology domain" and appear to be involved in rapid responses to extracellular signals. These proteins are normally complexed with repressor proteins in the cytoplasm and upon activation translocate to the nucleus where they bind specific target sequences as dimers. There is evidence that c-Rel may act as both an activator and as a repressor of a transcription (reviewed in GILMORE 1991; BOSE 1992; SARKAR and GILMORE 1993; KABRUN and ENRIETTO 1994). It is not yet clear whether it is the activation or repression of transcription by v-*rel* which leads to cell transformation.

The REV-T virus strain is morphologically and antigenically more closely related to the mammalian retroviruses than to avain retroviruses (MIZUTANI and TEMIN 1973; MALDONADO and BOSE 1973; MOELLING et al. 1975), presumably representing a jump in species during the evolution of the virus. REV-T is also unusual in that most clones transformed by the virus are immortalized (BEUG et al. 1981; LEWIS et al. 1981). Although it was recognized relatively early that the virus is extremely oncogenic and that it induces a lymphoid leukemia, considerable confusion remained concerning the actual transformation specificity of the virus, leading early reports to classify the cells as pre-B/pre-T lymphocytes (BEUG et al. 1981) or pre-B cells (LEWIS et al. 1981). One major source of the confusion appears to have arisen due to the influences of different helper viruses used for REV-T transformation. As is true for the majority of the actuely transforming avian retroviruses, the v-*rel* oncogene transduced by REV-T has replaced structural elements in its genome which are normally required for replication, including the envelope antigen (reviewed in GRAF and STÉHELIN 1982; LUCIW and LEUNG 1992). These must therefore be supplied *in trans* using a replication competent "helper" virus. Two distinct helper viruses have most commonly been used in studies with REV-T, namely, REV-A and chicken syncytial virus (CSV). Interestingly, the REV-A helper virus alone has a profound immunosupressive effect and leads to considerable atrophy of the thymus and bursa of Fabricius, the sites of primary T and B lymphopoiesis, respectively in avians. The immunosupressive effect in the bursa is probably due to destruction of both B cells and stromal elements required for normal B lymphocyte maturation (BARTH and HUMPHRIES 1988b). The end result is that when REV-T is combined with REV-A as a helper virus most infected animals die rapidly from T cell leukemia, probably due to elimination of potential B cell targets by the helper virus (MARMOR et al. 1993; BARTH and HUMPHIRES 1988a, b) T cell lines transformed by REV-T may express CD3, CD4, CD8, and α/β or γ/δ T cell receptors, indicating that most T cell subsets are targets for transformation (BARTH et al. MARMOR et al. 1993; McNAGNY, unpublished). Some clones obtained under these conditions have also been classified as "myeloid" based on the expression of MHC class II and the lack of B and T cell markers (BARTH et al. 1990). However, these cells fail to express markers typical of myelomonocytic cells and probably represent T lineage cells which have lost expression of T cell receptor (McNagny, unpublished). Analysis of these clones for gene rearrangements in T cell receptor loci using the recently cloned constant region probes may clarify this issue (see Chen, this issue).

In contrast to REV-A helper virus, co-infection of REV-T with CSV helper virus leaves the bursal architecture essentially intact. Under these conditions the primary targets for transformation are committed B lineage cells (Barth and Humphries 1988b). Transformed B cells cover the spectrum of phenotypes from cells expressing markers of committed B lineage cells but lacking immunoglobulin gene rearrangements to mature IgM- or IgG-positive cell lines (Zhang et al. 1991). Although the cells lacking Ig gene rearrangements fail to undergo VDJ recombination in vitro, those which have completed heavy and light chain rearrangements are capable of further diversifying their receptors by gene conversion of the light chain loci, showing that this activity can be preserved in the transformed clones (Kim et al. 1990; Zhang et al. 1989, 1991).

In another study a replication competent v-*rel* virus was constructed in the background of an avian sarcoma virus (Morrison et al. 1991). Infection of bone marrow with this virus leads to the outgrowth of transformed cells which demonstrates that the original helper virus is not essential for cell transformation. All transformed clones appear to express the T cell-specific CD3 complex and yet the majority of clones were found to express IgM. In addition, a minor subpopulation of cells were also reported to express erythroid-and myeloid-restricted antigens (Morrison et al. 1991). The classification of these cells to a specific lineage therefore is at variance with the results obtained with the helper dependent strains.

In an attempt to address the mechanism of v-*rel* function in transformation, inducible forms of v-*rel* have been constructed by fusion to the hormone binding domain of the estrogen receptor (ER). This results in proteins that are inhibited for DNA binding in the absence of estradiol and can be activated in its presence (Boehmelt et al. 1992; Capobianco and Gilmore 1993). One of these constructs was made in the context of a replication-competent avian sarcoma virus (RCAS) vector (Boehmelt et al. 1992) while another was packaged using a REV-A helper virus (Capobianco and Gilmore 1993). In both cases, DNA binding and transactivation is hormone-dependent, and it appears that the transactivation function is due in part to the estrogen-dependent transactivating domain in the ER fusion and not from the endogenous transactivation domain of the Rel protein. Both constructs transform hematopoietic cells from bone marrow or spleen in a factor-dependent fashion and while the phenotype of cells transformed by the repliciation defective strain were not determined, cells transformed by the RCAS/Rel-ER virus expressed MHC class II antigens and very low levels of surface IgM, suggesting affiliation to the B cell lineage (Boehmelt et al. 1995). Removal of estrogen and subsequent treatment with an estrogen antagonist resulted in cells resembling either dendritic cells or polymorphonuclear "granulocytes," depending on the culture conditions used. The nature of these cells is not entirely clear since the dendritic cells express at least one marker of nonmarrow-derived, follicular dendritic cells, and the polymorphonuclear cells lacked cytoplasmic granules characteristic of avian heterophils. Nevertheless, the data suggest that inactivation of v-*rel* can induce a trans-differentiation along different hematopoietic and non-hematopoietic lineages.

2.4 Myb, Ets Viruses

Cellular sequences encoding the Myb transcription factor have been transduced by two independent avian retroviruses, E26 and the avian myeloblastosis virus, AMV (reviewed in GRAF and STÉHELIN 1982). c-*myb* is expressed by precursors of most hematopoietic lineages as well as by a few nonhematopoietic cell types (GONDA et al. 1982; WESTIN et al. 1982; DUPREY and BOETTIGER 1985). Germ line inactivation of the c-*myb* gene in mice showed that it plays a crucial role in normal hematopoiesis since nullizygous embryos die at about day 14 of gestation from profound defects in definitive erythropoiesis and these mice probably lack the ability to produce most other hematopoietic lineages as well (MUCENSKI et al. 1991). In the E26 virus, v-*myb* is fused in frame to an oncogenic version of a second cellular transcription factor, c-*ets*1 (LEPRINCE et al. 1983; NUNN et al. 1983). v-*ets*, which acquired its name from the E-Twenty Six virus, is the founding member of an ever expanding family of transcription factors which share homology in their DNA binding domains. c-*ets*1 is most highly expressed by a variety of embryonic tissues, embryonic and adult thymocytes, and vascular endothelia (BHAT et al. 1987; VANDENBUNDER et al. 1989).

E26 was originally described as inducing a mixed erythroid and myeloid leukemia in vivo. Similarly, in vitro transformation assays suggested that the virus in capable of transforming both myelomonocytic cells and early erythroid cells (RADKE et al. 1982; MOSCOVICI et al. 1983). Although detailed analysis of the myeloid cells transformed by E26 revealed a clear resemblance to myeloblasts (granulocyte and macrophage precursors), the "erythroid" cells transformed in vitro have proven more difficult to classify. In the most recent analyses it has become clear that, although these cells can spontaneously differentiate at a low rate into erythroid cells, most of them express markers restricted to cells of the thrombocytic lineage (GRAF et al. 1992; MCNAGNY et al. 1992; FRAMPTON et al. 1995). This is based on morphological criteria, the expression of thrombocyte integrins $\alpha II\beta_3$ and $\alpha 2\beta_1$, and a number of other markers specific for this lineage (see Table 3). More surprisingly, it was found that when treated with phorbol esters these cells differentiate into eosinophils and myeloblasts (Fig. 1), suggesting that, despite their predominantly thrombocytic phenotype, they functionally correspond to multipotent hematopoietic progenitors (GRAF et al. 1992) We have therefore referred to these cells as "T-MEPs" for thrombocytic, Myb-Ets transformed progenitors (FRAMPTON et al. 1995 and see Fig. 1).

Comparison of the AMV "*myb*-only" virus with the virus expressing the E26 fusion protein or with viruses bearing temperature sensitive mutations in *myb* and *ets* (Table 2) have been helpful in determining the relative contributions of these

Table 3. Phenotypic markers of E26 transformed cells and differentiated derivatives

Cell type	Cell surface antigens	Cytoplasmic/nuclear antigens	Cellular transcription factors
T-MEP	11C3 ($\alpha IIb\beta_3$ integrin) MEP17 ($\alpha_2\beta 1$ integrin) MEP21 MEP26	Histone H5 TGFβ[a]	GATA-1 GATA-2[a] GATA-3[a] SCL[a]
E-MAP	MEP17 ($\alpha_2\beta_1$ integrin) MEP26 JS4	Histone H5 Hemoglobin	GATA-1 SCL[a]
Myeloblast	cMGF receptor MHC class II MYL51/2 4M12 MEP17 ($\alpha_2\beta$. integrin)	Mim-1	C/EBPβ (NF-M)
Promyelocyte	cMGF receptor IC3 MYL51/2 MHC class II 4M12	Chicken type lysozyme[a] Mim-1 Goose type lysozyme[a]	C/EBPβ (NF-M)
Macrophage	Fc receptor[a] cMGF receptor CSF-1 receptor (c-fms)[a] MYL51/2 MHC class II MEP17 ($\alpha_2\beta_1$ integrin)	Chicken type lysozyme[a]	C/EBPβ (NF-M)
Eosinophil	EOS47 (melanotransferrin) 4M12	Eosinophil peroxidase Mim-1	C/EBPβ (NF-M) GATA-1 GATA-2[a]
Thrombocyte	11C3 ($\alpha IIb\beta_3$ integrin) MEP17 ($\alpha_2\beta$. integrin) MEP21 MEP26	Serotonin granules TGFβ[a] Histone H5	GATA-1 GATA-2[a] GATA-3[a] SCL[a]
Erythrocyte	JS4	ALA-S[a] Carbonic anhydrase[a] Hemoglobin Histone H5	GATA-1 GATA-2[a] SCL[a]

References: McNagny et al. 1992; Graf et al. 1992; Kraut et al. 1994; Kulessa et al. 1995; Frampton et al. in press, and Frampton, Graf, Kulessa, McNagny unpublished observations.
TGF-β transforming growth factor-β; ALA-s, aminolevulinic acid synthase.
[a] Indicates the marker is expressed in the designated cell type but has not been examined in all other cell types listed.

two oncogenes to transformation. These experiments showed that, while the myeloid cells transformed by E26 resembled myeloblasts (granulocyte/macrophage precursors), the cells transformed by AMV resemble monoblasts (commited macrophage precursors; Pessano et al. 1979; Introna et al. 1990). Although at first glance this would suggest that coexpression of v-*myb* and v-*ets* allows transformation of earlier myeloid precursors than v-*myb* alone, this is clearly not the case, since E26 viruses in which *ets* is deleted also give myeloblasts (Introna

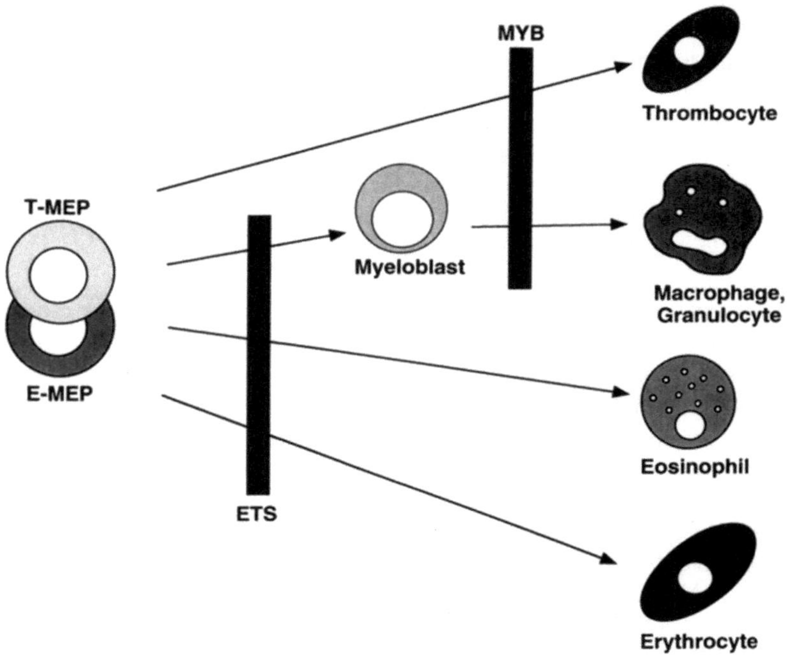

Fig. 1. Differentiation capacity of E26-transformed progenitors. The *bars* indicate the commitment events blocked by the Myb and Ets components of the E26 oncoprotein. For explanation of T-MEP and E-MEP, see accomanying text

et al. 1990). Thus differences in the 5' and 3' termini of the two oncogenic forms of *myb* plus the ten amino acid differences in the overlapping coding regions of these proteins must account for this effect. In experiments in which chimeric proteins were made by swapping subdomains of AMV *myb* with E26 *myb* it could indeed be shown that point mutations in v-*myb* can have a profound biological effect, leading to cells with a myeloblast, monoblast or promyelocyte phenotype (Introna et al. 1990).

The functional role of v-*myb* in the myeloid lineage was perhaps most directly addressed in experiments where it was ectopically expressed in v-*myc* transformed macrophages (Ness et al. 1987). These cells were induced to "dedifferentiate" and to acquire a myeloblast phenotype suggesting that the primary of v-*myb* is to prevent terminal differentiation and maintain the immature phenotype.

Although both AMV and E26 virus transform myeloid cells, only E26 is capable of transforming T-MEP cells (Graf et al. 1992, and unpublished). The role of v-*myb* and v-*ets* in the transformation of MEP cells has been analyzed extensively. Retroviral vectors constructed to allow expression of both v-*myb* and v-*ets* but as separate proteins (by introducing an intervening sequence) showed that these cells have an immature erythroid phenotype and spontaneously differentiate into erythocytes at a much higher rate than Myb-Ets transformed

MEP cells (Metz and Graf 1991a). Unlike Myb-Ets (fused) transformants, these cells do not differentiate into eosinophils or myeloblasts when treated with TPA, suggesting that fusion of the two proteins is required for transformation of multipotent progenitors (Metz and Graf 1991a). Fusion of the two proteins also appears to be crucial for the ability of the virus to induce leukemia since chickens injected with the *myb/ets* (separate) virus developed leukemia only after a much longer latency and, in each case, analysis of the retroviral genomes from leukemic cells revealed a "re-fusion" of the two transcription factors due to DNA rearrangements and the loss of the artificially introduced intervening sequence (Metz and Graf 1991b). Taken together, these data suggest that, while coexpression of separate Myb and Ets proteins leads to transformation in vitro, physical linkage of the two proteins is essential for rendering transformants multipotent and dramatically enhances their ability to induce leukemia. Recent evidence, however, has shown that v-*ets* is not unique in its ability to cooperate with v-*myb* in the transformation of multipotent cells; cells transformed by a recombinant virus expressing v-*myb* and the v-*erb*B tyrosine kinase type oncoprotein are phenotypically indistinguishable from E26 transformants and show a similar capacity for multilineage differentiation (Frampton et al. 1995, and in preparation).

To obtain insights about the role of Myb and Ets in maintaining the differentiation block of MEPs, temperature sensitive mutants in each of these two oncogenes were generated. The prototype of the category of mutants that carry a lesion in Ets is the ts1. 1 mutant. It contains a histidine to aspartic acid mutation in the DNA binding domain of Ets, leading to a complete inhibition of DNA binding at the non permissive temperature (Kraut et al. 1994). This variant was isolated in a biological screen of mutagenized E26 virus based on its ability to transform immature hematopoietic cells at 37°C which undergo a rapid erythroid differentiation after shift to 42°C (Golay et al. 1988). It was subsequently shown that the inactivation of the Ets DNA binding domain in MEP cells leads to maturation not only along the erythroid, but also along the myeloid and, in some clones, eosinophilic lineage (Kraut et al. 1994, and see Fig, 1). The finding that myeloid cells transformed by ts1.1 exhibited no phenotypic changes upon temperature shift again suggested that v-*myb* but not v-*ets* plays the primary role in myeloid cell transformation. Several observations, however, suggest that the ts1.1 point mutation is not a "null". The phenotypes of both myeloid cells and MEP cells transformed by ts1.1, differ from their wild-type counterparts even at the permissive temperature. Myeloid cells transformed by ts1.1 frequently contain cytoplasmic granules and more closely resemble promyelocytes than wild-type E26 transformants (Golay et al. 1988; Kraut et al. 1994). Ts1.1 MEP cells, by contrast, express the JS4 antigen (a marker of mature erythroid cells), significant levels of hemoglobin and no or lower amounts of thrombocytic antigens (Kraut et al. 1994). Like T-MEPs, they can be induced to differentiate along the myeloid or eosinophilic lineage following treatment with TPA. This transformed precursor has therefore been referred as E-MEP or "erythroid" MEP (see Table 3 and Fig. 1). The E-MEPs also show a relatively high rate of spontaneous erythroid and eosinophil differentiation at the permissive temperature. These differences in the phenotype of MEPs transformed by wild-type E26 or ts1.1 may either reflect a greatly reduced DNA binding affinity of ts 1.1 Ets at the permissive temperature (KRAUT et al. 1994) or an alteration in the ability of Ets to bind other protein(s). Regardless of the mechanism, it is clear that Ets is required for maintaining multipotency in E26-transformed progenitor cells as its inactivation by temperature shift allows multilineage differentiation.

The ability to inactivate v-Ets of E26 by temperature shift in progenitor cells provides an excellent system to search for downstream target genes and assess their role in blocking differentiation. Using a differential cDNA screening approach to identify genes expressed selectively in E-MEPs and which are down-regulated rapidly after temperature shift we have recently identified a candidate Ets-regulated gene. This gene, termed *rem*-1 (for regulated by Ets in MEPs), encodes a recoverin- and visinin-related calcium binding protein expressed in early hematopoietic cell lines, bone marrow, gut, and neural tissue, but not in most other tissues. When Ets is "reactivated" in MEPs by shifting them back to the permissive temperature *rem*-1 is re-expressed rapidly and in the absence of de novo protein synthesis, again suggesting direct regulation by Ets (KRAUT et al. 1995). Ectopic expression of Rem-1 in ts1.1 E-MEPs does not block temperature-induced differentiation which would suggest that its down-regulation is not required for cell maturation. The functional role of the Rem-1 protein in hematopoietic cell differentiation, therefore, remains to be clarified.

The ts21 mutant represents the prototype of the category of E26 mutants with a lesion in Myb (BEUG et al. 1984, 1987; FRYKBERG et al. 1988; FRAMPTON et al. 1995). This virus contains a threonine to arginine substitution in the DNA binding domain of Myb, rendering the protein unable to bind to DNA at the nonpermissive temperature. The mutation was identified in a biological screen for temperature-induced maturation of transformed myeloblasts into macrophages (BEUG et al. 1984; FRYKBERG et al. 1988). The fact that inactivation of Myb DNA binding results in production of postmitotic macrophages again suggests that Myb plays a primary role in bolcking terminal differentiation of myelomoncytic cells (BEUG et al. 1984; NESS et al. 1989). Interestingly, this temperature-induced differentiation can be reversed in a significant portion of the cells by "back shifting" the cells to the permissive temperature. As with ts1.1 this system was used in a differential screen to isolate a first Myb regulated gene, *mim*-1 (BEUG et al. 1987; NESS et al. 1989). This gene, which is specifically expressed in transformed and normal myeloblasts, promyelocytes and eosinophils, but not in MEPs, contains three Myb binding sites in its promoter region. A more detailed analysis of the regulation of *mim*-1 has shown that cooperation between Myb and a second transcription factor called C/EBPβ (also known as NF-M) governs the tissue specific expression of *mim*-1 (NESS et al. 1993; BURK et al. 1993; KATZ et al. 1993). Mim-1 is expressed in the granules of normal granulocytes and their precursors (NESS et al. 1989) but like Rem-1, its role in differentiation and/or transformation, if any, is not yet known.

In contrast to the MEP cells transformed by ts1.1, MEP cells transformed by the ts21 virus closely resemble wild-type E26-transformed MEPs in that they express thrombocytic markers and differentiate into eosinophils and myeloblasts upon TPA treatment (FRAMPTON et al. 1995). Interestingly, temperature shift of these cells to inactivate Myb results in terminal thrombocytic differentiation as determined by morphological changes and expression of a number of markers. These include the production of characteristic electron dense secretory granules, increased expression of cytoplasmic c-Src, serotonin release, production of TGFβ, and expression of the activated form of the platelet integrin $\alpha IIb\beta_3$ (FRAMPTON et al. 1995). As with ts21 transformed myeloblasts, this system should prove ideal for the identification of Myb regulated genes in T-MEPs.

Further experiments are required to answer the question whether thromboblasts transformed by E26 faithfully represent the phenotype of normal multipotent hematopoietic precursors. Here, the isolation and analysis of normal cells which serve as targets for transformation by E26 virus are required. Preliminary experiments suggest that E26 targets indeed express the thrombocyte-specific MEP 21 antigen and are capable of forming multilineage colonies (MCNAGNY, unpublished). However, the fact that ts1.1 transformed E-MEPs and wild-type T-MEPs can both give rise to same set of hematopoietic lineages and yet exhibit distinct lineage specific markers would argue that the fusion protein is capable (at least to some extent) of modulating the phenotype of infected target cell.

3 Concluding Remarks

The avian retrovirus models described have proven to be useful tools in dissecting the machinery involved in hematopoietic growth control and neoplasia by providing a way of selectively expanding and maintaining relatively rare hematopoietic precursors in sufficient numbers for biochemical and molecular analyses. The remarkable transformation specificity of different viral oncogenes (receptor tyrosine kinase-type oncogenes for erythroid lineage cells, v-*rel* for lymphoid lineages, v-*myc* for macrophages, and v-*myb* for myeloid precursors) suggests that they act as dominant alleles of genes which are involved in differentiation and lineage-specific growth control. Because avian cells tolerate larger temperature shifts than their mammalian counterparts, it has been relatively easy to select for temperature sensitive mutations in transforming proteins and address their function by inactivating and reactivating them at will. Such simple inactivation experiments have provided profound insights into the role of these oncoproteins in disregulated growth.

Avian retroviruses have in addition provided some of the first models for studying the cooperativity of oncogenes in leukemogenesis. This began with the observation that a surprising number of avian retroviruses carry pairs of oncogenes (AEV, E26, and MH2). The subsequent assessment of the individual contributions of these genes to transformation and the observation that in each case both are required for the formation of aggressive leukemias provided some of the first examples to demonstrate the multistep nature of tunorigenesis. The construction of new viruses which express novel combinations of oncogene pairs is now being used as a method to assess oncogene cooperativity as a more rapid and cost effective alternative to transgenic mice carrying combinations of oncogenes.

The utility of transformed hematopoietic cells as a model for normal processes lies in the degree to which they resemble normal hematopoietic cells. In light of their capacity to grow in the absence of a normal hematopoietic microenvironment, it is remarkable that with few exceptions avian transformants appear to closely resemble their normal counterparts. Myb-Ets induced myeloblasts, thromboblasts and eosinophils, for example, are phenotypically similar to their bone marrow counterparts (McNagny et al. 1992, Table 3, and unpublished). Thus, these in vitro systems can be exploited to gain insights into normal hematopoietic cell differentiation such as by searching for target genes for both Myb and Ets (Ness et al. 1989; Kraut et al. 1995). Such genes might provide important clues for the regulation of growth of normal multipotent progenitors and the mechanisms of decision making during lineage comittment. In the case of the Myb regulated *mim-1* gene, analysis of its promoter has already provided a much clearer picture of how a combinatorial overlap of transcription factors governs myeloid-specific gene expression (Burk et al. 1993; Katz et al. 1993; Ness et al. 1993 reviewed in Ness and Engel 1994). Similarly, analysis of v-*rel* transformants has provided insights into the mechanisms of generating diversity in immunoglobulin and T cell receptor loci (see C-L. Chen, this issue; Kim et al. 1990; Zhang et al. 1989, 1991, Chen, this issue, and Thompson et al., this issue).

Perhaps one of the most exciting new applications of avian retrovirus technology is the use of viral vectors to overexpress genes in embryonic tissues (Morgan et al. 1992). Because of the easy accessibility of chick embryos this promises to become an important alternative to the expression of transgenes in mice using tissue-specific promotors. The technique might also be applicable to the expression of genes, such as those encoding transcription factors with a known or suspected role in hematopoiesis, in the bone marrow of chick embroys. In addition, since avian embryos are accessible for manipulation within hours of fertilization they will continue to provide a valuable addition to mice and other species as a system to analyze normal and abnormal developmental processes, including hematopoiesis.

Acknowledgments. We would like to thank Drs. Jon Frampton, Holger Kulessa, Filip Lim, Claus Nerlov, Fabio Rossi, Michael Sieweke, and Melissa Woodrow for discussions and helpful suggestions. We also thank Hilary Davies-Rück and Servane Leillard their help in preparing this manuscript. KMM was supported by NRSA fellowship #F32 HL08736 from the National Heart, Lung and Blood Institute of the NIH.

References

Barettino D, Bugge TH, Bartunek P, Vivanco-Ruiz MD, Sonntag-Buck V, Beug H, Zenke M, Stunnenberg HG (1993) Unligated T3R, but not its oncogenic variant, v-*erbA*, suppresses RAR-dependent transactivation by titrating out RXR. EMBO J 12: 1343–1354

Barth CF, Humphries EH (1988a) Expression of v-*rel* induces B-cell lines that reflect the diversity of avian immunoglobulin heavy-and light-chain rearrangements. Mol Cell Biol 8: 5358–5368

Barth CF, Humphries EH (1988b) A nonimmunosupressive helper virus allows high efficiency induction of B cell lymphomas by reticuloendotheliosis virus strain T.J Exp Med 167: 89–108

Barth CF, Ewert DL, Olson W C, Humphries EH (1990) Reticuloendotheliosis virus REV-T(REV-A)-induced Neoplasia: development of tumors within the T-lymphoid and myeloid lineages. J Virol 64: 6054–6062

Beug H, and Graf T (1989). Co-operation between viral oncogenes in avian erythroid and myeloid leukemia. Eur J Clin Invest 19: 491–502

Beug H, von Kirchbach A, Döderlein G, Conscience J.-F, Graf T (1979) Chicken hematopoietic cells transformed by seven strains of defective avian leukemia viruses display three distinct phenotypes of differentiation. Cell 18: 375–390

Beug H, Müller H, Grieser S, Doederlein G, Graf T (1981) Hematopoietic cells transformed in vitro by REV-T avian reticuloendotheliosis virus express characteristics of very immature lymphoid cells. Virology 115: 295–309

Beug H, Palmieri S, Freudenstein C, Hanswalter Z, Graf T (1982) Hormone-dependent terminal differentiation in vitro of chicken erythroleukemia cells transformed by ts mutants of avian erythroblastosis virus. Cell 28: 907–919

Beug H, Leutz A, Kahn P, Graf T (1984 Ts mutants of E26 leukemia virus allow transformed myeloblasts, but not erythroblasts or fibroblasts to differentiate at the nonpermissive temperature. Cell 39: 579–588

Beug H, Hayman MJ, Graf T, Benedict SH, Wallbank AM, Vogt PK (1985a) S13, a rapidly oncogenic replication-defective avian retrovirus. Virology 145: 141–153

Beug H, Kahn P, Doederlein G, Hayman MJ, Graf T (1985b) Characterization of hematopoietic cells transformed in vitro by AEV-H, a v-*erb*B-containing avian erythroblastosis virus. Springer Berlin Heidelberg New York

Beug H, Blundell P, Graf T (1987) Reversibility of differentiation and proliferative capacity in avian myelomonocytic cells transformed by tsE26 leukemia virus. Genes Dev 1: 277–286

Beug H, Müllner EW, Hayman MJ (1994) Insights into erythroid differentiation obtained from studies on avian erythroblastosis virus. Cur Opin Cell Biol 6: 816–824

Bhat NK, Fisher R, Fujiwara S, Ascione R, Papas T (1987) Temporal and tissue-specific expression of mouse *ets* genes. Proc Natl Acad Sci USA 84: 3161–3165

Bister K, Loliger, H.-C, Duesberg PH (1979) Oligonucleotide map or and protein of CMII:detection of conserved and nonconserved genetic elements in avian acute leukemia viruses CMII, MC29, and MH2. J Virol 32: 208–219

Boehmelt G, Madruga J, Dörfler P, Briegel K, Schwarz H, Enrietto P, Zenke M (1995) Dendritic cell progenitor is transformed by a conditional v-Rel estrogen receptor fusion protein v-RelER. Cell 80: l–20

Boehmelt G,Walker A, Kabrun N, Mellitzer G, Beug H, Zenke M, Enrietto P (1992) Hormone-regulated v-*rel* estrogen receptor fusion protein: reversible induction of cell transformation and cellular gene expression. EMBO J 11: 4641–4652

Bose HR (1992) The Rel family: models for transcriptional regulation and oncogenic transformation. Bioch Bioph Acta 1114: 1–17

Burk O, Mink S, Ringwald M, Klempnauer K.-H (1993) Synergistic activation of the chicken *mim-1* gene by v-Myb and C/EBP transcription factors. EMBO J 12: 2027–2038

Capobianco A, Gilmore T (1993). A conditional mutant of *v-Rel* containing sequences from the human eostrogen receptor. Virology 193: 160–170

Choi O.-R, Trainor C, Graf T, Beug H, Engel JD (1986). A single amino acid substitution in *v-erbB* confers a thermolabile phenotype to ts 167 avian erythoblastosis virus-transformed erythroid cells. Mol Cell Bio 6: 1751–1759.

Coll J,Righi M, deTaisne C, Dissous C, Gegonne G, Stehelin D (1983) Molecular cloning of the avian acute transforming retrovirus MH2 reveals a novel cell-derived sequence (v-*mill*) in addition to the *myc* oncogene. EMBO 2: 2189–2194.

Damm K, Thompson CC, Evans RM (1989) Protein encoded by v-erbA functions as a thyroid-hormone receptor antagonist. Nature 339: 593–597

Denhez F, Heimann B, d'Auriol L, Graf T, Coquillaud M, Coll J, Galibert F, Moelling K, Stehelin D, Ghysdael J (1988) Replacement of lys 622 in the ATP binding domain of P100$^{gag-mil}$ abolishes the in vitro autophosporylation of the protein and the biological properties of the v-mil oncogene of MH2 virus. EMBO J 7: 541–546

Disela C, Glineur C, Bugge T, Sap J, Stengl G, Dodgson J, Stunnenberg H, Beug H, Zenke M (1991) *v-erbA* overexpression is required to extinguish *c-erbA* function in erythroid cell differentiation and regulation of the *erbA* target gene CAII. Genes Dev 5: 2033–2047

Downward J, Yarden Y, Mayes E, Scrace G, Totty N, Stockwell P, Ulrich A, Schlessinger J, Waterfield MD (1984) Close similarity of epidermal growth factor receptor and *v-erbB* oncogene protein sequences. Nature 307: 521–527

Duprey SP, Boettiger D (1985) Developmental regulation of *c-myb* in normal myeloid progenitor cells. Proc Natl Acad Sci USA 1985: 6937–6941

Frampton J, McNagny KM, Sieweke M, Philip A, Smith G, Graf T (1995) Myb DNA binding is required to block thrombocytic differentiation of Myb-Ets transformed multipotent haematopoietic progenitors. EMBO J 14: 2866–2875

Frykberg L, Palmieri S, Beug H, Graf T, Hayman MJ, Vennström B (1983) Transforming Capacities of avian erythroblastosis virus mutants deleted in the *erbA* or *erbB* oncogenes. Cell 32: 227–238

Frykberg L, Metz T, Brady G, Introna M, Beug H, Vennstrom B, Graf T (1988) A point mutation in the DNA binding of the v-myb oncogene of E26 virus confers temperature sensitivity for transformation of myelomonocytic cells Oncogene Res 3: 3131–322

Gandrillon O, Jurdic P, Pain B, Desbois C, Madjar JJ, Moscovici MG, Moscovivi C, Samarut J (1989) Expression of the *v-erbA* product, an altered nuclear hormone receptor, is sufficient to transform erythrocytic cells in vitro. Cell 58: 115–121

Gazzolo L, Moscovici C, Moscovici MG, Samarut J (1979) Response of hemopoietic cells to avian acute leukemia viruses: effects on the differentiation of the target cells. Cell 16: 627–638

Gilmore T D (1991) Malignant transformation by mutant Rel proteins. Trends Genet 7: 312–322

Golay J, Introna M, Graf T (1988) A single point mutation in the *v-ets* oncogene affects both erythroid and myelomoncytic cell differentiation. Cell 55: 1147–1158

Gonda TJ, Sheiness DK, Bishop JM (1982) Transcripts from the cellular homologs of retroviral oncogenes:distribution among chicken tissues. Mol Cell Biol 2: 617–624

Graf T, Beug H (1978) Avian leukemia viruses:interaction with their target cells in vitro and in vivo. Biochim Biophys Acta 516: 269–299

Graf T, Stéhelin D (1982) Avian leukemia viruses: oncogenes and genome structure. Biochim Biophys Acta 651: 245–271

Graf T, Ade N, Beug H (1978) Temperature-sensitive mutant of avian erytroblastosis virus suggests a block of differentiation as mechanism of leukaemongenesis. Nature247: 496–501

Graf T, von Kirchbach A, Beug H (1981) Characterization of the hematopoietic targetcells of AEV, MC29 and AMV avian acute leukemia viruses. Exp Cell Res 131: 331–343

Graf T, von Weizsecker F, Grieser S, Coll J, Stehelin D, Patschinsky T, Bister K., Bechade C, Calothy G, Leutz A (1986) v-*mill* induces autocrine growth and enhanced tumorigenicity in v-*myc*-transformed avian macrophages. Cell 45: 357–364

Graf T, McNagny K, Brady G, Frampton J (1992) Chicken 'erythroid cells transformed by the gag-myb-ets-encoding E26 leukemia virus are multipotent. Cell 70: 201–213

Hayman MJ, Meyer S, Martin F, Steinlein P, Beug H (1993) Self-renewal and differentiation of normal avian erythroid progenitor cells: regulatory roles of the TGF/α/c-*erbB* and SCF/c-*Kit* receptors. Cell 74: 157–169

Hermann T, Hoffman B, Piedrafita FJ, Zhang XK, Pfahl M (1993) *v-erbA* requires auxiliary proteins for dominant negative activity. Oncogene 8: 55–65

Hrdlickova R, Nehyba J, Humphries EH (1944) In vivo evolution of c-*rel* oncogenic potential. J Virol 68: 2371–2382

Hu W-S, Temin HM (1990) Retroviral replication and reverse transcription. Science 250: 1227–1233

Huff JL, Jelinek MA, Borgman CA, Lansing TJ, Parsons JT (1993) The protooncogene c-*sea* encodes a transmembrane protein-tyrosine kinase related to the Met/hepatocyte growth factor/scatter factor receptor. Proc Natl Acad Sci USA 90: 6140–6144

Introna M, Golay J, Frampton J, Nakano T, Ness S, Graf T (1990) Mutations in *v-myb* alter the differentiation of myelomonocytic cells transformed by the oncogene. Cell 63: 1287–1297

Jansen HW, Ruckert B, Lurz R, Bister K (1983) Two unrelated cell-derived sequences in the genome of avian leukemia and carcinoma inducing retrovirus MH2. EMBO J 2: 1969–1975

Kabrun N, Enrietto PJ (1994) The Rel family of proteins in oncogenesis and differentiation. Semin Cancer Biol. 5: 103–112

Kahn P, Leutz A, Graf T (1986) Individual and combined effects of viral oncogenes in hematopoietic cells. Berlin Heidelberg New York

Kan NC, Flordellis CS, Garon GF, Deusberg PH, Papas TS (1983) Avian carcinoma virus MH2 contains a transformation specific sequence, *mht* and shares the *myc* sequence with MC29, CMII, and OK10 viruses. Proc Natl Acad Sci USA 80: 6566–6570

Katz S, Kowenz-Leutz E, Müller C, Meese K, Ness SA, Leutz A (1993) The NF-M transcription factor is related to C/EBPβ and plays a role in signal transduction, differentiation, and leukemogenesis of avian myelomonocytic cells. EMBO J 12: 1321–1332

Kim S, Humphries EH, Tjoelker L, Carlson L, Thompson CB (1990) Ongoing diversification of the rearranged immunoglobulin light-chain gene in a bursal lymphoma cell line. Mol Cell Biol 10: 3224–3231

Knight J, Zenke M, Disela C, Kowenz E, Vogt P, Engel JD, Hayman MJ, Beug H (1988) Temperature-sensitive v-*sea* transformed erythroblasts: a model system to study gene expression during erythroid differentiation. Genes Dev 2: 247–258

Kraut N, Frampton J, McNagny KM, Graf T (1994) A functional Ets DNA-binding domian is required to maintain multipotency of hematopoietic progenitors transformed by Myb-Ets. Genes Dev 8: 33–44

Kraut N, Frampton J, Graf T (1995) Rem-1, a putative direct target gene of the Myb-Ets fusion oncoprotein in haematopoietic progenitors, is a member of the recoverin family. Oncogene 10: 1027–1036

Kleessa H, Frampton J, Graf T (1995) GATA-1 reprograms avian myelomonocytic cell lines into eosinophils, thromboblasts, and erythroblasts, Genes Dev 9: 1250–1262

Lax I, Johnson A, Howk R, Sap J, Bellot F, Winkler M, Ulrich A, Vennström B, Schlessinger J, Givol D (1988) Chicken epidermal growth factor (EGF) receptor: cDNA cloning, expression in mouse cells and differential binding of EGF and transforming growth factor-alpha. Mol Cell Biol 8: 1970–1978.

Leprince D, Gegonne A, Coll J, de Taisne C, Schneeberger A, Lagrou C, Stehelin D (1983) A putative second cell-derived oncogene of the avian leukemia retrovirus E26. Nature 306: 395–397

Leutz A, Beug H, Graf T (1984) Purification and characterization of cMGF, a novel chicken myelomoncytic growth factor. EMBO J 3: 3191–3197

Leutz A, Damm K, Sterneck E, Kowenz E, Ness S, Frank R, Gausephol H, Pany YCE, Smart J, Hayman M, Graf T (1989) Molecular cloning of the chicken myelomoncytic growth factor. EMBO J 8: 175–181

Lewis RBJM, Rub B, Nielsel DW, Garry RF, Hoelzer JD, Nazerian K, Bose HR (1981) Avian reticuloendotheliosis virus:identification of the hematopoietic target cell for transformation. Cell 25: 421–431

Luciw PA, Leung LN (1992) Mechanisms of retrovirus replication. Plenum, New York

Maldonado RL, Bose HR (1973) Relationship of reticuloendotheliosis virus to the avian tumor viruses:nucleic acid and polypeptide composition. J Virol 11: 741–747

Marmor MD, Benater T, Ratcliffe MJH (1993) Retroviral transformation in vitro of chicken T cells expressing either α/β or γ/δT cell receptors by reticulendotheiliosis virus strain T.J Exp Med 177: 647–656

McNagny KM, Lim F, Grieser S, Graf T (1992) Cell surface proteins of chicken hematopoietic progenitors, thrombocytes and eosinophils detected by novel monoclonal antibodies. Leukemia 6: 975–984

Metcalf D (1984) The hematopoietic stimulating factors. Elsevier New York

Metcalf D (1988) The molecular control of blood cells. Harvard University Press, Cambridge

Metz T (1994) Oncogenes and erythroid differentiation. Semin Cancer Biol 5: 125–135

Metz T, Graf T (1991a) Fusion of the nuclear oncoproteins *v-myb* and *v-ets* is required for the leukemogenicity of E26 virus. Cell 66: 95–105

Metz T, Graf T (1991b) *v-myb* and v-*ets* transform chicken erythroid cells and cooperate both in trans and in cis to induce distinct differentiation phenotypes. Genes Dev.5: 369–380

Mizutani S, Temin HM (1973) Lack of serological relationship among DNA polymerases of avian leukosis-sarcoma viruses, reticuloendotheliosis virus, and chicken cells. J Virol. 12: 440–448

Moelling K, Gelderblom H, Pauli G, Friis R, Bauer H (1975) A comparative study of the avian reticuloendotheliosis virus:relationship to murine leukemia virus and viruses of the avian sarcoma-leukosis complex. Virology 65: 546–557.

Moelling K, Heimann B, Beimling P, Rapp UR, Sander T (1984) Serine and threonine-specific protein protein kinase activities of purified gag-*mil* and gag-*raf* proteins. Nature 312: 558–561

Morgan BA, Izpisua-Belmonte J-C, Duboule D, Tabin CJ (1992) Targeted misexpression of Hox-4.6 in the avian limb bud causes apparent homeotic transformations. Nature 358: 236–239
Morrison LE, Boehmelt G, Beug H, Enrietto P (1991) Expression of v-*rel* in a replication competent virus:transformation and biochemical characterization. Oncogene 6: 1657–1666
Moscovici C, Gazzolo L (1982) Transformation of hemopoietic cells with avian leukemia viruses. In: Klein G (ed) Advances in viral oncology. Raven, New York; pp 83–106
Moscovici MG, Jurdic P, Samarut J, Gazzolo L, Mura CV, Moscovici C (1983) Characteristics of the hemopoietic target cells for the avian leukemia virus E26. Virology 129: 65–78
Mucenski MI, Mclain K, Kier AB, Swerdlow SH, Schreiner CM, Miller TA, Pietryga DW, Scott WJ, Potter SS (1991) A functional *c-myb* gene is required for normal mouse fetal hepatic hematopoiesis. Cell 65: 677–689
Ness SA, Engel JD (1994) Vintage reds and whites:combinatorial transcription factor utilization in hematopoietic differentiation. Cur Opin Gen Dev 4: 718–724
Ness S, Beug H, Graf T (1987) *v-myb* dominance over v-*myc* in doulby transformed chick myelomoncytic cells. Cell 51: 41–50
Ness SA, Marknell Å, Graf T (1989) The *v-myb* oncogene product binds to and activates the promyelocyte-specific *mim-1* gene. Cell 59: 1115–1125
Ness SA, Kowenz-Leutz E, Casini T, Graf T, Leutz A (1993) Myb and NF-M:combinatorial activators of myeloid genes in heterologous cell types. Genes Dev.7: 749–759
Nunn M, Seeburg PH, Moscovici C, Duesberg PH (1983) Tripartite structure of the avian erythroblastosis virus E26 transforming gene. Nature 306: 391–395
Pain B, Melet F, Jurdic P, Samarut J (1990) The carbonic anhydrase II gene, a gene regulated by thyroid hormone and erythropoietin, is repressed by the v-erbA oncogene in erythrocytic cells. New Biol 2: 284–294
Pain B, Woods CM, Saez J, Flickinger T, Raines M, Peyrol S, Moscovici C, Moscovici MG, Kung HJ, Jurdic P, Lazarides E, Samarut J (1991) EGF-R as a hematopoietic growth factor receptor: the c-*erbB* product is present in chicken erythrocytic progenitors and controls their self-renewal. Cell 65: 37–46
Pathak VK, Temin HM (1990a) Broad spectrum of in vivo forward mutations, hypermutations, and mutational hotspots in a retroviral shutte vectro after a single replication cycle: deletions and deletions with insertions. Proc Natl Acad Sci USA 87: 6024–6028
Pathak VK, Temin HM (1990b). Broad spectrum of in vivo forward mutations, hypermutations, and mutational hotspots in a retroviral shuttle vectro after a single replication cycle: substitutions, frameshifts, and hypermutations. Proc Natl Acad Sci USA 87: 6019–6023
Payne LN (1992). Biology of avian retroviruses. Plenum, New York
Pessano S, Gazzolo L, Moscivici C (1979). The effect of tumor promoter on avian leukemic cells. Microbiologica 2: 379–392
Quesenberry PJ (1992). Stroma-dependent hematolympoietic stem cells. Springer, Berlin Heidelberg New York
Radke K, Beug H, Kornfeld S, Graf T (1982). Transformation of both erythroid and myeloid cells by E26, an avian leukemia virus that contains the *myb* gene. Cell 31: 643–653
Roussel M, Saule S, Lagrou C, Rommens C, Beug H, Graf T, Stehelin D (1979). Three new types of viral oncogenes of cellular origin specific haematopoietic cell transfromation. Nature 281: 452–455
Samarut J, Gazzolo L (1982). Target cells infected by avian erythroblastosis virus differentiate and become transformed. Cell 28: 921–929
Sap J, Munoz A, Damm K, Goldberg Y, Ghysdael J, Leutz A, Beug H, Vennström B (1986). The *c-erbA* protein is a high affinity receptor for thyroid hormone. Nature 324: 635–640
Sap J, Munoz A, Schmitt J, Stunnenberg H, Vennström B (1989). Repression of transcription mediated at a thyroid hormone response element by the *v-erbA* oncogene product. Nature 340: 242–244
Sarkar S, Gilmore TD (1993). Transformation by the vRel oncoprotein requires sequences carboxy-terminal to the Rel homology domain. Oncogene 8: 2245–2252
Schroedere C, Raynoschek C, Fuhrmann U, Damm K, Vennström B, Beug H (1990). The *v-erbA* oncogene causes repression of erythrocyte-specific genes and an immature, aberrant differentiation in normal erythroid progenitors. Oncogene 5: 1445–1453
Schroeder C, Gibson L, Zenke M, Beug H (1992). Modulation of normal erythroid differentiation by the endogenous thyroid hormone and retinoic acid receptors: a possible target for v-*erbA* oncogen action, Oncogene 7: 217–227
Schroeder C, Gibson L, Nordstrom C, Beug H (1993). The estrogen receptor cooperates with the TGFa receptor (c-*erbB*) in the regulation of chicken erythroid progenitor self-renewal. EMBO J 12: 951–960

Steinlein P, Wessely O, Meyer S, Deiner E-M, Hayman MJ, Beug H (1995). Triggering the development of self-renewing erythroid progenitors. Curr Biol 5: 191–204

Stephens RM, Rice NR, Hiebsch RR, Bose HR, Gilden RV (1983). Nucleotide sequence of *v-rel*: the oncogen of reticuloendotheliosis virus. Proc Natl Acad Sci USA 80: 6229–6233

Sutrave P, Bonner TI, Rapp UR, Jansen HW, Patschinsky T, Bister K (1984). Nucleotide sequence of avian retroviral oncogene *v-mil*: homologue of murine retroviral oncogene *v-raf*. Nature 309: 85–88.

Vandenbunder B, Pardanaud L, Jaffredo T, Mirabel MA, Stehelin D (1989). Complementary patterns of c-*ets* 1 c-*myb* and c-*myc* in the blood forming system of the chick embryo. Development 106: 265–274

Vennström B, Bishop JM (1982). Isolation and characterization of chicken DNA homologous to the two putative oncogenes of avian erythroblastosis virus. Cell 28: 135–143

Weinberger C, Thompson CC, Ong ES, Lebo R, Gruol DJ, Evand RM (1986). The c-*erbA* gene encodes a throid hormone receptor. Nature 324: 641–646

Westin EH, Gallo RC, Arya SK, Eva A, Souza LM, Baluda MA, Aaronson SA, Wong-Staal F (1982) Differential expression of the AMV gene in human hematopoietic cells. Proc Natl Acad Sci USA 79: 2194–2198

White DW, Roy A, Gilmore TD (1995). The v-Rel oncoprotein blocks apoptosis and proteolysis of IκB-α in transformed chicken spleen cells. Oncogene 9: (in press)

Yamamoto T, Nishida T, Mitajimi N, Kawai S, Ooi T, Toyoshima K (1983). The erbB gene of avian erythroblastosis virus is a member of the *src* gene family. Cell 35: 71–78

Zenke M, Kahn P, Disela C, Vennström B, Leutz A, Keegan K, Hayman MJ, Choi HR, Yew N, Engle JD, Beug H (1988). v-*erbA* specifically supresses transcription of the avian erythrocyte anion transporter (band 3) gene. Cell 52: 107–119

Zenke M, Munoz A, Sap J, Vennström B, Beug H (1990). V-*erbA* oncogene activation entails loss of hormone-dependent regulator activity of c-*erbA*. Cell 61: 1035–1049

Zhang J, Termin HM (1993). Rate and mechanism of nonhomologous recombination during a single cycle of retroviral replication. Science 259: 234–238

Zhang J, Bargmann W, Bose HRJ (1989). Rearrangement and diversification of immunoglobulin light-chain genes in lymphoid cells transformed by reticuloendotheliosis virus. Mol Cell Biol 9: 4970–4976

Zhang J, Olson W, Ewert D, Bargmann W, Bose HRJ (1991). The *v-rel* oncogene of avian reticuloendotheliosis virus transforms immature and mature lymphoid cells of the B cell lineage in vitro. Virology 183: 457–466

Control of the Differentiation Commitment by Nuclear Hormone Receptors in Chicken Erythrocytic Progenitor Cells

J. Samarut

1 Introduction

Hematopoiesis is a complex process which involves specification of multiple lineages from a common stem cell and then differentiation and maturation into fully functional blood cells within each lineage. It is clear now that, before reaching the final mature step, cells must first implement a specific morphogenetic program, then express this program to fulfill their physiological functions.

The role of environmental factors of either cellular or biochemical origin is now clearly established. However, major questions remain unanswered concerning the intracellular signals induced by these factors and, more specifically, the mechanisms through which they control the gene expression programms involved at the successive steps of the differentiation pathways.

Laboratoire de Biologie Moléculaire et Cellulaire de l' Ecole Normale Supérieure de Lyon, UMR49 CNRS, Laboratoire Associé INRA, 46 Allée d'Italie, 69364 Lyon Cedex 07, France

The avian hematopoietic system has long been a fruitfull model for investigating the mechanisms of differentiation of blood cells. The easy access to the embryo has allowed pioneering studies on the ontogeny of the system (review in Dieterlen-Lièvre 1988). As an example, it should be noted that, in contrast to most common laboratory strains of mice, the chicken embryo produces a fetal hemoglobin similar to that of humans.

Another great advantage of the avian model is the availability of many acute leukemia retroviruses which alter the differentiation program at various steps within each hematopoietic lineage. Each retrovirus carries a specific oncogene whose product alters cellular biochemical functions (Bishop 1985). These retroviruses can thus be considered as genetic tools to investigate the molecular bases of the morphogenetic programming of hematopoietic progenitor cells. In this review we will concentrate on the erythrocytic differentiation pathway.

2 The Chicken Erythropoietic System

As in mammals, hematopoietic differentiation in chicken proceeds from self-renewing pluripotent progenitor cells. Although indirect data suggest that a chicken pluripotent hematopoietic stem cell should exist (Graf et al. 1992), there is still no direct functional assay for such cells. Many progenitors committed to specific hematopoietic cell lineages have been identified in semisolid culture assays (review in Dieterlen-Lièvre 1988).

The erythrocytic progenitors identified so far include BFU-Es, the earliest precursors and CFU-Es, the latest, precursors of erythroblasts (Samarut and Bouabdelli 1980).

One very specific feature of BFU-Es in the chicken is that they are endowed with high self-renewing potentialities when stimulated by transforming growth factor (TGFα). These cells harbor at their plasma membrane TGFα receptors which are identical to the epidermal growth factor (*EGF) receptors encoded by the *c-erbB* proto-oncogene (Pain et al. 1991). Schroeder et al. (1993) showed that estradiol and stem cell factor (SCF) can also stimulate self-renewal of early erythrocytic precursors. It is not yet clear whether the same progenitors activated by TGFα are also activated by estradiol and SCF. There are some indications that they represent different subsets of early erythrocytic precursors (see Beug, this volume).

Since early erythrocytic progenitors can be maintained as self-renewing cells in culture, they provide an invaluable tool for biochemical investigations. Moreover, the discovery that these cells express the *c-erbB* membrane receptor was quite unexpected and provided the first demonstration that the TGFα- and EGF-receptor is also expressed in hematopoietic cells.

3 Commitment of Erythrocytic Progenitors to Final Differentiation

Self-renewing BFU-Es can be triggered out of their replication program by treatment with either thyroid hormone (T3) or *all-trans*-retinoic acid (RA) (Gandrillon et al. 1994a). Treatment for 24 h is sufficient to commit the cells to differentiation. Interestingly, this length of time corresponds to the generation time of the progenitors, which suggests that commitment takes places during a single cell cycle. In the presence of insulin and erythropoietin, the committed cells undergo final maturation into erythrocytes after six to eight cell divisions. In the absence of these growth factors the committed cells undergo apoptosis. We might anticipate that commitment by T3 or RA induces some genetic unstability in the cells and that apoptosis is an emergency pathway for those cells which cannot correctly achieve differentiation. The self-renewing BFU-Es express the nuclear receptors for T3, T3Rα, and those for RA, RARα, RARγ, and, in response to RA, RARβ. It is not yet known if all RARs play identical redundant functions in the commitment process induced by RA.

4 General Scheme of Red Cell Differentiation

We can now devise a general scheme for the differentiation of hematopoietic stem cells into erythrocytes (Fig. 1). The pluripotent stem cells undergo a first commitment process which drives the cells into the erythrocytic lineage. We shall call this step lineage commitment. This step is important since it leads to a strong restriction of the morphogenetic potentialities of the stem cells. Important genetic events should then take place at this step to select the erythrocytic morphogenetic program.

The self-renewing BFU-Es then undergo a second commitment process which triggers them into final maturation. We shall call this second commitment maturation commitment. During this process a strong change occurs between two kinetic and morphogenetic programs. In the early self-renewing progenitors, the programs of cell division and cell differentiation are totally uncoupled to each other since the cells can replicate without progressing toward the final differentiation. Beyond the maturation commitment point, the programs of cell division and cell maturation become strictly coupled to each other since the cells cannot achieve their differentiation without undergoing a fixed few number of divisions, and they cannot replicate without progressing toward the erythrocyte stage. At maturation commitment, biochemical events should then take place to limit the proliferation potential and to activate genes involved in final maturation.

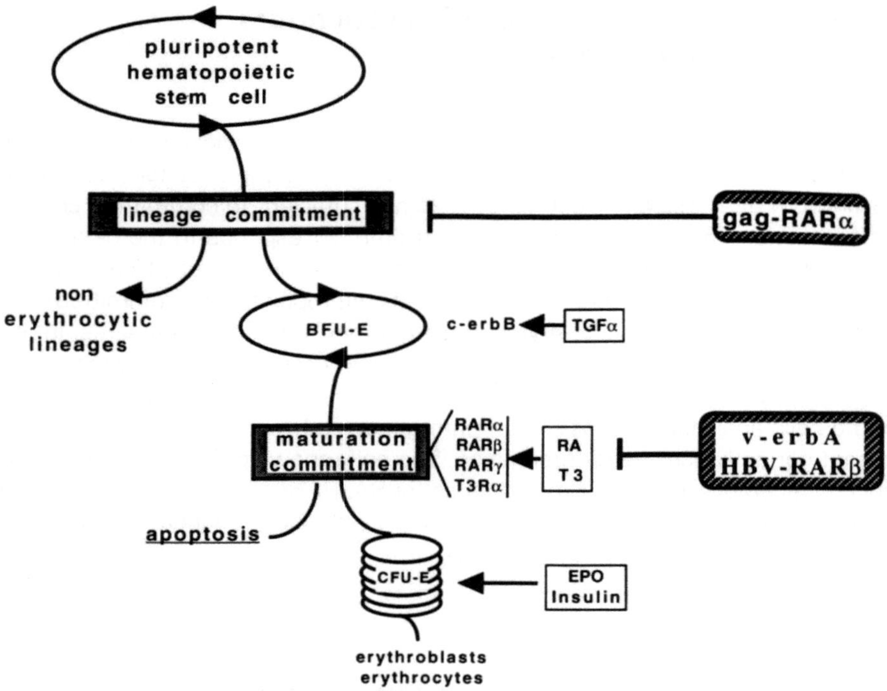

Fig. 1. Pathway of erythrocytic differentiation in chicken. Only the pathway to erythrocyte differentiation from pluripotent hematopoietic stem cell is detailed. *Horizontal arrows* represent positive effects; *horizontal lines ended by vertical bars* represent inhibitory effects. *RAR*, retinoic acid receptor; EPO erythropoietin; HBV, hepatitis B virus

5 Transformation of Erythrocytic Progenitors by the Avian Erythroblastosis Virus

The avian erythroblastosis virus (AEV) is an acute leukemia retrovirus which induces erythroleukemias in chicken in vivo and transforms erythrocytic progenitors in culture. Interestingly, the virus blocks differentiation specifically at the early CFU-E step (Gazzolo et al. 1980; Beug et al. 1982). Whereas the virus can infect earlier hematopoietic precursors, it will block the cells only when they proceed from the BFU-E to the CFU-E stage (Samarut and Gazzolo 1982). This step is now identified as the maturation commitment step. The virus should then alter the implementation of the final morphogenetic program.

The virus contains two oncogenes, *v-erbA* and *v-erbB*, that are derived from the respective proto-oncogenes *c-erbA* and *c-erbB* (review in Graf and Beug 1983). These two viral oncogenes play different but complementary functions in this transformation process. The *v-erbA* oncogene is necessary and sufficient for blocking the differentiation program of the leukemic cells (Frykberg et al. 1983;

GANDRILLON et al. 1989). The *v-erbB* oncogene promotes growth factor-independent cell proliferation (FRYKBERG et al. 1983; PAIN et al. 1991). That *v-erbA* blocks differentiation was quite interesting because it provided one way to assess the biochemical mechanisms of maturation commitment in the erythrocytic progenitor cells.

6 The *v-erbA* Oncogene Encodes an Altered Form of the Nuclear Receptor for Thyroid Hormone

The *c-erbA* proto-oncogene encodes the nuclear receptor for the thyroid hormone triiodothyronine, T3 (SAP et al. 1986; WEINBERGER et al. 1986). This receptor belongs to the subfamily of nuclear hormone receptors which also includes the receptors for RA (RARs), the receptors for the *9-cis*-retinoic acid isomer (RXRs), the vitamin D3 receptor (VDR) and the receptors for peroxisome proliferators (PPARs) (EVANS 1988). All these receptors share the same overall structure, with a zinc finger DNA binding domain localized in the middle or at the NH_2-terminal of the protein. The COOH-terminal of the protein contains the ligand binding domain and a dimerization domain (GREEN and CHAMBON 1988).

All these receptors are transcription factors which function as dimers. The RXRs seem to be the preferential common partner for making heterodimers with the other receptors. However, it is not excluded that homodimers and heterodimers between the other receptors are also active in the cells (CARLBERG et al. 1993, review in LEID et al. 1992).

The v-ErbA product is a rearranged version of the α form of the T3 receptor. The NH_2-terminal of the normal receptor has been replaced by a peptide encoded by the residual viral *gag* coding sequence. Thirteen point mutations are scattered all over the receptor sequence, with many of them in the ligand binding domain. Moreover, a nine amino acid deletion is present in the ligand binding domain. As a consequence of all these mutations the v-ErbA protein is unable to bind T3 and has a reduced affinity for DNA; but it can still heterodimerize with reduced affinity to RXRs (review in GANDRILLON et al. 1994b).

7 The v-ErbA Oncoprotein Reverses the Pattern of Gene Expression

As transcription factors, T3Rs and RARs activate the transcription of specific target genes through the binding to specific recognition motifs in the regulatory domains of these genes. In most of the documented cases, the ligand-activated receptors activate the transcription of the target genes. However, some genes are repressed by the receptors in a ligand-dependent fashion. The receptors can also control gene expression indirectly by functionally interfering with the AP-1 transcription factor. Ligand-activated T3Rs or RARs inactivate AP-1 and thereby indirectly down-modulate AP-1-regulated genes (Desbois et al. 1991, review in Pfahl 1993). This dual function of the receptors accounts for their ability to simultaneously activate some sets of genes and down-modulate others (Fig. 2). As a result of down-modulating AP-1, the ligand-activated receptors could slow down cell proliferation.

The v-ErbA oncoprotein is a transdominant repressor of T3Rs and RARs (Sap et al. 1989; Damm et al. 1989; Sharif and Privalsky 1991). As a matter of fact, the oncoprotein inactivates the genes that are normally acativated by T3Rs and RARs. Moreover, v-ErbA abrogates the receptor-mediated down-modulation of AP-1-dependent genes and thereby maintains these genes fully active (Desbois et al. 1991). Therefore, v-ErbA totally reverses the pattern of gene expression in the cells (Fig. 2). By maintaining AP-1 activity, *v-erbA* promotes growth of the cells (Gandrillon et al. 1987; Desbois et al. 1991; Khazaie et al. 1991).

It cannot be excluded that the v-ErbA product deregulates cellular genes that are not directly controlled by T3R and RARs. In yeast *v-erbA* can act as a transcriptional activator (Privalsky et al. 1990). Because of point mutations in the DNA binding domain, the v-ErbA protein might bind to DNA motifs different from those recognized by T3R and RARs (Forman and Samuels 1990). For a review on the biological effects of *v-erbA* see Gandrillon et al. (1994b).

8 v-ErbA Alters Gene Expression at the Maturation Commitment Step of Erythrocytic Progenitors

The *v-erbA* oncogene blocks maturation commitment of BFU-Es. It also inhibits entry into apoptosis (Gandrillon et al. 1994a). Both effects are strongly dependent upon the occurrence of the single amino acid difference between v-ErbA and c-ErbA at the 61st amino acid residue in the DNA binding domain. A glycine residue in c-ErbA has been changed into a serine in the natural v-ErbA oncoprotein. Reversion of this serine residue into glycine in the mutant S61G of v-ErbA totally abrogates its ability to block differentiation and to protect from apoptosis (Sharif and Privalsky 1991; Gandrillon et al. 1994a). The mutant is unable to inhibit transcription of genes regulated directly by RARs but still able to block genes regulated directly by T3R.

During maturation commitment induced by T3, the gene encoding carbonic anhydrase II (CAII) is turned on by the liganded T3Rs (Pain et al. 1990; Disela et al. 1991; Rascle et al. 1994). In *v-erbA*-expressing erythrocytic progenitors the transcriptional activation of the CAII gene is annihilated (Pain et al. 1990; Disela et al. 1991; Rascle et al. 1994).

The mutant S61G of *v-erbA* is unable to abrogate the inactivation of AP-1 by T3R or RAR (Desbois et al., unpublished) and is also unable to inhibit apoptosis

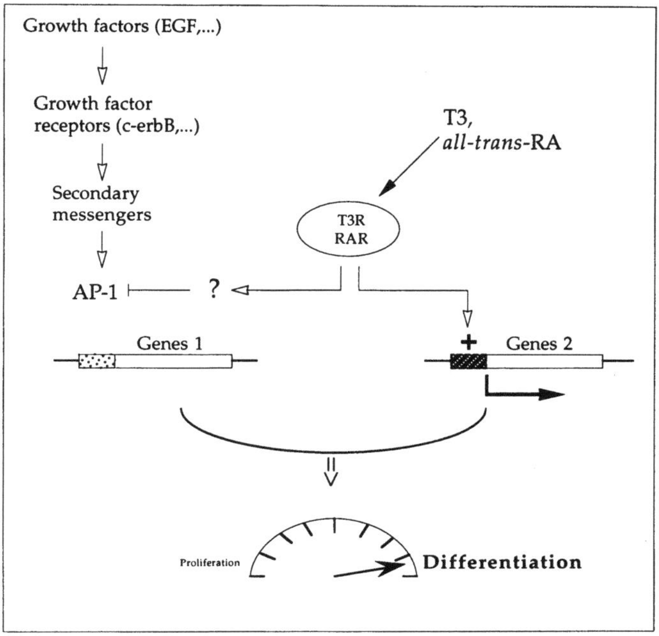

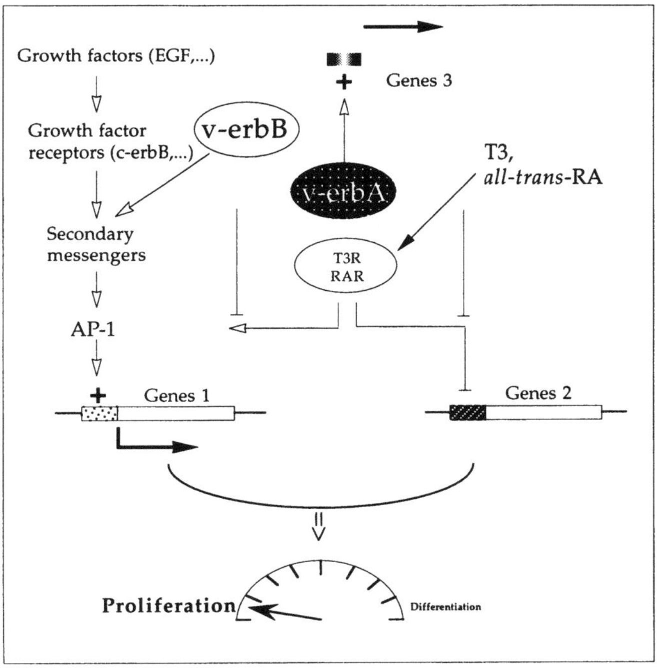

Fig. 2. Control of gene expression by c-erbA and retinoic acid receptors (RARs) and its deregulation by v-ErbA. *Arrows* and *lines ended by vertical bars* (—) schematize activation and repression effects, respectively. Upper panel, normal pattern of gene regulation in cells treated with T3 or RA. Genes 1 represent AP-1-dependent genes. Genes 2 are hormone-regulated genes directly under the control of T3R or RAR. *Lower panel*, pattern of gene expression in v-erbA-expressing cells. Genes 3 represent hypothetical genes which might be directly activated by v-ErbA

induced by either T3 or RA (GANDRILLON et al. 1994a). However, this mutant still inhibits transcriptional activation of the CAII gene by T3 or RA (RASCLE et al., unpublished). Taken together, these observations suggest that apoptosis induced by T3 might be controlled through an AP-1 pathway or a related pathway. We may speculate that apoptosis results from down-modulation by T3R or RAR of AP-1-controlled genes.

There is thus, a strong correlation between the ability of *v-erbA* to block maturation commitment by T3 and RA and its ability to alter the control of gene expression by T3R and RAR. This is consistent with the hypothesis that these receptors govern the commitment of the cells by controlling a specific gene expression program necessary for final maturation or apoptosis.

9 Block of Maturation Commitment by an Altered Retinoic Acid Receptor-β

An altered form of the RARβ has been isolated from a human case of hepatocarcinoma occurring consecutive to hepatitis induced by the hepatitis B virus (HBV, DEJEAN et al. 1986). This mutant form (HBV-RARβ) was encoded by a rearranged RARβ gene resulting from insertion of part of the HBV genome into the gene. In the mutant receptor, 74 amino acids at the NH_2-terminal are replaced by 30 residues of the HBV pre-S1 peptide. To check if this rearranged receptor is oncogenic, we inserted it in place of *v-erbA* in AEV-derived retrovectors, with or without the *v-erbB* oncogene. Even those viruses devoid of *v-erbB* could block differentiation of erythrocytic progenitors at the maturation commitment step, similar to *v-erbA*-containing viruses. Interestingly, the HBV-RARβ form was also sufficient to activate proliferation of the blocked cells in the absence of TGFα or any other specific growth facator (GARCIA et al. 1993). Thus, a rearranged form of RARβ can alter the morphogenetic program at the maturation commitment step, similar to *v-erbA*. This is strongly consistent with the view that RARs control this commitment event.

10 Block of Differentiation by a Mutated Retinoic Acid Receptor-α

To test the oncogenic potentialities of mutated RARα, we constructed a retrovirus vector in which residual *gag* sequences were fused to a RARα truncated of its NH_2-terminal domain (ALTABEF et al., submitted). This virus induces an acute leukemia in chickens in vivo and transforms bone marrow cell in vitro. Interestingly, these leukemic cells exhibit membrane antigens specific for multipotent hematopoietic progenitors (GRAF et al. 1992). This observation suggests that an altered RARα can block the lineage commitment process. Furthermore, the data also suggest that commitment of pluripotent hematopoietic stem cells into the erythrocytic lineage involves the function of nuclear hormone receptors of the RAR family.

11 Avian Erthroblastosis Virus: A Paradigm for Oncogene Activation and Cooperation

It is quite interesting to see that the two oncogenes *v-erbA* and *v-erbB* that were selected by AEV both originate from proto-oncogenes which play specific and complementary functions in the erythrocytic progenitors. We may then speculate that the two oncogenes were transduced by an ancestor retrovirus during infection of these particular cells. It is likely that the two oncogenes were not transduced simultaneously into the parental retrovirus. Indeed, a natural variant of AEV, AEV-H, which contains only the v-erbB oncogene, has been isolated from chicken infected with nontransforming helper retroviruses (YAMAMOTO et al. 1983). AEV-H induces an erythroblastosis in which erythroblats are not strictly blocked in their differentiation. So far, no natural retrovirus containing only the *v-erbA* oncogene has ever been isolated. Taken together, these observations suggest that the transduction of the *v-erbB* oncogene might have been the initial step in generating AEV. The *v-erbA* oncogene would have been transduced as a second event. It would be interesting to pass AEV-H into successive host chickens to check whether the virus could further transduce a *v-erbA* or related oncogene.

Oncogenic activation of the *c-erbA* and *c-erbB* genes by AEV is also a good illustration of the two possible mechanisms of oncogene activation. The v-ErbB product represents an overactivated membrane receptor. In contrast, the v-ErbA product acquires its oncogenic activity due to loss of function of the parental nuclear receptor.

12 Conclusions and Future Prospects

The avian hematopoietic system provides a unique model to investigate the multistep programming of hematopoietic stem cells into differentiated blood cells. Two majors advantages are provided by this model. First, the occurrence of self-renewing committed erythrocytic progenitors which is so far unique in vertebrates. Second, the availability of acute leukemia retroviruses and in the present case AEV. This virus, a natural vector of a rearranged nuclear hormone receptor, provided the first demonstration of the oncogenicity of such receptors.

It also allowed demonstration of the role of these receptors in the commitment process of erythrocytic progenitors.

Preliminary data with rearranged RARα suggest that these receptors, or closely related receptors are involved in lineage commitment.

Taken together, these data suggest that different nuclear receptors of the RAR/erbA family might be involved all along the erythrocytic differentiation pathway. Constructed retrovirus vectors carrying various altered nuclear hormone receptors (like VDR and PPAR) will be useful in testing the putative role of such receptors in the differentiation of other hematopoietic lineages.

An important task now will be to identify the target genes of these nuclear receptors at the commitment steps. Here again, the avian system is quite useful due to the possibility of making thermosensitive mutants of the blocking retroviruses whose transforming activity can be controlled by temperature shift.

References

Beug H, Palmieri S, Freudenstein C, Zentgraf H, Graf T (1982) Hormone-dependent terminal differentiation in vitro of chicken erythroleukemia cells transformed by *ts* mutants of avian erythroblastosis virus. Cell 28: 907–919

Bishop JM (1985) Viral oncogenes. Cell 42: 23–38

Carlberg C, Bendik I, Wyss A, Meier E, Sturzenbecker LJ, Grippo JF, Hunziker W (1993) Two nuclear signalling pathways for vitamin D. Nature 361: 657–660

Damm K, Thompson CC, Evans RM (1989): Protein encoded by v-erbA functions as a thyroid-hormone receptor antagonist. Nature 339: 593–597

Dejean A, Bouguelerey L, Grzeschik KH, Tiollais P (1986) Hepatitis B virus DNA integration in a sequence homologous to v-erbA and steroid receptor genes in a hepatocellular carcinoma. Nature 322: 70–72

Desbois C, Aubert D, Legrand C, Pain B, Samarut J (1991) A novel mechanism of action for v-ErbA: abrogation of the inactivation of transcription factor AP-1 by retinoice acid and thyroid hormone receptors. Cell 67: 731–740

Dieterlen-Lièvre F (1988) Birds. In: (eds) Vertebrate blood cells. Rowley AF, Ratcliffe NA, Cambridge University Press, CAmbridge

Disela C, Glineur C, Bugge T, Sap. J, Stengl G, Dodgson J, Stunnenberg H, Beug H, Zenke M (1991) v-erbA overexpression is required to extinguish c-erbA function in erythroid cell differentiation and regulation of the erb A target CAII. Genes Dev 5: 2033–2047

Evans RM (1988) The steroid and thyroid hormone receptor family. Sciene 240: 889–895

Forman BM, Samuels HH (1990) Interaction among a subfamily of nuclear receptors: the regulatory zipper mode. Mol Endocrinol 4: 1293–1301

Frykberg L, Palmieri S, Beug H, Graf T, Hayman MJ, Vennström B (1983) Transforming capacities of avian erythroblastosis virus mutants delected in the erbA or erbB oncogenes. Cell 32: 227–238

Gandrillon O, Jurdic P, Benchaibi M, Xiao J-H, Ghysdael J, Samarut J (1987) Expression of the v-erbA oncogene in chicken embryo fibroblasts stimulates their multiplication in vitro and enhances tumor growth in vivo. Cell 49: 687–697

Gandrillon O, Jurdic P, Pain C, Desbois B, Madjar JJ, Moscovici MG, Moscovici C, Samarut J (1989) Expression of v-erbA, an altered nuclear hormone receptor, is sufficient to transform erythrocytic cells in vitro. Cell 58: 115–121

Gandrillon O, Ferrand N, Michaille JJ, Roze L, Zile MH, Samarut J (1994a) c-erbAa/T3R and RARs control commitment of hematopoietic self-renewing progenitor cells to apoptosis or differentiation and are antagonized by the v-erbA oncogene. Oncogene 9: 749–758

Gandrillon O, Rascle A, Samarut J (1994b) The v-erbA oncogene: a superb tool for dissecting the involvement of nuclear hormone receptors in differentiation and neoplasia. Int J Oncol 6: 215–231

Garcia M, de Thè H, Tiollais P, Samarut J, Dejean A (1993) A hepatitis B virus preS-retinoic receptor b chimera transforms erythrocytic progenitor cells in vitro. Proc Natl Acad Sci USA 90: 89–93

Gazzola L, Samarut J, Bouabdelli M, Blanchet JP (1980) Early precursors in the erythroid lineage are the specific target cells of the avian erythroblastosis virus in vitro. Cell 22: 683–691

Graf T, Beug H (1983) Role of the v-erbA and v-erbB oncogenes of avian erythroblastosis virus in erythroid cell transformation. Cell 34: 7–9

Graf T, Mc Nagny K, Brady G, Frampton J (1992) Chicken "erythroid" cells transformed by the gag-myb-ets-encoding E26 leukemia virus are multipotent. Cell 70: 201–213

Green S, Chambon P (1988) Nuclear receptors enhance our understanding of transcriptional regulation. TIG 4: 309–314

Khazaie K, Panayotou G, Aguzzi A, Samarut J, Gazzolo L, Jurdic P (1991) EGF promotes in vivo tumorigenic growth of primary chicken embryo fibroblasts expressing v-myc and enhances in vitro transformation by the v-erbA oncogene. Oncogene 6: 21–28

Leid M, Kastner P, Chambon P (1992) Multiplicity generates diversity in the retinoic acid signalling pathways. TIBS 17: 427–433

Pain B, Melet F, Jurdic F, Samarut J (1990) The carbonic anhydrase II gene, a gene regulated by thyroid hormone and erythropoietin, is repressed by the v-erbA oncogene in erythrocytic cells. New Biol 2: 284–294

Pain B, Woods CM, Saez J, Flickinger T, Raines M, Kaung H-J, Peyrol S, Moscovici C, Moscovici GM, Jurdic P, Lazarides E, Samarut J (1991) EGF-R as a hemopoietic growth factor: the c-erbB product is present in normal chicken erythrocytic progenitor cells and controls their self-renewal. Cell 65: 37–46

Pfahl M (1993) Nuclear receptor/AP-1 interaction. Endocr Rev 14: 651–658

Privalsky ML, Sharif M, Yamamoto KR (1990) The viral erbA oncogene protein, a constitutive repressor in animal cells, is a hormone-regulated activator in yeast. Cell 63: 1277–1286

Rascle A, Ghysdael J, Samarut J (1994) c-ErbA, but not v-ErbA, competes with a putative erythroid repressor for binding to the carbonic anhydrase II promoter. Oncogene 9: 2853–2867

Samarut J, Bouabdelli M (1980) In vitro development of CFU-E and BFU-E in cultures of embryonic and postembryonic chicken hematopoietic cells. J Cell Physiol 105: 553–563

Samarut J, Gazzolo L (1982) Target cells infected by avian erythroblastosis virus differentiate and become transformed. Cell 28: 921–929

Sap J, Munoz A, Damm K, Goldberg Y, Ghysdael J, Leutz A, Beug H, Vennström B (1986) The c-erbA protein is a high-affinity receptor for thyroid hormone. Nature 324: 635–640

Sap J, Munoz A, Schmitt J, Stunnenberg H, Vennström B (1989) Repression of transcription mediated at thyroid hormone response element by the v-erbA oncogene product. Nature 340: 242–244

Schroeder C, Gibson L, Nordstrom C, Beug H (1993) The estrogen receptor cooperates with the TGFalpha receptor (c-erbB) in regulation of chicken erythroid progenitor self-renewal. EMBO J 12: 951–960

Sharif M, Privalsky ML (1991) v-erbA oncogene function in neoplasia correlates with its ability to repress retinoic acid receptor action. Cell 66: 885–893

Weinberger C, Thompson CC, Ong ES, Lebo R, Gruol DJ, Evans RM (1986) The c-erb-A gene encodes a thyroid hormone receptor. Nature 324: 641–646

Yamamoto T, Hihara H, Nishida T, Kawai S, Toyoshima K (1983) A new avian erythroblastosis virus, AEV-H, carries erbB gene responsible for the induction of both erytrhoblastosis and sarcomas. Cell 34: 225–232

Cell Cycle Regulation and Erythroid Differentiation

E.W. Müllner[1], H. Dolznig[1], and H. Beug[2]

1 Introduction

The regulation of cell cycle progression is one of the most intensively studied fields in contemporary cell biology. This already becomes evident from the exploding number of papers and reviews on this issue. More important, of course, is the multitude of biologically significant problems that now can be addressed (at least in part) by applying recently developed molecular tools. To name but a few, these problems include characterization of the basic machinery driving the cycle, determination of how signals transduced via growth factor receptors become integrated into the cell's decision to proceed through G1 into S phase or – in case of conflicts – become apoptotic, and analysis of how orderly withdrawal from the cell cycle is achieved during terminal differentiation. One specialized case of cell cycle regulation during such a terminal differentiation process, namely, the events occurring during progression of committed hematopoietic precursor cells, via erythroblasts, to mature erythrocytes, will form the core of this text.

The cell cycle of a continuously growing cell consists of four discernible phases. During G1 (for "gap 1") the cells accumulate proteins, grow in size and

[1]Institute of Molecular Biology, Vienna Biocenter Dr. Bohr-Gasse 9, 1030 Vienna, Austria
[2]Institute of Molecular Pathology, Vienna Biocenter, Dr. Bohr-Gasse, 1030 Vienna, Austria

decide from exogenous and endogenous clues whether to progress into S (for "synthesis") phase or become arrested (see the monograph by MURRAY and HUNT 1993 for a recent introduction to many concepts of cell cycle biology mentioned in this article). Once a positive decision is reached, the cycle will be completed without any requirement for further extracellular stimulation. Among the most distinctive features of S phase are the duplication of cellular DNA content and synthesis of histones to package the newly synthesized DNA into chromatin. The G2 period ("gap 2") preceedes cell division which takes place during M phase (for "mitosis"). An important function of G1 and G2 is to block entry into S or M phases in the case of DNA damage. The existence of G2 also assures that DNA replication is completed before chromosome segregation takes place (for comprehensive reviews on these checkpoint controls see for example HARTWELL and WEINERT 1989 and MURRAY 1992). In a typical animal cell cycle of about 24 h, G1, S, G2 and M last for about 12, 6, 6 and 1 h, respectively. Shorter cell cycle phases are observed in special cases, e.g., in some tumor cells or during early embryonic development, while longer intervals between divisions are observed for instance in senescent fibroblasts. Here we will discuss another example of a drastically shortened cell cycle, the "terminal differentiation divisions" of maturing normal erythroid progenitor cells.

2 Regulators of Cell Cycle Progression

The purpose of the next two sections is to outline the main routes of interplay between major components of the cycle clock that are required for the progression of every continously growing normal or primary cell from one division to the next. This may seem a redundant task to any specialist in the cell cycle field, given the abundance of timely and excellent reviews on the topic (see below). Nevertheless, we hope to provide some useful information to scientists outside the cell cycle field, namely insiders in cell biology or hematology.

As briefly mentioned above, many of the cell biological characteristics of cell cycle progression were already layed out 25 years ago. The study of underlying molecular details became possible by a combination of yeast genetics and the vigorous advent of molecular biology some 10 years later. The complementation of a large number of temperature sensitive cdc (for "cell division cycle") mutants in *Saccharomyces cerevisiae* (HARTWELL et al. 1970) and *Schizosaccharomyces pombe* (Nurse et al. 1976; LEE and NURSE 1988), restoring their ability to grow, allowed the isolation and subsequent molecular cloning of many regulatory factors from these organisms. Up to this day of the highlights in the field was the unraveling of the regulatory functions of cdc 2 in *S. pombe* (cdc 28 in *S. cerevisiae;* BEACH et al. 1982), a 34 kDa serine/threonine protein kinase whose activity is essential for traversal of G1 and G2. The enzymatic activity of this catalytic subunit is modulated by the binding of regulatory components, the so-called cyclins. This name refers to the variations in synthesis and degradation rates that these proteins undergo during distinct phases of the cell cycle (Fig. 1; EVANS et al. 1983). In addition, phosphorylation and dephosphorylation play an important role in regulating the activity of complexes between the catalytic kinase subunit, the so called cyclin-dependent kinase (cdk) and its regulatory partners, the various cyclins (e.g., GOULD and NURSE 1989). Due to space limitations, these few remarks have to suffice for the situation in yeasts (for an excellent recent monograph see MURRAY and HUNT 1993). In the last several years, scores of animal homologs for cyclins and cdks have been identified, frequently by their ability to compensate the defects of the corresponding yeast mutants.

The picture outlined in the following paragraphs (and illustrated in Fig. 2), although quite complex, is still a drastically simplified summary of what is known today on the expression of cyclins and cdks in the progression of logarithmically growing animal cells through the cell cycle (for more detailed information the interested reader is referred to a series or recent reviews, e.g., those by NURSE 1994; SHERR 1994; HEICHMAN and ROBERTS 1994; KING et al. 1994; HUNTER and PINES 1994; HUNT and SHERR 1994).

As in yeast, the cdc2 of higher eukaryotes (more systematically termed cdk1) is complexed with the mitotic cyclin B (LABBÉ et al. 1989; MEIJER et al. 1991). Activation of this complex brings about M phase. A similar complex between cdk2 and cyclin A is activated at the beginning of DNA replication (GIRARD et al. 1991; ZINDY et al. 1992). Cyclin A is required again at a second stage in the cycle prior to mitosis, this time in association with cdc2, at a point close to but preceeding the requirement for cdc2/cyclin B (MINSHULL et al. 1990; PINES and HUNTER 1990). At least two more types of cyclins as well as additional cdks are observed and thought to function during the progression through G1 (Sherr 1993). The resulting complexes are proving to be integrators of growth factor mediated signals which

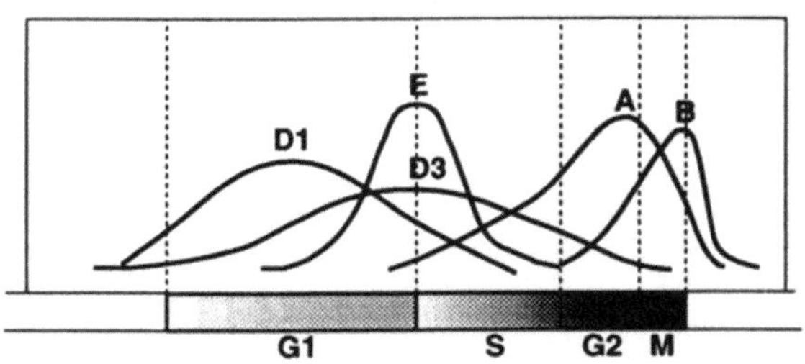

Fig. 1. Periodic expression of cyclins during the various cell cycle phases of logarithmically growing cultures. The levels indicated on the *y-axis* do not represent absolute amounts of protein and can, therefore, not be compared from one cyclin to another. Expression patterns look different for growth stimulated cells, where especially D-type cylins start to appear later in G1. In contrast, a rise of D1 levels can already be observed in the G2 phase of the previous cycle in synchronized chicken erythroid cells (H. DOLZNIG, unpublished; see also MOTOKURA and ARNOLD 1993). In accordance with their assumed role in growth factor signaling, maximal expression of D cyclins may occur at different times in G1 for different cell types. Down-regulation of most cyclin genes is facilitated by so-called destruction boxes or PEST regions (for example, see GLOTZER et al. 1991) which result in short half-life times of the proteins

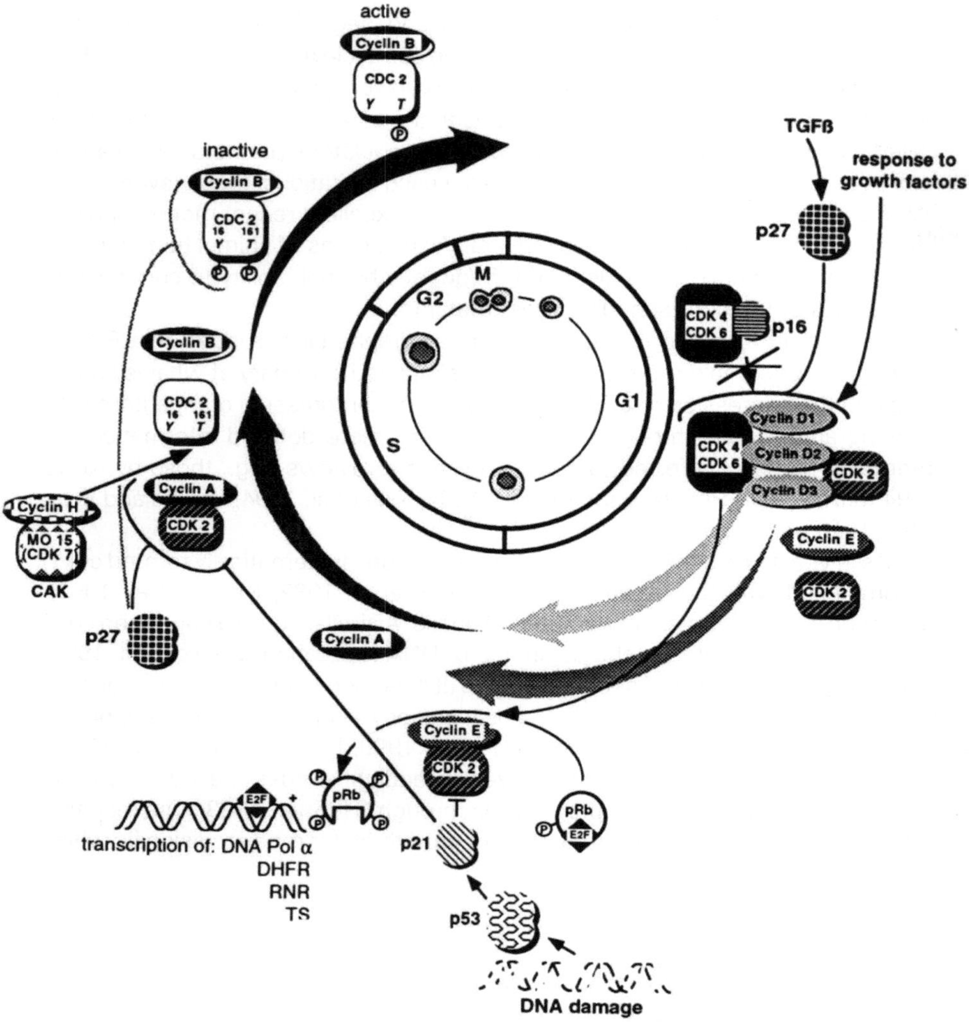

Fig. 2. Interactions between cyclins, cdks and cdk inhibitors during the cell cycle of higher eukaryotes. The *inner circle* shows the cell cycle phases, the length of the *bars* indicating approximate duration. Further outside, the *arrows* illustrate the timing of expression and oscillating nature of cyclins. Combinations of various cyclins with cdks are indicated by *boxes*. In addition, some of the pathways (action of TGFβ, DNA damage) that induce the binding of cdk inhibitors (p16, p21, p27) are outlined schematically. At the end of G1, the active forms of cyclin D and E containing complexes have the ability to hyperphosphorylate pRb, resulting in the release of transcription factor E2F, which in turn is essential for the expression of many S phase specific genes (*DNA pol α*, DNA polymerase α *DHFR*, dihydrofolate reductase; *RNR*, ribonucleotide reductase; *TS*, thymidylate synthase). As one example for the regulatory potential of phosphorylation, the possible action of cdk7/cyclin H (CAK) on cdc2 and the subsequent activation of cyclin B/cdc2 complexes by selective dephosphorylation on Tyr 16 is depicted

are essential to drive the cell cycle engine through another complete turn. Cyclins D2 and D3 and especially D1 (XIONG et al. 1991; LEW et al. 1991; MATSUSHIME et al. 1991a,b) appear rather early in G1, although in some cell types their levels may only fluctuate minimally (Fig. 1). Unlike other family members, D type cyclins exhibit a pattern of differential expression in various cell lineages. For example, mammalian D1 is not expressed in normal lymphoid or myeloid cells (WITHERS et al. 1991; INABA et al. 1992), whereas D2 levels are highest in T lymphocytes (MATSUSHIME et al. 1991a). These features underscore the function of D cyclins as growth factor sensors which, therefore, do not really represent integral components of the cell cycle clock. In vivo, mainly cdks 2, 4 and 6 seem to associate with these regulatory subunits (MEYERSON and HARLOW 1994 and references therein). Last, but nevertheless important to mention, is cyclin E, which is also essential for cell cycle progression (KOFF et al. 1991). Cyclin E expression rises throughout G1 and reaches a maximum later than D type cyclins, just prior to the onset of S. Cyclin E associates mainly with cdk2 (KOFF et al. 1991, 1992).

A significant advance in the understanding of cell cycle regulation came with the recent identification of a family of cyclin-cdk inhibitory proteins (CDIs; see Fig. 1), including p16 (and the very similar p15; KAMB et al. 1994; NOBORI et al. 1994; HANNON and BEACH 1994), p21 (reviewed in HUNTER 1993) and p27 (POLYAK et al. 1994a; TOYOSHIMA and HUNTER 1994) that bind to and inactivate cdks (for the sake of clarity we refrain from using the multitude of synonyms under which these CDIs have become known in the last 2 years). Since one or the other of these inhibitors are absent or inactivated in many transformed cells, they appear to have a potential as tumor suppressors (for recent reviews see HUNTER and PINES 1994 or PETER and HERSKOWITZ 1994). p16 competes with D cyclins for binding to cdk4 (SERRANO et al. 1993) or cdk6 (HANNON and BEACH 1994). Upon DNA damage, the tumor suppressor gene product p53 increases the levels of CDI p21, leading to inhibition of G1-specific cyclin/cdk complexes, in turn preventing the cells from entering S phase (DILEONARDO et al. 1994). Like p21, p27 also inhibits the formation of a wide variety of cdk interactions, resulting in a block to G1 progression. One of the triggers to this process is transforming growth factor β (TGFβ; HANNON and BEACH 1994; POLYAK et al. 1994b).

Conversely, there are kinases which in turn activate the various cdks (hence their designation as "CAKs" for "cylin-dependent kinase" activating kinases). Unphosphorylated cdk's produced in bacteria or insect cells cannot assemble into active complexes (e.g. KATO et al. 1994), and differential phosphorylation accompanies the transition between cell cycle phases (KREK and NIGG 1991), emphasizing the regulatory potential of this post-translational modification. One example of a mammalian CAK, acting on complexes of cdc2 and cdk2 with a variety of cyclins, has recently been characterized in some detail (FISHER and MORGAN 1994; MÄKELÄ et al. 1994; MATSUOKA et al. 1994; for a review see MORGAN 1995). It consists of two subunits, one of them has significant homology to other cyclins and was consequently termed cyclin H, the second contains the kinase domain and therefore was called cdk7 (also MO15). The regulatory potential of these CAKs still remains to be determined.

Although much is known on the assembly of complexes between G1 cyclins and their kinase partners, so far only a few physiologically relevant substrates of these cell cycle kinases have been identified. One such substrate is pRb, the gene product of the retinoblastoma tumor suppressor gene, on which most in vitro studies on G1-specific cdks rely. In its unphosphorylated / underphosphorylated form, pRb (and its more recently characterized homologs p107, p130, etc.; COBRINIK et al. 1993; HANNON et al. 1993; LI et al. 1993 and references therein) exerts its growth inhibitory properties in part via binding and inactivation of the E2F transcription factor gene family, the members of which, in turn, appear essential for the expression of many S phase-specific genes, including dihydrofolate reductase, thymidylate synthase, DNA polymerase or ribonucleotide reductase (HENGSTSCHLÄGER et al. 1994; for reviews on transcriptional regulation by pRb and E2F see KOUZARIDES 1993; HELIN and HARLOW 1993; LATHANGUE 1994; a comprehensive summary on substrates for G2/M specific cdks can be found in NIGG 1993). This provides at least one example of how the cell cycle machinery is connected to ultimate targets in the genome.

The scenario outlined above for cell cycle progression in continuously growing normal cells does not really apply for cultures withdrawing from the cycle during terminal differentiation or following restimulation/activation of resting cells. In addition, it should be emphasized that deviations from this scheme are frequently observed upon oncogenic transformation. Obviously, gaps in our knowledge on the cell cycle machinery remain to be filled and many of the facts mentioned here may be outdated and subject to revision in the light of more recent insights (e.g., the identification of new players as well as additional members of known gene families). There are also issues for which a molecular solution is not even in sight. One of these is the question how cells maintain their constant size during consecutive cell cycles, another one is if and how intimately withdrawal from the cell cycle is coupled to terminal differentiation. Below, we will describe a novel system that may help to overcome the difficulty in resolving some of these issues.

3 Cell Cycle Regulators in Hematopoiesis

The model for cell cycle regulation described so far is mainly (if at all) applicable to continuously cycling cells. Most of the molecules mentioned will be absent or inactive in resting or terminally differentiated cells. Thus the mechanism for orderly withdrawal from the cycle during terminal differentiation and the consequences of disturbances in this process have found wide interest. Conversely, under some circumstances (e.g., during T cell activation) committed primary cells must be able to re-enter the cell cycle upon proper stimulation. In this article, we can only briefly touch upon some observations made during the activation or maturation of normal hematopoietic progenitors.

Resting human T cells express only low levels of cyclin D2. Stimulation with 12-0- tetradecanoylphorpbol-13-acetate (TPA) or phytohemagglutin (PHA) results in rapid induction of D2 mRNA in early G1, followed by an up-regulation of D3 transcripts in late G1. D1 was not detected in T cells under any condition tested (AJCHENBAUM et al. 1993). Similar observations were described by several other groups working on T cells and early myeloid precursors (KATO and SHERR 1993; KIYOKAWA et al. 1992b; MATSUSHIME et al. 1991a; MEYERSON and HARLOW 1994). In macrophages, however, expression of D1 mRNA and protein is part of the delayed early response after growth stimulation by colony stimulating factor-1 (CSF-1). In these cells, cyclin D1 preferentially activates cdk4, whereas D3 is undetectable and D2 is synthesized at very low levels. Also, in another study, the main phosphorylating activity for pRb in macrophages was identified as cdk4 (MATSUSHIME et al. 1994; phosphorylation of pRb is believed to down-regulate its growth inhibitory activities; CHELLAPAN et al. 1991; HELIN et al. 1992). Contrary to the situation in macrophages, in primary T lymphocytes the major kinase associated with G1-specific cyclins appears to be cdk6 (MEYERSON and HARLOW 1994). Activation of cdk6 in PHA-stimulated human T cells occurs in mid-G1, prior to the activation of cdk2. Again, the preferred substrate is pRb; thus, cdk6 is a good candidate for the initial pRb kinase in T cells. These results from various lineages suggest that either cdk6 or its homolog cdk4, in different cell type-specific combinations with D type cyclins, can link growth factor simulation to the onset of cell cycle progression.

Two more papers describe interesting data on cell cycle regulation during an "opposite" process, namely, terminal differentiation accompained by growth arrest of myeloid precursors along the granulocyte lineage. In the murine interleukin (IL)-3-dependent myeloid cell line 32Dcl3, both cyclins D2 and D3 are expressed in proliferating cells while cyclin D1 is undetectable. Overexpression of transfected D2 and D3 genes did not affect cell viability in the presence of (IL-3) but delayed apoptotic cell death in the absence of this growth factor. In addition, the transfectants exhibited an increased fraction of cells in S phase, apparently due to a compensatory shortening of G1 (KATO and SHERR1993; ANDO et al. 1993). When the cells were induced to differentiate into mature neutrophils by the addition of granulocyte-CSF (G-CSF), D cyclin and cdk4 levels were reduced while cdk2 expression was sustained. Transfectants ectopically expressing D2 or D3 (but not D1) were unable to differentiate and died in G-CSF (KATO and SHERR 1993).

Also, the role of CDIs in hematopoiesis has recently begun to be addressed. When peripheral human T lymphocytes are subjected to antigen receptor stimulation they respond by synthesis of cyclins and cdks. The activity of complexes containing cdk2 stays repressed, however, most likely by p27, until the cells receive a second signal from the mitogenic lymphokine IL-2. Another CDI, p21, could not be detected in growth arrested lymphocytes (FIRPO et al. 1994). This may relate to the recent observation that p21 is induced in differentiating myoblasts by the basic helix-loop-helix protein MyoD upon terminal differentiation (HALEVY et al. 1995; SKAPEK et al. 1995; PARKER et al. 1995), since the nuclei in myotubes are irreversibly withdrawn from the cell cycle, whereas resting lymphocytes still have the potential to proliferate upon stimulation.

4 Lessons Learned from Erythroleukemic Cell Systems

Although a large body of literature exists on erythroid cell proliferation and differentiation, both in experimental and clinical settings, only a small number of reports has so far addressed erythroid-specific features of cell cycle regulation and the underlying molecular principles. One explanation may be the difficulty in isolating substantial amounts of pure and untransformed erythroid precursors (SAWADA et al. 1987, 1990) as compared to the ease with which high amounts of primary lymphoid cells or macrophages are obtained that can subsequently be stimulated to re-enter the cell cycle. Research on erythroid differentiation up to now relied on leukemic model systems that would faithfully reproduce parts of the erythroid differentiation program but are consistently unable to show all aspects of normal erythroid differentation. A well characterized and extensively used system of this type are the Friend virus infected murine erythroleukemia (MEL) cells, which consequently have also been employed for cell cycle studies. Due to their altered responsiveness to erythroid growth/differentiation factors and their inability to form mature erythrocytes upon differentiation induction these cells can only be viewed as the second best choice next to primary cells.

Friend erythroleukemia cells emerge in adult mice after infection with the Friend leukemia virus complex (for review, see BEN-DAVID and BERNSTEIN 1991). This retrovirus is composed of a replication-defective spleen focus forming virus (SFFV) and a replication-competent helper virus (F-MuL V). Several of the molecular events that constitute the different steps of disease development have been identified. The initial polyclonal expansion of erythroid progenitors results from the constitutive activation of the erythropoietin receptor (EpoR) as the result of the intracellular binding of the 55 aD envelope glycoprotein of SFFV (LI et al. 1990; HOATLIN et al. 1990). The emergence of leukemic clones is associated with at least two further genetic alternations. The first is overexpression of Spi-1/Pu-1, a member of the Ets family of transcriptional regulators caused by an SFFV proviral insertion at the Spi-1 locus (MOREAU-GACHELIN et al. 1988; PAUL et al. 1991). The second alternation is inactivation of the p53 tumor suppressor gene by deletion or mutation (BEN-DAVID et al. 1990; MUNROE et al. 1990). Due to these alterations, most Friend cell lines either exhibit an aberrant response or fail to respond at all to erythropoietin; differentiation is induced instead by chemicals (hexamethylene bisacetamide, HMBA; dimethyl sulfoxide, DMSO) in a poorly understood fashion. Although these terminally differentiating cells share several properties with mature erythrocytes, they fail to enucleate and undergo apoptosis instead. A few selected lines progress quite far towards mature erythrocytes, a fraction even enucleates under specific conditions (PATEL and LODISH 1987). Such lines have unfortunately not yet been employed for cell cycle studies.

In spite of these caveats, a whole series of interesting results on cell cycle regulation of MEL cells during chemically induced differentiation has been obtained over the last several years. To start with, mouse cyclin D2 was originally cloned from MEL cells by its homology with mouse cyclin D1, then referred to as

CYL1. A broad peak of expression was observed through G1 and S followed by a decrease in G2/M in cultures synchronized by centrifugal elutriation (KIYOKAWA et al. 1992a). The findings on differentiating cells were more revealing: the commitment to this process is stochastic and requires the presence of the inducing agent in G1 (KIYOKAWA et al. 1992b). Addition of HMBA also resulted in an initial prolongation of the G1 phase, associated with a decrease in cyclin A levels and underphosphorylation of pRb (KIYOKAWA et al. 1993). In subsequent studies with the same cell system, a rapid decline in the level of cdk4 due to decreased stability of the protein was detected (KIYOKAWA et al. 1994), which appears to be the reason for the repression of cdk4-associated pRb kinase activity observed previously. Cyclin D3, one of the binding partners for cdk4, was increased by HMBA treatment and found in complexes containing the transcription factor E2F and pRb. At later stages of differentiation, growth arrest also caused a general decline in the expression of cyclin A and cdk2. Whereas overexpression of cdk4 rendered transfected MEL cells insensitive to induction by HMBA, no such effect was seen with cdk2, suggesting that cdk4 down-regulation may be a critical event during terminal erythroid differentiation (KIYOKAWA et al. 1994). This resembles the situation described in myeloid precursors, in which ectopic overexpression of cyclins D2 or D3 (binding partners for cdk4) inhibited differentiation along the granulocyte lineage (KATO and SHERR 1993).

Although MEL cells seem to execute many functions of maturing erythroblasts during induced differentiation, the results may be conflicting with respect to the expression of some cell cycle regulators. For example, levels of endogenous pRb have been reported to either rise several fold upon HMBA treatment (RICHON et al. 1992) or decrease under the same regimem (KHOCHBIN et al. 1992). In addition, expression of transfected pRb in a nonproducer Friend cell line led to apoptosis and hemoglobin synthesis. This phenotype could be partially rescued by the addition of erythropoietin (a physiological inducer of erythropoiesis) with a concommitant decline in globin production (JOHNSON et al. 1993). These data clearly indicate the need to study erythroid maturation in normal, nonleukemic progenitors in order to elucidate the proper execution of all regulatory pathways involved.

It may be emphazied at this point, that oncogenic transformation of many cell types is frequently associated with aberrant expression or activity of cell cycle regulators (reviewed in HARTWELL and KASTAN 1994; HUNTER and PINES 1991, 1994). Although it may be premature to postulate a distinct causative role of D cyclins in tumor development, several cyclins (D1 is still a candidate for the BCL1 B cell lymphoma oncogene on human chromosome 11) can cooperate with proto-oncogenes, e.g., *ras* or *myc*, both in cell transformation in vitro and in transgenic animals (BODRUG et al. 1994; LOVEC et al. 1994). Cyclin E may even have potential as a prognostic marker for breast cancer (KEYOMARSI et al. 1994). The tumor suppressor activity of proteins such as p53 or members of the Rb family has been appreciated for some time, a similar importance for tumorigenesis has recently been ascribed to the CDIs (cdk inhibitors) p16, p21 or p27 (for concise recent summaries see HUNTER and PINES 1994; ELLEDGE and HARPER 1994).

One important conclusion from the above discussion is that neoplastic transformation as well as immortalization crucially alter regulation of cell cycle progression in cells. Therefore, results concerning differentiation-related cell cycle regulation employing such cell lines need to be interpreted with great caution. The obvious alternative is the use of primary erythroid progenitors. However, since such progenitors, even if purified to homogeneity (SAWADA et al. 1987, 1990; DALYOT et al. 1993), undergo only a few cell divisions until terminal differentiation, cell numbers large enough for biochemical studies are difficult to obtain. In the avian system, erythroid progenitors transformed with temperature-sensitive mutants of tyrosine kinase oncogenes have been successfully used (BEUG et al. 1982; KNIGHT et al. 1988). In these cells, however, residual activity of the oncogenic tyrosine kinases after inactivation of the state of the art was only obtained after it became possible to cultivate self-renewing, normal erythroid progenitors in vitro.

5 Culture and Differentiation of Normal Chicken Erythroid Progenitors

The critical clues leading to successful culture of self-renewing, normal erythroid progenitors were provided by avian retroviruses causing lethal erythroleukemia. The avian retrovirus AEV (avian erythroblastosis virus) induced proliferation of transformed erythroblast for 25–40 generations, thereafter, the cells underwent cellular senescence. Leukemic transformation by AEV is caused by the protein products of its two cooperating oncogenes. V-ErbB represents a mutated avian epidermal growth factor receptor (c-ErbB; DOWNWARD et al. 1984), while v-ErbA is a mutated, nuclear thyroid hormone receptor (SAP et al. 1986).

The v-*erb*B oncogene of AEV suffices to transform erythroblasts (FRYKBERG et al. 1983). Initially it was suspected that v-*erb*B functioned ectopically in erythroblasts, i.e. that c-*erb*B, the cellular gene from which it was derived, would not normally be expressed in erythroid cells (GRAF and BEUG 1983). However, a more detailed examination of the possible role of c-ErbB in erythroid growth and differentiation (KHAZAIE et al. 1988) revealed that normal progenitors did indeed express endogenous c-ErbB, which was able to employ mammalian TGFα instead of the unknown avian ligand of c-ErbB (LAX et al. 1988). Furthermore, this receptor erythroblast self-renewal in colony assays (PAIN et al. 1991) when stimulated by TGFα. Later, it became clear, however, that long-term growth of these progenitors in suspension required cooperation of signals derived from the endogenous c-ErbB with the ligand-activated, endogenous estrogen receptor (HAYMAN et al. 1993; SCHROEDER et al. 1993).

These self-renewing progenitors, referred to as SCF/TGFα progenitors, clearly differ from normal avian CFU-E/BFU-E (termed SCF progenitors, HAYMAN et al. 1993). SCF/ TGFα progenitors are much less abundant than SCF progenitors

(1 in 15,000 as compared to 1 in 300). SCF/TGFα progenitors express both c-erbB and c-Kit and respond to both TGFα and the c-Kit ligand stem cell factor (SCF). In contrast, SCF progenitors express c-Kit but fail to express c-ErbB; and SCF induces only transient self-renewal (8–10 cell divisions) in them (Fig. 3). Despite these differences, both cell types differentiate according to the same program when exposed to anemic serum and insulin. They both divide five times within 60–70 h and drastically decrease in size while maturing into erythrocytes (Fig. 3, HAYMAN et al. 1993).

The above studies, identifying a novel, self-renewing progenitor, raised the question how this new cell type was related to the known progenitors, CFU-E and

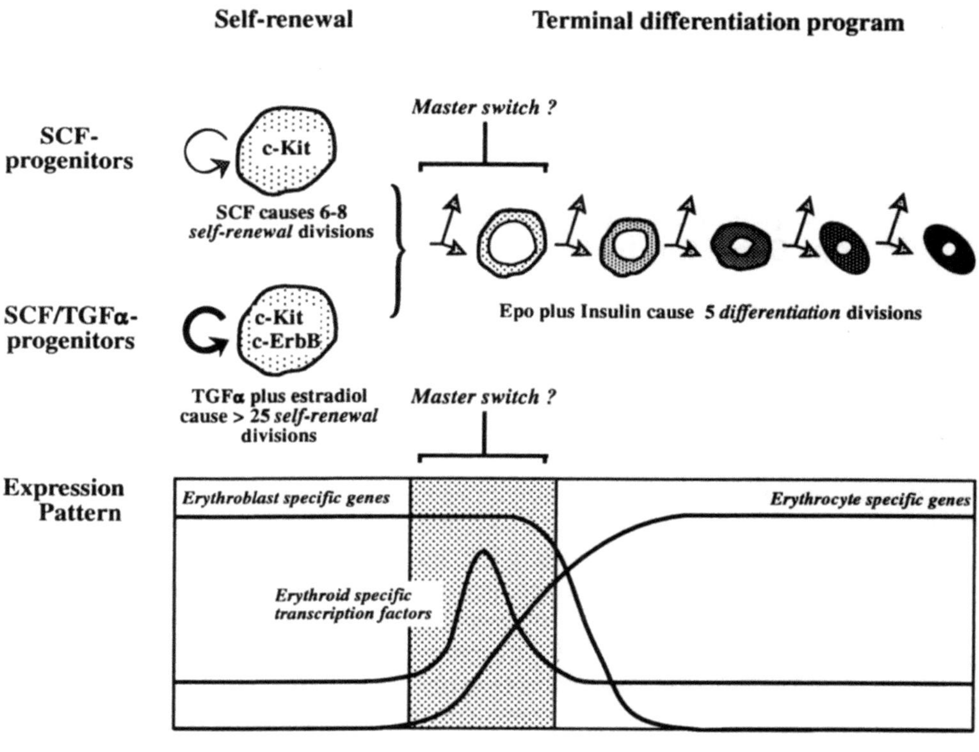

Fig. 3. Self-renewal and differentiation of normal chicken erythroid progenitors in response to different exogenous signals. c-Kit expressing SCF progenitors are capable of transient growth in the presence of SCF, the c-kit ligand. In contrast, the c-Kit/c-ErbB expressing SCF/ TGFα progenitors show prolonged self-renewal when exposed to TGFα and estradiol. When these factors are removed and replaced by the differentiation inducing factors erythropoietin (*Epo*) and insulin, the cells are re-programmed to terminally differentiate. This involves massive changes in: (1) the gene expression program (i.e., down-regulation of erythroblasts specific genes, transient induction of erythroid specific transcription factors, followed by the rise in levels of erythrocyte-specific genes), (2) cell morphology (round erythroblasts turn into oval erythrocytes) and size (from more than 300 fl to less than 80 fl), and (3) cell cycle regulation (transient acceleration of growth by shortening the G1 phase of the cell cycle from about 11 to 5 h). The possibility that gene expression and cell cycle regulation are subject to concerted alteration by "master switch" transcription factors is indicated.

BFU-E. Recently, these SCF/TGFα progenitors were found to develop from normal CFU-E/BFU-E when costimulated with SCF, TGFα, estradiol and an unknown activity in chicken serum (STEINLEIN et al. 1995). This developmental switch from SCF to SCF/TGFα progenitors was remarkable in that an "unorthodox" combination of three, if not four, growth factors and hormones had to be present for at least 6–8 days to induce efficient switching, suggesting that the cells had to undergo several cell divisions in the presence of these factors (STEINLEIN et al. 1995).

Thus, examining the role of v-ErbB in leukemia has revealed the existence of a new erythroid progenitor capable of extended self-renewal, resulting from the cooperation between signals evoked by the c-ErbB ligand and steroid hormones. This work has also provided us with in vitro systems to grow large amounts of differentiating normal progenitors and to engage in a detailed molecular analysis of how self-renewal and differentiation are controlled in these cells.

6 Erythroid Differentiation Coordinately Reprograms Gene Expression Including Specific Alterations in Cells Cycle Control

A major advantage of the normal SCF or SCF/TGFα progenitors was that cell numbers large enough for extensive biochemical analysis could be produced in the presence of self-renewal factors. These cells could then be induced to synchronously differentiate into mature erythrocytes by the same differentiation factors as required by human CFU-E (erythropoietin containing anemic chicken serum and insulin, SAWADA et al. 1989). Using this system, several interesting new aspects of erythropoiesis were uncovered. For example, the cells in such mass cultures underwent five cell divisions during the first 3 days of differentiation and then arrested in G1. The final steps to fully mature erythrocytes occurred in the absence of further cell divisions. Two important events occurred in the first 16–24 h after differentiation induction. Firstly, the gene expression pattern of these cells was extensively reprogrammed at this time point (Fig. 3). Genes characteristic of self renewing cells (c-*myb*, c-*kit*, estrogen receptor) abruptly ceased to be expressed, whereas erythroid-specific transcription factors such as GATA-1, GATA-2, SCL and NF-E2 were strongly up-regulated, followed slightly later by activation of numerous late erythrocyte genes. Secondly, the cell proliferation rate increased drastically, as indicated by a reduction of cell cycle length from about 20 to 12 h. This was almost entirely due to a shortening of the G1 period from 12 to 5 h, while S and G2 phase lengths were not affected. Contraction of the cell cycle and shortening of G1 commenced 16–24 h after differentiation induction, exactly coinciding with the general reprogramming of the progenitors' gene expression program. Furthermore, cell cycle and G1 phase shortening was accompained by a decrease in cell volume, from about 300 to less than 70 femtoliters. Measurements of ^{14}C-labeled aminoacid incorporation versus cell size showed that the protein synthesis rate normalized to cell volume remained constant. This indicated that size reduction is a necessary consequence of G1 shortening, since it could not be compensated for by a higher rate of protein synthesis. The tentative conclusion from these experiments was, therefore, that control of cell size at the restriction point before entering S seems to be lost or altered (DOLZNIG et al. 1995).

Thus, erythroid differentiation, like myoblast differentiation (ALEMA and TATO 1994), may involve control by a master switch, which not only completely reprograms the pattern of gene expression, but also profoundly alters cell cycle control (Fig. 3). This immediately raises the question which molecular components may be involved in such a bypass mechanism. As described above, progression of cells through G1 and into S is mainly mediated by activation of cdks (cyclin-dependent kinases) via binding to their cyclin partners, most notably the D type cyclin/cdk4 and cyclin E/cdk2 complexes, whose expression peaks in G1 and declines rapidly thereafter (see Fig. 2). While D type cyclins seem to act in response to signals from growth factor receptors, cyclins A, B, and E appear to be intrinsic parts of the cell cycle clock. Therefore, one could speculate that alterations in D type cyclin/cdk4 function or expression could be held responsible for rapid progression beyond the restriction point in maturing chicken erythroblasts. Indeed, constitutive or inducible overexpression of cyclin D1 or E altered the control of G1-S transition in certain mammalian cells. Two papers report a prolongation of S compensating the observed shortening of G1 and unaltered control of cell size (RESNITZKY et al. 1994; OHTSUBO and ROBERTS 1993), other authors reported a shortening of G1 only, without alternations in S as well as with a concomitant decrease of cell size (QUELLE et al. 1993). Although D1 is not expressed in erythroid or myeloid cells of mammalian origin (WITHERS et al. 1991), cyclin D2 or D3 fulfills similar functions (ANDO et al. 1993). It has to be kept in mind, however, that all previous studies were done in immortalized cell lines, mostly of nonhematopoietic origin, and that the regulation in primary, differentiating cells may be totally different.

7 Outlook

We have recently begun to study the particular molecular mechanisms of cell cycle progression and differentiation in the primary chicken erythroid progenitor cell system described above. There are two major questions we will have to address: (1) What is the basis for the initial growth acceleration during the erythroid differentiation divisions? (2) How is the phase of rapid growth ultimately terminated within a period as short as a single cell cycle? In connection with these problems, we would like to learn something about size control, how cell volume is precisely maintained from one doubling to the next during self-renewing growth and reduced in a highly regulated manner during terminal maturation.

To approach these questions, the primary tools required will be complete cDNAs of chicken origin encoding many, if not all, of the cell cycle clock components mentioned in Fig. 2. Several of these molecular tools were already generated previously (cdc2: KREK and HIGG 1989; cyclin A: MARIDOR et al. 1993; cyclin B2: GALLANT and NIGG 1992; cyclin B3: GALLANT and NIGG 1994; pRb: BOEHMELT et al. 1994; p53: SOUSSI 1988), some were isolated in the author's labs in the course of the last months and additional ones are still missing (see Fig. 4 for an inventory of reagents already available). Starting from known sequences of mammalian and amphibian cDNAs, degenerate oligonucleotide primers were used to amplify partial cDNAs for G1/S cell cycle regulators by PCR from either embryonic, erythroleukemic or normal erythroblast cDNA libraries. In this way, probes specific for chicken cyclins D1, D2, E and cdk4 were obtained which show high homology with their respective mammalian counterparts. We next want to isolate the avian equivalents of p21, cdk2, cdk6 and, eventually, p27. Another major task will be to raise antibodies against the respective chicken proteins or to identify existing antibodies cross-reactive with the chicken proteins.

Besides these cloning efforts, we have employed the probes already at hand in a first series of experiments using RNAs isolated from differentiating primary chicken erythroblasts. These studies have yielded some interesting results. We observed a rapid down-regulation of cdk4 and D2 mRNAs at the phase of accelerated growth while expression of cyclins A and B as well as cdc2 stayed up until the cells became postmitotic. This suggests that during these final divisions the cells no longer regulate cell cycle progression via growth factor signaling, that is, integrated via cyclin D2/cdk4. Thus, cells undergoing "differentiation divisions" may proceed through the cycle guided exclusively by internal clues. Our findings add some new facets to the possibility (see also above) that D2/cdk4 are not intrinsic parts of the machinery required for keeping the cycle turning. Similar observations were recently made for chemically induced MEL cells (KIYOKAWA et al. 1994) and myeloid cells (KATO and SHERR 1993; ANDO et al. 1993), although no initial increase in growth rates comparable to what we observed in the primary erythroid progenitors was reported.

Contrary to the situation in mammalian hematopoietic progenitors (WITHERS et al. 1991; INABA et al. 1992), the normal chicken erythroblasts cells also express cyclin D1 mRNA in parallel with transcripts for D2. At present we cannot even speculate on the consequences of this apparently species-specific variation, although the possibility that cyclin D1 is absent in mammalian erythroid cell lines as a consequence of immortalization and ensuring genetic instability cannot be ruled out. Interestingly, so far we were not able to pick any clone resembling D3 from any of the chicken libraries tested. The same may be true for *Xenopus laevis* (M. COCKERILL, personal communication) indicating a possible evolutional divergence in the tissue specificity of D type cyclin expression.

Future experiments may include attempts to determine, whether ectopic overexpression of one or the other cell cycle regulator can inhibit erythroid

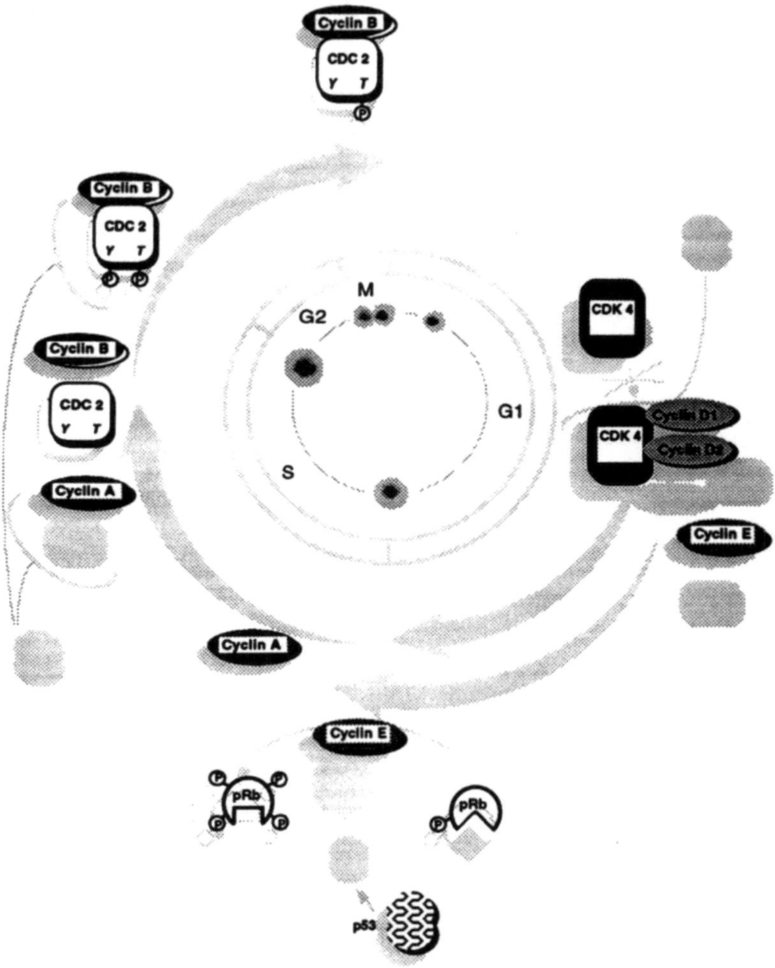

Fig. 4. A summary of the molecular tools currently available to study the regulation of cell progression in avian cells. Complementary DNAs for several mitotic (cyclins A, B, cdc2) and G1-specific functions (cyclins D1, D2, E, cdk4; pRb, p53) are already available, corresponding antibodies remain to be tested/ generated in many cases. Future efforts will concentrate on isolation of probes specific for cdk2, cdk6, p21 and, eventually p27 (outlined in *gray*, see Fig. 2)

differentiation of the primary cells and whether such a block enables the cells to undergo self-renewal. Precedents for a D cyclin-induced developmental block were found in myeloid and erythoid cell lines (KATO and SHERR 1993; KIYOKAWA et al. 1994). Another series of experiments may concentrate on how terminally maturing cells execute their highly synchronous and abrupt withdrawal from the cell cycle. Our efforts will concentrate on the role of p21 in this process. A series of recent papers has elegantly demonstrated the relation between the myogenic basic helix-loop-helix transcription factor MyoD and elevated p21 expression during muscle cell differentiation (HALEVY et al. 1995; SKAPEK et al. 1995; PARKER et al. 1995). A similar mechanism – this time involving one of the erythroid-specific transcription factors – may be in place to achieve terminal growth arrest of mature erythrocytes.

Obviously, it would be highly desirable to extend our studies to normal hematopoietic progenitors of human origin (see BERARDI et al. 1995). Trials to apply the principles governing self-renewal of avian erythroid progenitors (HAYMAN et al. 1993; STEINLEIN et al. 1995) to human erythroid progenitors from bone marrow, umbilical cord blood or even peripheral blood have recently met with some success. Using factor combinations similar, but not identical, to those used for avian cells allowed growth of cell populations highly enriched in proerythroblasts for 14–16 days, achieving a >1000-fold increase in the number of immature erythroblasts. We are currently trying to develop this system to a stage where it can be used to perform studies similar to those described here.

Acknowledgments. Most importantly, we have to thank numerous group members as well as collaborators and colleagues in the field. Without their persistent dedication, patience and support the work described in this contribution would not have been possible. We apologize for selecting single references when a whole series could have been cited and thus to all scientists whose work we could not mention. Research in the labs of the authors was supported by the "Herzfeld Family Foundation" (Austria), the "Fonds zur Förderung der Wissenschaftlichen Forschung" (Austria) and the "Austrian Industrial Research Promotion Fund."

References

Ajchenbaum F, Ando K, DeCaprio JA, Griffin JD (1993) Independent regulation of human D-type cyclin gene expression during G1 phase in primary human T lymphocytes. J Biol Chem 268: 4113–4119

Alema S, Tatò F (1994) Oncogenes and muscle differentiation: multiple mechanisms of interference. Semin Cancer Biol 5: 147–156

Ando K, Ajchenbaum-Cymbalista F, Griffin JD (1993) Regulation of G1/S transition by cyclins D2 and D3 in hematopoietic cells. Pro Natl Acad Sci USA 90: 9571–9575

Beach D, Durkacz B, Nurse P (1982) Functionally homologous cell cycle control genes in budding and fission yeast. Nature 300: 706–709

Ben-David Y, Bernstein A (1991) Friend virus-induced erythroleukemia and the multistage nature of cancer. Cell 66: 831–834

Ben-David Y, Lavigueur A, Cheong G, Bernstein A (1990) Insertional inactivation of the p53 gene during friend leukemia: a new strategy for identifying tumor suppressor genes. New Biol 2: 1015–1023

Berardi AC, Wang A, Levine JD, Lopez P, Scadden DT (1995) Functional isolation and characterization of human hematopoietic stem cells. Science 267: 104–108

Beug H, Palmieri S, Freudenstein C, Zentgraf H, Graf T (1982) Hormone-dependent terminal differentiation in vitro of chicken erythroleukemia cells transformed by ts mutants of avian erythroblastosis virus. Cell 28: 907–919

Bodrug SE, Warner BJ, Bath ML, Lindeman GJ, Harris AW, Adams JM (1994) Cyclin D1 transgene impedes lymphocyte maturation and collaborates in lymphomagenesis with *myc* gene. EMBO J 13: 2124–2130

Boehmelt G, Ulrich E, Kurzbauer R, Mellitzer G, Bird A, Zenke M (1994) Structure and expression of the chicken retinoblastoma gene. Cell Growth Differ 5: 221–230

Chellapan S, Hiebert S, Mudryj M, Horowitz J, Nevins J (1991) The E2F transcription factor is a cellular target for the RB protein. Cell 65: 1053–1061

Cobrinik D, Whyte P, Peeper DS, Jacks T, Weinberg RA (1993) Cell cycle-specific association of E2F with the p130 E1A-binding protein. Genes Dev 7: 2392–2404
Dalyot N, Fibach E, Ronchi A, Rachmilewitz EA, Ottolenghi S, Oppenheim A (1993) Erythropoietin triggers a burst of GATA-1 in normal human erythroid cells differentiating in tissue culture. Nucleic Acids Res 21: 4031–4037
DiLeonardo A, Linke SP, Clarkin K, Wahl GM (1994) DNA damage triggers a prolonged p53-dependent G1 arrest and long-term induction of Cip1 in normal human fibroblasts. Genes Dev 8: 2540–2551
Dolznig H, Bortunek P, Nasmyth K, Müllner E, Beug H (1995) Terminal differentiation of normal chicken erythroid progenitors: Shortening of G1 correlates with loss of D cyclin/cdk 4 expression and altered size control. Cell Growth Diff 6: 1341–1352
Downward J, Yarden Y, Mayes E, Scrace G, Totty N, Stockwell P, Ullrich A, Schlessinger J, Waterfield MD (1984) Close similarity of epidermal growth factor receptor and v-erbB oncogene protein sequences. Nature 307: 521–527
Elledge SJ, Harper JW (1994) Cdk inhibitors: on the threshold of checkpoints and development. Curr Opin Cell Biol 6: 847–852
Evans T, Rosenthal ET, Youngblom J, Distel D, Hunt T (1983) Cyclin: a protein specified by maternal mRNA in sea urchin eggs that is destroyed at each cleavage division. Cell 33: 389–396
Firpo EJ, Koff A, Solomon MJ, Roberts JM (1994) Inactivation of a cdk2 inhibitor during interleukin 2-induced proliferation of human T lymphocytes. Mol Cell Biol 14: 4889–4901
Fisher RP. Morgan DO (1994) A novel cyclin associates with MO15/cdk7 to form the cdk-activating kinase. Cell 78: 713–724
Frykberg L, Palmieri S, Beug H, Graf T, Hayman MJ, Vennström B (1983) Transforming capacities of avian erythroblastosis virus mutants deleted in the erbA or erbB oncogenes. Cell 32: 227–238
Gallant P, Nigg EA (1992) Cyclin B2 undergoes cell cycle-dependent nuclear translocation and, when expressed as a non-detructible mutant, causes mitotic arrest in HeLa cells. J Cell Biol 117: 213–224
Gallant P, Nigg EA (1994) Identification of a novel vertebrate cyclin: cyclin B3 shares properties with both A and B-type cyclins. EMBO J 13: 595–605
Girard F, Strausfeld U, Fernandez A, Lamb NJC (1991) Cyclin A is required for the onset of DNA replication in mammalian fibroblasts. Cell 67: 1169–1179
Glotzer M, Murray AW, Kirschner MW (1991) Cyclin in degraded by the ubiquitin pathway. Nature 349: 132–138
Gould KL, Nurse P (1989) Tyrosine phosphorylation of the fission yeast $cdc2^+$ protein kinase regulates entry into mitosis. Nature 342: 39–45
Graf T, Beug H (1983) Role of the v-erbA and v-erbB oncogenes of avian erythroblastosis virus in erythroid cell transformation. Cell 34: 7–9
Halevy O, Novitch BG, Spicer DB, Skapek SX, Rhee J, Hannon GJ, Beach D, Lassar AB (1995) Correlation of terminal cell cycle arrest of skeletal muscle with induction of p21 by MyoD. Science 267: 1018–1021
Hannon GJ, Beach D (1994) $p15^{INK4B}$ is a potential effector of TGFβ-induced cell cycle arrest. Nature 371: 257–261
Hannon GJ, Demetrick D, Beach D (1993) Isolation of the Rb-related p130 through its interacation with CDK2 and cyclins. Genes Dev 7: 2378–2391
Hartwell LH, Kastan MB (1994) Cell cycle control and cancer. Sciene 266: 1821–1828
Hartwell LH, Weinert TA (1989) Checkpoints: controls that ensure the order of cell cycle events. Science 246: 629–634
Hartwell LH, Culotti J, Reid B (1970) Genetic control of cell division in yeast, I. Detection of mutants. Proc Natl Acad Sci USA 66: 352–359
Hayman MJ, Meyer S, Martin F, Steinlein P, Beug H (1993) Self renewal and differentiation of normal avian erythroid progenitor cells: regulatory roles of the c-erbB/TGFα receptor and the c-kit/SCF receptor. Cell 74: 157–169
Heichman KA, Roberts JM (1994) Rules to replicate by. Cell 79: 557–562
Helin K, Harlow E (1993) The retinoblastoma protein as a transcriptional repressor. Trends Cell Biol 3: 43–49
Helin K, Lees JA, Vidal M, Dyson N, Harlow E, Fattaey A (1992) A cDNA encoding a pRb-binding protein with properties of the transcription factor E2F. Cell 70: 337–350
Hengstschläger M, Mudrak I, Wintersberger E, Wawra E (1994) A common regulation of genes encoding enzymes of the deoxynucleotide metabolism is lost after neoplastic transformation. Cell Growth Differ 5: 1389–1394

Hoatlin M, Kozak S, Lilly F, Chakraborti A, Kozak C, Kabat D (1990) Activation of erythropoietin receptors by Friend viral gp55 and by erythropoietin and down-modulation by the murine Fv-2r resistance gene. Proc Natl Acad Sci USA 87: 9985–9989

Hunt T, Sherr CJ (eds) (1994) Cell multiplication. Curr Opin Cell Biol 6/(6): 833–882

Hunter T (1993) Braking the cycle. Cell 75: 839–841

Hunter T, Pines J (1991) Cyclins and cancer. Cell 66: 1071–1074

Hunter T, Pines J (1994) Cyclin and cancer II: cyclin D and cdk inhibitors come of age. Cell 79: 573–582

Inaba T, Matsushime H, Valentine M, Roussel M, Sherr CJ, Look T (1992) Genomic organization, chromosomal localization, and independent expression of human cyclin D genes. Genomics 13: 565–574

Johnson P, Chung S, Benchimol S (1993) Growth suppresssion of Friend virus-transformed erythroleukemia cells by p53 protein is accompanied by hemoglobin production and is sensitive to erythropoietin. Mol Cell Biol 13: 1456–1463

Kamb A, Gruis NA, Weaver-Feldhaus J, Liu Q, Harshman K, Tavtigian SV, Stockert E, Day RSI Johnson BE, Skolnick MH (1994) A cell cycle regulator potentially involved in genesis of many tumor types. Science 264: 436–440

Kato J-Y, Sherr CJ (1993) Inhibition of granulocyte differentiation by G1 cyclins. Proc Natl Acad Sci USA 90: 11513–11517

Kato J-Y, Matsuoka M, Strom DK, Sherr CJ (1994) Regulation of cyclin D-dependent kinase 4 (cdk4) by cdk4-activating kinase. Mol Cell Biol 14: 2713–2721

Keyomarsi K, O'Leary N, Molnar G, Lees E, Fingert HJ, Pardee AB (1994) Cyclin E, a potential prognostic marker for breast cancer. Cancer Res 54: 380–385

Khazaie K, Dull TJ, Graf T, Schlessinger J, Ullrich A, Beug H, Vennstrom B (1988) Truncation of the human EGF receptor leads to differential transforming potentials in primary avian fibroblasts and erythroblasts . EMBO J 7: 3061–3071

Khochbin S, Brocard MP, Grunwald D, Larence JJ (1992) Antisense RNA and p53 regulation in induced murine cell differentiation. Ann NY Acad Sci 660: 77–87

King RW, Jackson PK, Kirschner MW (1994) Mitosis in transition. Cell 79: 563–571

Kiyokawa H, Busquests X, Powell CT, Ngo L, Rifkind RA, Marks PA (1992a) Cloning of a D-type cyclin from murine erythroleukemia cells. Proc Natl Acad Sci USA 89: 2444–2447

Kiyokawa H, Ngo L, Kurosaki T, Rifkind RA, Marks PA (1992b) Changes in $p34^{cdc2}$ kinase activity and cyclin A during induced differentiation of murine erythroleukemia cells. Cell Growth Differ 3: 377–383

Kiyokawa H, Richon VM, Venta-Perez G, Rifkind RA, Marks PA (1993) Hexamethylenbisacetamide-induced erythroleukemia cell differentiation involves modulation of events required for cell cycle progression through G1. Proc Natl Acad Sci USA 90: 6746–6750

Kiyokawa H, Richon VM, Rifkind RA, Marks PA (1994) Suppression of cyclin-dependent kinase 4 during induced differentiation of erythroleukemia cells. Mol Cell Biol 14: 7195–7203

Knight J, Zenke M, Disela C, Kowenz E, Vogt P, Engel JD, Hayman MJ, Beug H (1988) Temperature-sensitive v-sea transformed erythroblasts: a model system to study gene expression during erythroid differentiation. Genes Dev 2: 247–58.

Koff A, Cross F, Fisher A, Schumacher J, Leguellec K, Philippe M, Roberts JM (1991) Human cyclin E, a new cyclic that interacts with two members of the cdc2 gene family. Cell 66: 1217–1228

Koff A, Giordano A, Desai D, Yamashita K, Harper JW, Elledge S, Nishimoto T, Morgan DO, Franza BR, Roberts JM (1992) Formation and activation of a cyclinE-cdk2 complex during the G1 phase of the human cell cycle. Science 257: 1689–1694

Kouzarides T (1993) Transcriptional regulation by the retinoblastoma protein. Trends Cell Biol 3: 211–213

Krek W, Nigg EA (1989) Structure and developmental expression of the chicken cdc2 kinase: evidence for regulation at multiple levels. EMBO J 8: 3071–3078

Krek W, Nigg EA (1991) Differential phosphorylation of vertebrate $p34^{cdc2}$ kinase at the G1/S and G2/M Transitions of the cell cycle: identification of major phosphorylation sites. EMBO J. 10: 305–316

Labbé J-C, Capony J-P, Caput D, Cavadore JC, Derancourt J, Kaghad M, Lelias J-M, Picard A, Dorée M (1989) MPF from starfish oocytes at first meiotic metaphase is a heterodimer containing one molecule of cdc2 and one molecule of cyclin B. EMBO J 8: 3053–3058

LaThangue NB (1994) DRTF1/E2F: an expanding family of heterodimeric transcription factors implicated in cell-cycle control. Trends Biochem Si 19: 108–114

Lax I, Johnson A, Howk R, Sap J, Bellot F, Winkler M, Ullrich A, Vennstrom B, Schlessinger J, Givol D (1988) chicken epidermal growth factor (EGF) receptor: cDNA cloning, expression in mouse cells, and differential binding of EGF and transforming growth factor alpha. Mol Cell Biol 8: 1970–1978

Lee MG, Nurse P (1988) Cell cycle control genes in fission yeast and mammalian cells. Trends Genet 4: 287–290

Lew DJ, Dulic V, Reed SI (1991) Isolation of three novel human cyclins by rescue of G1 cyclin (Cln) function in yeast. Cell 66: 1197–1206

Li J, D'Andrea A, Lodish H, Baltimore D (1990) Activation of cell growth by binding of Friend spleen focus-forming virus gp55 glycoprotein to the erythropoietin receptor. Nature 343: 762–644

Li Y, Graham C, Lacy S, Duncan AMV, Whyte P (1993) The adenovirus E1A-associated 130-kD protein is encoded by a member of the retinoblastoma gene family and physically interacts with cyclin A and E. Genes Dev. 7: 2366–2377

Lovec H, Grzeschiczek A, Kowalski M-B, Möröy T (1994) Cyclin D1/*bcl1* cooperates with the *myc* genes in the generation of B-cell lymphoma in transgenic mice. EMBO J 13: 3487–3495

Mäkelä TP, Tassan J-P, Nigg EA, Frutiger S, Hughes GH, Weinberg RA (1994) A cyclin associated with the CDK-activating kinase MO15. Nature 371: 254–257

Maridor G, Gallant P, Golsteyn R, Nigg EA (1993) Nuclear localization of vertebrate cyclin A correlates with its ability to form complexes with cdk catalytic subunits. J Cell Sci 106: 535–544

Matsuoka M, Kato J-Y, Fisher RP, Morgan DO, Sherr CJ (1994) Activation of cyclin-dependent kinase 4 (cdk4) by mouse MO15-associated kinase. Mol Cell Biol 14: 7265–7275

Matsushime H, Roussel M, Ashmun R, sherr CJ (1991a) Colony-stimulating factor 1 regulates novel cyclins during the G1 phase of the cell cycle. Cell 65: 701–713

Matsushime H, Roussel MF, Sherr CJ (1991b) Novel mammlian cyclin (CYL) genes expressed during G1. CSH Symp Quant Biol 56, 69–74

Matsushime H, Quelle D, Shurtleff SA, Shibuya M, Sherr CJ, Kato J-Y (1994) D-type cyclin-dependent kinase activity in mammalian cells. Mol Cell Biol 14: 2066–2076

Meijer L, Azzi I, Wang JYJ (1991) Cyclin targets $p34^{cdc2}$ for tyrosine phosphorylation. EMBO J 10: 1545–1554

Meyerson M, Harlow E (1994) Identification of G1 kinase activity for cdk6, a novel cyclin D partner. Mol Cell Biol 14: 2077–2086

Minshull J, Golsteyn R, Hill CS, Hunt T (1990) The A- and B- type cyclin associated cdc2 kinases in *Xenopus* turn on and off at different times in the cell cycle. EMBO J 9: 2865–2875

Moreau-Gachelin F, Tavitian A, Tambourin P (1988) Spi-1 is a putative oncogene in virally induced murine erythroleukaemias. Nature 331: 277–280

Morgan DO (1995) Principles of CDK regulation. Nature 374: 131–134

Motokura T, Arnold A (1993) Cyclins and oncogenesis. Biochim Biophys Acta 1155: 63–78

Munroe D, Peacock J, Benchimol S (1990) Inactivation of the cellular p53 gene is a common feature of Friend virus–induced erythroleukemia: relationship of inactivation to dominant transforming alleles. Mol Cell Biol 10: 3307–3313

Murray AW (1992) Creative blocks: cell cycle checkpoints and feedback controls. Nature 359: 599–604

Murray AW, Hunt T (1993) The cell cycle: an introduction. Oxford University Press, Oxford

Nigg EA (1993) Cellular substrates of p34cdc2 and its companion cyclin-dependent kinases. Trends Cell Biol 3: 296–301

Nobori T, Miura K, Wu DJ, Lois A, Takabayashi K, Carson DA (1994) Deletions of the cyclin-dependent kinase-4 inhibitor gene in multiple human cancers. Nature 368: 753–756

Nurse P (1994) Ordering S phase and M phase in the cell cycle. Cell 79: 547–550

Nurse P, Thuriaux P, Nasmyth KA (1976) Genetic control of th cell division cycle in the fission yeast *Schizosaccharomyces pombe*. Mol Gen Genet 146: 167–178

Ohtsubo M, Roberts JM (1993) Cyclin-dependent regulation of G1 in mammalian fibroblasts. Science 259: 1908–1912

Pain B, Melet F, Pain S J., Flickinger T. Raines M, Peyrol S, Moscovici C, Moscovici MG, Kung H-J, Jurdic P. Lazarides F, Samarut J (1991) EGF-R as a hemopoietic growth factor receptor: the c-crbB product is present in erythrocytic cells. Cell 65: 37–46

Parker SB, Eichele G, Zhang P, Rawis A, Sands AT, Bradley A, Olson EN, Harper JW, Elledge SJ (1995) p53-independent expression of $p21^{Cip1}$ in muscle and other terminally differentiating cells. Science 267: 1024–1027

Patel VP, Lodish HF (1987) A fibronectin matrix is required for differentiation of murine erythroleukemia cells into reticulocytes. J Cell Biol 105: 3105–3118

Paul R, Schuetze S, Korzak S, Kozak C, Kozat D (1991) The Sfpi-1 proviral integration site of Friend erythroleukemia encodes the ets-related transcription factor Pu. 1. J Virol 65: 464–467

Peter M, Herskowitz I (1994) joining the complex: cyclin-dependent kinase inhibitory proteins and the cell cycle. Cell 79: 181–184

Pines J, Hunter T (1990) Human cyclin A is adenovirus E1A-asociated protein p60 and behaves differently from cyclin B. Nature 346: 760–763

Polyak K, Lee M-H, Erdjument-Bromage H, Tempst P, Massague J (1994a) Cloning of p27^{Kip1}, a cyclin-dependent kinase inhibitor and a potential mediator of extracellular antimitogenic signals. Cell 78: 59–66

Polyak K, Kato Y-Y, Solomon Mj, Sherr CJ, Massague J, Roberts JM, Koff A (1994b) p27^{Kip1}, a cyclin-cdk inhibitor, links transforming growth factor-β and contact inhibition to cell cycle arrest, Genes Dev 8: 9–22

Quelle DE, Ashmun RA, Shurtleff SA, Kato J-Y, Bar-Sagi D, Roussel MF, Sherr CJ (1993) Overexpression of mouse D-type cyclins accelerates G1 phase in rodent fibroblasts. Genes Dev 7: 1559–1571

Resnitzky D, Gossen M, Bujard H, Reed SI (1994) Acceleration of the G1/S phase transition by expression of Cyclins D1 and E with an inducible system. Mol Cell Biol 14: 1669–1679

Richon VM, Rifkind RA, Marks PA (1992). Expression and phosphorylation of the retinoblastoma protein during induced differentiation of murine erythroleukemia cells. Cell Growth Differ 3: 413–420

Sap J, Munoz A, Damm K, Goldberg Y, Ghysdael J, Leutz A, Beug H, Vennstrom B (1986) The c-erb-A protein is a high-affinity receptor for thyroid hormone. Nature 324: 635–640

Sawada K, Krantz SB, Kans JS, Dessypris EN, Sawyer ST, Glick AD, Civin CI (1987) Purification of human erythroid colony forming units and demonstration of specific binding of erythropoietin. J Clin Invest 80: 357–366

Sawada K, Krantz SB, Dessypris EN, Koury ST, Sawyer ST (1989) Human colony forming units-erythroid do not require accessory cells, but do require direct interaction with insulin-like growth factor I and/or insulin for erythroid development. J Clin Invest 83: 1701–1709

Sawada K, Krantz SB, Dai CH, Koury ST, Horn ST, Glick AD, Civin CI (1990) Purification of human blood burst-forming units erythroid and demonstration of the evolution of erythropoietin receptors. J Cell Physiol 142: 219–230

Schroeder C, Gibson L, Nordström C, Beug H (1993) The estrogen receptor cooperates with the TGFα receptor (c-erbB) in regulation of chicken erythroid progenitor self renewal. EMBO J 12: 951–960

Serrano M, Hannon GJ, Beach D (1993) A new regulatory motif in cell-cycle control causing specific inhibition of cyclin D/CDK4. Nature 366: 704–707

Sherr CJ (1993) Mammalian G1 cyclins. Cell 73: 1059–1065

Sherr CJ (1994) G1 phase progression: cycling on cue. cell 79: 551–555

Skapek SX, Rhee J, Spicer DB, Lassar AB (1995) Inhibition of myogenic differentiation in proliferating myoblasts by cyclin D1-dependent kinase. Science 267: 1022–1024

Soussi T (1988) Nucleotide sequence of a cDNA encoding the chicken p53 nuclear oncoprotein. Nucleic Acids Res 16: 11383

Steinlein P, Wessely O, Meyer S, Deiner EM, Hayman MJ, Beug H (1995) Primary, self-renewing erythroid progenitors develop from BFU-E through activation of both tyrosine kinase and steroid hormone receptors. Curr Biol 5: 191–204

Toyoshima H, Hunter T (1994) p27, a novel inhibitor of G1 cyclin-Cdk protein kinase activity, is related to p21. Cell 78: 67–74

Withers D, Harvey R, Faust J, Melnyk O, Carey K, Meeker T (1991) Characterization of a candidate bcl-1 gene, Mol Cell Bil 11: 4846–4853

Xiong Y, Connolly T, Futcher B, Beach D (1991) Human D-type cyclin. Cell 65: 691–699

Zindy F, Lamas E, Chenivesse X, Sobczak J, Wang J, Fesquet D, Henglein B, Brechot C (1992) Cyclin A is required in S phase in normal epithelial cells. Biochem Biophys Res Commun 182: 1144–1154

Chick Stem Cells

C.D. Stern

1 Introduction

A stem cell is defined as a proliferating cell with the ability to renew itself. In the adult organism, such cells are required to maintain a continuous supply of cells to compensate for cell loss throughout the lifetime of the individual and therefore include the basal proliferating cells that renew the epidermis, specialized cells that maintain the inner lining of the digestive system, progenitors of the spermatozoa in the male, hematopoietic progenitors responsible for ensuring a supply of blood cells, and probably the olfactory epithelium, which produces sensory neurons throughout life. The fertilized egg itself could be considered as a stem cell, since subsequent divisions generate all the somatic cells of the organism as well as germ cells.

In the embryo, there are surprisingly few cases in which the existence of stem cells has been unambiguously demonstrated, other than the progenitors of cells known to have this property in the adult. And yet, embryonic development can in some ways be likened to the maintenance of the hematopoietic (including the immune) system: both must generate many cells and ensure a correct balance of cell diversity to fulfill appropriate functions. Stem cells are therefore

Department of Genetics and Development, College of Physicians and Surgeons of Columbia University, 701 W 168th Street, New York, NY 10032, USA

centrally implied by the title of this volume, which aims to stress the similarities between immunology and embryology as well as to underline the value of the avian embryo in both disciplines.

2 Cell Fate and Asymmetry of Cell Division

A common misconception is that stem cells must be multipotent; that is, they must give rise to different cell types and therefore they cannot be committed to any particular fate. However, this is not necessarily the case. In the adult, some of the best examples of stem cells, such as the basal cells of the skin, appear to be committed to a single fate: that of generating epidermal cells. Their division generates some progeny that continue to divide and other progeny that begin to differentiate and lose the potential to proliferate. Other examples, notably in the hematopoietic system, do include cells that are multipotent, and the initiation of the commitment of their progeny to differentiate is accompanied, at some stage, by selection from a specific subset of fates.

Is the decision between proliferation, differentiation and self-renewal made at the level of single cells or at the level of cell populations? These two modes imply very different mechanisms to ensure a continued supply of stem cells. In the former, each cell division is asymmetric: one daughter must retain the stem cell property while the other daughter enters into a pathway of differentiation. This mode is probably best exemplified by the basal layer of the epidermis (Potten and Morris 1988; Parkinson 1992), where at each cell division one daughter remains in the basal layer and continues to proliferate, retaining its stem cell properties and representing the founder cell of a single epidermal proliferative unit (EPU; see Potten and Morris 1988). In the case in which the decisions are made at the level of cell populations, cell divisions are not necessarily asymmetric; when a stem cell divides, it is conceivable that both daughters retain the stem cell character. This mode is advantageous when a decision from among several possible fates must be made and is probably the mechanism by which the hematopoietic system maintains itself (Sigal et al. 1992; Zipori 1992; Janssen 1993). However, there is little or no direct evidence for this.

The second mechanism requires some signal, external to the cell itself, ensuring that at least one cell retains its multipotency and stem cell character. One possible mechanism, favored by many workers on the bone marrow, is that there are specialized physical niches within the marrow which maintain cells in this special state (Parakh and Kannan 1993; Tamayo et al. 1993; Rosendaal et al. 1994). Cells divide and retain their multipotency and ability to self-renew as long as they remain in this niche, but when progeny leave it, they may lose one or both of these characteristics. In a sense, therefore, both modes of stem cell self-renewal are compatible with the existence of special spatial niches that maintain this characteristic, and the basal layer of the skin is probably one such niche; however, the second mechanism is absolutely dependent on external stimuli.

Perhaps the clearest, and therefore the most extreme, example of stem cells operating at the level of a cell population is that of the so-called embryonic stem cells (ES cells). When an early mammalian embryo (at a stage before the appearance of the blastocoele) is dissociated into single cells and these are placed in culture on a confluent monolayer of feeder cells, they continue to divide without differentiating for a very long time, greatly expanding the cell population. Treatment of such cultures with specific factors, such as retinoic acid, can induce their differentiation into many different recognizable cell types. Under some conditions, cells begin to aggregate into nodules,known as embryoid bodies. If cells from one such body, or indeed subconfluent ES cells, are injected into a young blastocyst, the cultured cells can contribute to any cell lineage in the embryo, including the germ cells. Indeed, it is this property that is exploited in the production of transgenic mice (see JOYNER 1991; ROBERTSON et al. 1992). The case of ES cells represents a clear demonstration that stimuli external to the cells can direct their continued self-renewal or the initiation of cell differentiation and the loss of the stem cell character.

Formally, therefore, a cell can only be defined as a stem cell if it always produces more stem cells. According to this strict definition, asymmetry of cell division is required; ES cells are therefore not "true" stem cells because their divisions are not necessarily asymmetric and because the control of their self-renewal is external to the cell being considered.

In the remainder of this review, I will concentrate first on asymmetric cell divisions and then briefly discuss several candidate cell populations that may have stem cell properties in the avian embryo, concentrating on a specific population of putative stem cells in the "organizer" region of the gastrulating chick embryo (Hensen's node) as a possible example of these. Finally, I will consider whether it will be possible to generate the avian equivalent of ES cells and the future prospects for constructing transgenic chicks.

3 Asymmetric Cell Divisions in Early Development

The "standard" view of cell division is of a cell that divides to generate two identical daughters. However, the nematode *Caenorhabditis elegans*, an organism in which the complete cell lineage during development is known, illustrates the fallacy of this view: out of the 949 somatic cell divisions that produce this animal, no fewer than 807 (85%) are asymmetric, giving rise to daughters that differ from one another in their fate. In this organism at least, asymmetry of cell division is the rule rather than the exception. One example of this is in the progeny of the N neuroblast. This cell divides to give rise to a daughter (A) that differentiates into a neuron and to another neuroblast which, at the next cell division, generates another neuron (B) and whose other daughter undergoes programmed cell death (Fig. 1).

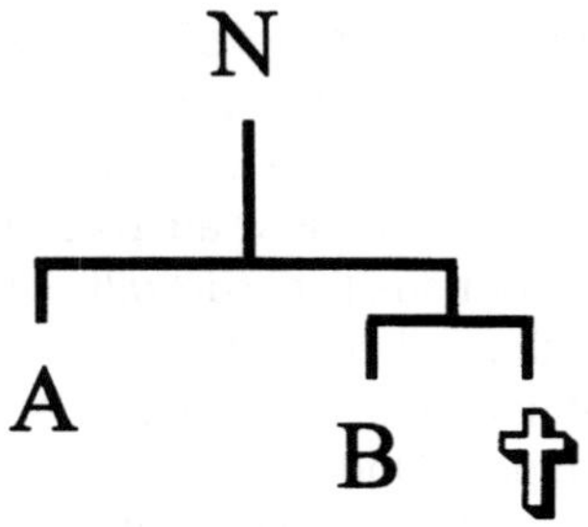

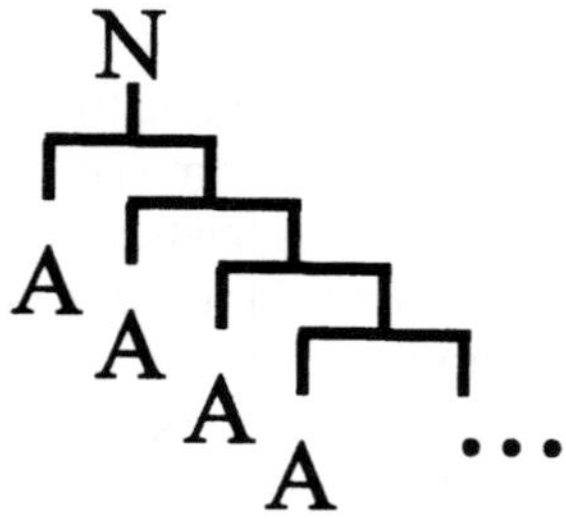

Fig. 1. Progeny of the N neuroblast

Fig. 2. Division of the N neuroblast in *unc–86* mutants

The asymmetry of division of the N neuroblast requires the expression of a POU domain containing transcription factor, *unc-86* (FINNEY and RUVKUN 1990). In *unc-86* mutants, rather than generating the above scheme, the N neuroblast acquires stem cell-like characteristics, so that at each of many subsequent divisions it gives rise to a neuron and another neuroblast (Fig. 2).

These results suggest that the *unc-86* gene product is required to prevent the self-renewal of the N progenitor; that is, it suppresses the stem cell-like character of the cell division. The *unc-86* gene is expressed in 57/302 neurons in different lineages, but always in just one of the two daughters of a cell division and not in the other daughter or the mother cell. However, it is not expressed in the remaining 750 asymmetric cell divisions in the animal, suggesting that other genes are also required to define the extent of this asymmetry.

Genes homologous to *unc-86* exist in vertebrates. One example is *brn-3*, with 85% homology within the POU domain. It is expressed in various neural organs in the chick, mouse, rabbit, monkey and human (XIANG et al. 1993; GERRERO et al. 1993). Another vertebrate POU domain gene with homology to *unc-86* is *Oct-2*, which, in addition to the nervous system, is expressed in B cells and has been implicated in the regulation of immunoglobulin gene transcription and B cell development (FELDHAUS et al. 1994). However, it is not known whether the cells expressing either gene, or their progenitors, divide asymmetrically.

4 Stem Cells of the Vertebrate Nervous System

In the avian embryo, as in mammalian systems, there are cells that have the ability to self-renew in the skin and hematopoietic system as well as during bone development (osteocytes). There is also some evidence that skeletal muscle cell development includes a population wih the capacity to self-renew (QUINN et al. 1985). However, work from several groups indicates that, during early development of the avian and mammalian nervous systems, some of the cells there may also possess stem cell properties. In the optic tectum of the chick, for example, cell lineage analysis reveals that one of the last divisions of a multipotent stem cell-like precursor gives rise to radial glia and that these cells may themselves have stem cell-like properties (GRAY and SANES 1992). Similarly, in the neural tube, some cells are believed to be equivalent to the invertebrate neuroblasts and also be capable of self-renewal (SENDA et al. 1992), although direct evidence for this self-renewal is still lacking. The only exception is perhaps a recent study (DAVIS and TEMPLE 1994) revealing the existence of such a self-renewing population of neuroblasts in the rat forebrain.

Perhaps the best evidence for stem cells in the developing nervous system comes from studies on the neural crest. BAROFFIO et al. (1991) and DUPIN et al. (1993) first suggested, based on clonal analysis of avian neural crest cells, that these cells are multipotent and able to renew themselves, giving rise to multipotent descendants. Similarly in the mammalian neural crest, STEMPLE and ANDERSON (1992) have succeeded in isolating a stem cell-like cell population which is multipotent and whose descendants in vitro are also multipotent, because they give rise to both neurons and glia. Moreover, neural crest derivatives such as sensory neurons in the dorsal root ganglia can respond to stem cell factor, which induces their growth (CARNAHAN et al. 1994).

In all these studies, however, cells have been followed for only a few divisions, and it is therefore unknown whether the progenitors continue to have stem cell properties for a long time. In this sense, neural crest cells appear to be more akin to the ES cells of the mouse, where the stem cell property is only retained in vitro for either a few cell divisions or for as long as the culture conditions remain permissive. In one in vivo study, the multipotentiality of the avian neural crest has been demonstrated by short-term single cell lineage analysis (BRONNER-FRASER and FRASER 1988), but there has been as yet no direct demonstration of self-renewal in vivo for neural crest cells.

5 Putative Stem Cell Progenitors of the Somites and Notochord of the Avian Embryo

Cells destined to form the somites and notochord of the embryo come from a small region in the anterior part of the primitive streak and Hensen's node (SELLECK and STERN 1991; SCHOENWOLF 1992). How do they generate the large number of cells required to make all the somites and the whole length of the notochord? At the time of its formation, a somite comprises about 2000 cells (MENKES and SANDOR 1969), and the cells divide about every 10h (KEYNES and STERN 1988; PRIMMETT 1988; PRIMMETT et al. 1989). Since about 50 somites form from each side of the embryo, formation of the somitic mesoderm involves some 18 million cells. How is this vast number of cells generated over a few days from such a small region of the early gastrula stage embryo, which also gives rise to other structures at the same time? And what mechanisms ensure a continuous supply of cells, such that segmentation can proceed smoothly, with new somites forming every 100 min? One possibility is that the presumptive somite cells of Hensen's node have stem cell properties; that is, they renew themselves at the same time as they give rise to committed progeny. And, given the large number of progeny that they must generate, if they are indeed stem cells, they must retain this property for a considerable time.

Since cells destined to form somites and notochord leave Hensen's node in a continuous fashion, some mechanism must punctuate the stream of cells so that groups of prospective somitic cells adhere together to form a somite. It is thought that the cell division cycle is involved in this punctuation on the basis of heat shock studies, measurements of cell cycle length, pharmacological experiments and analysis of expression of heat shock proteins (PRIMMETT et al. 1988, 1989; STERN et al. 1988). The cell cycle was also suggested to be involved in some way in the elongation of the notochord (SELLECK and STERN 1992b), but there is no direct evidence for this.

SELLECK and STERN (1991) mapped the descendants of cells in Hensen's node (the anterior tip of the primitive streak) of the gastrulating embryo, using carbocyanine dyes (DiI, DiO) to mark small groups of cells and lysinated rhodamine-dextran to follow the progeny of single cells. One unexpected finding was that the node contains cells that contribute only to the medial halves of the somites. Their lateral halves were found to come from progenitors located about 100 μm further posteriorly in the primitive streak (Fig. 3). At the time, there was no obvious reason to expect this on functional or embryological grounds. But at about the same time an independent study by ORDAHL and LE DOUARIN (1992) reported that the medial halves of somites normally contribute only to the axial musculature, while their lateral halves contribute to muscles of the body wall and limbs.

In addition to progenitor cells for the medial halves of the somites, Hensen's node also contains prospective notochord cells, situated in the anterior, median quadrant. Progenitors of the gut endoderm and of the floor plate of the neural tube are also present in the node, but more widely distributed. The intermediate region of the node situated between the medial somite precursors and the prospective notochord area contains cells that contribute to both notochord and somites, as

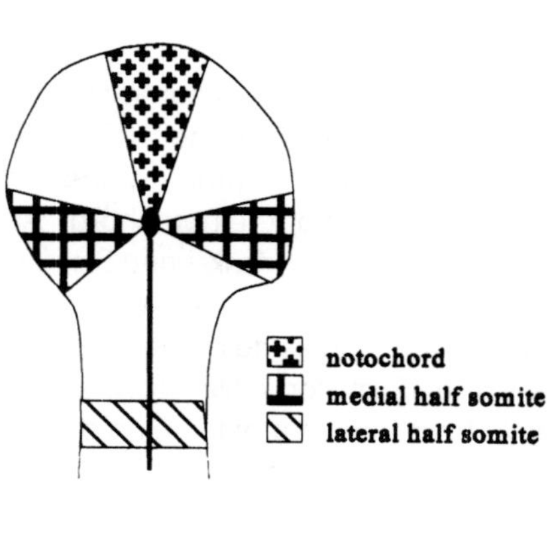

Fig. 3. A summary fate map of Hensen's node of the chick embryo at stage 4. Adapted from the results of SELLECK and STERN (1991)

revealed by analysis of descendants of single marked cells (SELLECK and STERN 1991). These are therefore pluripotent precursor cells. It was suggested (SELLECK and STERN 1992a,b; STERN et al. 1992) that these pluripotent precursors give rise to committed progenitors of medial half somite and notochord, situated in the adjacent regions. These findings beg the question: do the pluripotent precursor cells in the intermediate region of the node have stem cell properties; that is, are they able to renew themselves? Such a mechanism would ensure the maintenance of a progenitor population of constant size which would be responsible for generating the large number of cells required to make notochord and somites spanning the entire length of the embryo.

5.1 Evidence for Stem Cells from Heat Shock Experiments and Cell Lineage Analysis

When chick embryos are given a single, short heat shock, an unexpected result is obtained: discrete anomalies of somite development are seen at regular intervals along the axis of the embryo, appearing every seven or so segments (PRIMMETT et al. 1988). The anomalies consist of an abnormal (either large or small) number of cells being allocated to the defective somites. Because somite formation occurs sequentially, this suggests that heat shock affects some repeated cyclic process. A pair of somites forms every 100 min or so (MENKES and SANDOR 1969), therefore the time interval between these anomalies corresponds to groups of cells that are about 10h apart. An obvious candidate for the repeated process is the cell division cycle. This was measured by [^{3}H] thymidine pulse and chase, which confirmed that presumptive somite cells divide every 10h (PRIMMETT et al. 1989).

These findings suggest that cells that segment at the same time as each other divide relatively synchronously. This is consistent with measurements of the mitotic index of presumptive somite cells in the segmental plate (STERN and BELLAIRS 1984): a large peak of cells in the M-phase of the cycle is seen just before segmentation, at the anterior tip of the segmental plate, another in the middle and a third one at the posterior end of the plates. Since the segmental plate contains 13 presumptive somites (JACOBSON and MEIER 1986), these peaks of mitosis are separated by about seven somites.

Thus, the cell cycle is probably involved somehow in punctuating the continuous stream of cells into a discrete pattern of somites (STERN et al. 1988; PRIMMETT et al. 1989). Further evidence for this comes from single cell lineage analysis of presumptive somite cells at earlier stages of development in Hensen's node (SELLECK and STERN 1991, 1992b). When a single cell in the presumptive medial half somite region of the node is marked by intracellular injection of lysinated rhodamine-dextran (LRD), its progeny appears clustered in small groups, separated by about five to seven somites (Fig. 4). Interestingly, when the injected cell is in a region that contributes to the notochord, repeated clusters of labeled cells are also seen, but these are much closer together: 1.5–2 somite-lengths apart. This

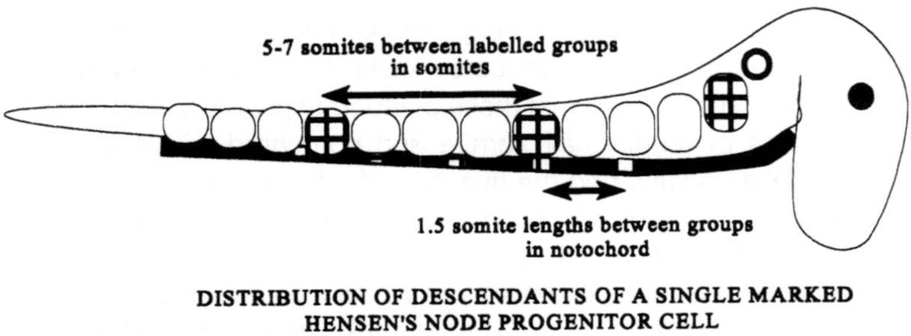

Fig. 4. After injecting the lineage tracer LRD into a single cell in the intermediate region of Hensen's node, its precursors are sometimes found in regularly spaced clusters in the somites and/or notochord. The spacing is 5–7 somite lengths in the somitic mesoderm and 1.5 somite equivalents in the notochord. Based on the results of SELLECK and STERN (1992b)

suggests that the notochord precursors divide about 2.5 times faster than the somite progenitors and predicts a doubling time of about 3.5–4h.

Since, as discussed above, there are cells in the intermediate region of the node that contribute to both notochord and somite, then it is expected that they will divide also at the faster rate of 3.5–4h and that the descendants slow down to a cycle time of about 10h if they become committed to a somite fate.

5.2 Evidence for Asymmetric Cell Division

Are the divisions of the putative stem cells in Hensen's node asymmetric? Three modes of cell division can be envisaged (in each of the three schemes shown in Fig. 5; the interrupted line shows the limits of the node where the dividing progenitor cells are located. The "progenitor" is considered at the last cell division before its daughters begin to leave the node).

In the first mode (Fig. 5A), the progenitor cell in Hensen's node behaves as a stem cell because it renews itself, but when it stops doing so it gives rise to two differentiating daughters which leave the node region to start the segmentation process. This scheme would not lead to periodic arrangements of the daughters but to one or at most two clusters of labeled cells at the site at which the daughters emerge. It is important to realize that in the period between the division shown and the next one of the progenitor cell, other progenitors (not labeled) also divide. This accounts for the continuous supply and the movement of cells from the node to more anterior regions of the embryo as segmentation proceeds.

In the second mode (Fig. 5B), the progenitor cell does not behave as a true stem cell. Even if both daughters do continue to divide, as in the first scheme, this mechanism will produce, at most, two clusters of labeled cells. In order to account for periodic labeled clusters in this and the previous scheme, unlabeled cells must become interspersed with the daughters of those emerging in a very precise way.

The third mode (Fig. 5C) is consistent with the results of the experiments described above, and also uniquely and simply accounts for the existence of the

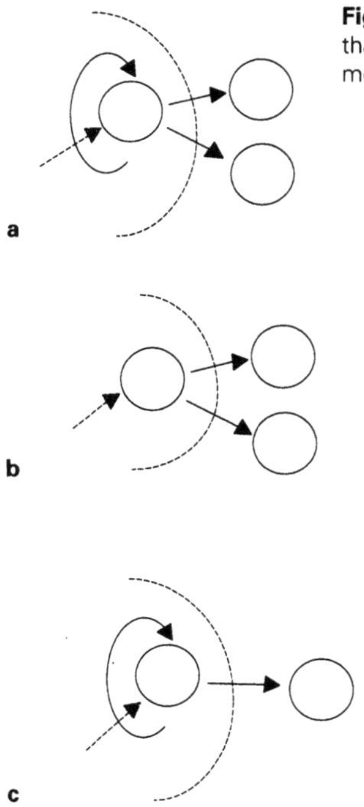

Fig. 5a–c. Three modes of cell division for progenitors in Hensen's node that can generate periodically spaced clusters of cells destined for mesodermal tissues of the embryo

stem cells as postulated. At each cell division of the progenitor cell, only one daughter (which itself continues to divide) leaves the node region and becomes a founder cell for just one cluster of labeled cells. The next division of the progenitor, still situated in the node as it regresses, will yield a similar cell, which will generate the next cluster. This scheme is the only one of the three presented which is, in a sense, inexhaustible, since it can continue to produce clusters of labeled cells as long as the progenitor continues to divide. The hypothesis proposes that the spacing between consecutive clusters is related directly to the cell division cycle of the progenitor cell.

Only two of the above schemes (the first and third) are compatible with a stem cell character of the progenitor cell. All three could, in principle, generate multiple periodic clusters of labeled cells from a single precursor as seen (SELLECK and STERN 1991), but the third scheme seems the simplest because it does not require other neighboring cells to intersperse with dividing cells outside the node in a very precise and predictable way.

5.3 Hensen's Node Stem Cells and Mesoderm Induction

The discovery of pluripotent precursor cells in Hensen's node indicates that individual cells at the end of the primitive streak stage still contribute progeny to both ectodermal (floor plate, neural tube) and mesodermal (somites, notochord) derivatives (SELLECK and STERN 1991). How do cells decide between these fates? From studies in amphibians, it is generally believed that responsiveness to mesodermal induction is lost at the beginning of gastrulation (see GREEN and SMITH 1991 for review). The situation in Hensen's node of the amniote embryo is therefore either different from *Xenopus*, or the appropriate region of the frog (dorsal lip of the blastopore) has not been adequately explored.

In amphibians, there is good evidence for the involvement of the peptide growth factor activin in mesodermal induction. Some of our recent results (STERN et al. 1995) suggest that activin could play a role in allocating cells to mesodermal fates during the later phases of gastrulation in the amniote embryo. First, we have cloned two activin receptors homologous to the amphibian ActRIIA and IIB. Transcripts of both genes are first expressed when the primitive streak appears; after this, cActRIIA is concentrated in Hensen's node of the full-length streak. Moreover, when different tissues are treated with activin and allowed to differentiate in culture, the primitive streak is found to be the most responsive tissue, giving rise to all mesodermal derivatives, including the most axial/dorsal types, in a concentration-dependent manner, as has been found in the frog blastula. These results argue that the role of activin-related signaling pathways in normal development may be confined to rather late stages of mesoderm formation. A study of the behavior of single cells in the node and their progeny in an amniote embryo will provide valuable information complementing present knowledge based on amphibian mesoderm induction and will be essential to provide direction for future research on the molecular bases of mesoderm induction in amniotes.

6 Chick Embryonic Stem Cells? Towards the Production of Transgenic Birds

Despite the obvious desirability of an avian equivalent of the mouse ES cell system for the production of targeted mutations (see SHUMAN 1991), and many attempts to generate them, it has so far proved impossible to produce stable cell lines that remain multipotent and which, upon further passages in culture, retain the ability to populate an entire host embryo including the germ line. The reasons for this remain unclear. However, recent progress has been made in culturing cells on feeder layers which, when supplied with various cytokine supplements to the medium, continue to divide and resemble mammalian ES cells both morphologically and in terms of the expression of many molecular markers. When treated with substances like retinoic acid, they can be induced to differentiate and then appear to give rise to most if not all possible recognizable cell types (PAIN et al. 1995).

Although previous studies have found relatively limited colonization of early embryos by implanted cells (WATANABE et al. 1992), spectacular advances have been made recently, dramatically improving the efficiency of production of chimeric embryos made with donor cells and a host preprimitive streak stage blastoderm. The method consists of treating the host embryo with X-rays to compromise the early blastodermal cells (ETCHES et al. 1993; FRASER et al. 1993). These two developments promise, in the not too distant future, the possibility of producing transgenic birds in almost routine manner. One of the great attractions of the avian system is that it allows sophisticated lineage analysis and manipulation of early and later embryonic cells. It is to be hoped that future studies will benefit from a combination of targeted mutations produced by the new technology and well-thought-out embryonic manipulations. With this combination, the avian embryo will no doubt become an even more powerful system to study early developmental mechanisms than many others in current use.

Acknowledgments. Our current research on this topic is supported by a grant from the National Institutes of Health.

References

Baroffio A, Dupin E, Le Douarin NM (1991) Common precursors for neural and mesectodermal derivatives in the cephalic neural crest. Development 112: 301–305
Bronner-Fraser M, Fraser SE (1988) Cell lineage analysis reveals multipotency of some avian neural crest cells. Nature 335: 161–164
Carnahan JF, Patel DR, Miller JA (1994) Stem cell factor is a neurotrophic factor for neural crest derived chick sensory neurons. J Neurosci 14: 1433–1440
Davis AA, Temple S (1994) A self-renewing multipotential stem cell in embryonic rat cerebral cortex. Nature 372: 263–266
Dupin E, Sextier-Sainte-Claire Deville F, Nataf V, Le Douarin NM (1993) The ontogeny of the neural crest. C R Acad Sci [III] 316: 1062–1081
Etches RJ, Carsience RS, Clark ME, Fraser RA, Toner A, Verrinder Gibbins AM (1993) Chimeric chickens and their use in manipulation of the avian genome. Poultry Sci 72: 882–889
Feldhaus AL, Klug CA, Arvin KL, Singh H (1994) Targeted disruption of the Oct-2 locus in a B cell provides genetic evidence for two distinct cell type specific pathways of octamer element mediated gene activation. EMBO J 12: 2763–2772
Finney M, Ruvkun G (1990) The unc-86 gene product couples cell lineage and cell identity in C. elegans. Cell 63: 895–905
Fraser RA, Carsience RS, Clark ME, Etches RJ (1993) Efficient incorporation of transfected blastodermal cells into chimeric embryos. Int J Dev Biol 37: 381–385
Gerrero MR, McEvilly RJ, Turner E, Lin CR, O'Connell S, Jenne KJ, Hobbs MV, Rosenfeld MG (1993) Brn-3.0: a POU domain protein expressed in the sensory, immune and endocrine systems that functions on elements distinct from known octamer motifs. Proc Natl Acad Sci USA 90: 10841–10845
Gray GE, Sanes JR (1992) Lineage of radial glia in the chicken optic tectum. Development 114: 271–283
Green JBA, Smith JC (1991) Growth factors as morphogens: do gradients and thresholds establish body plan? Trends Genet 7: 245–250
Jacobson A, Meier S (1986) Somitomeres: the primordial body segments. In: Bellairs R, Ede DA, Lash JW (eds) Somites in developing embryos. Plenum, New York, pp 1–16
Janssen WE (1993) Peripheral blood and bone marrow hematopoietic stem cells: are they the same? Sem in Oncol 20: 19–27
Joyner AL (1991) Gene targeting and gene trap screens using embryonic stem cells: new approaches to mammalian development. Bioessays 13: 649–656
Keynes RJ, Stern CD (1988) Mechanisms of vertebrate segmentation. Development 103: 413–429
Menkes B, Sandor S (1969) Researches on the development of axial organs. Rev Roum Embryol Cytol 6: 65–88

Ordahl CP, Le Douarin NM (1992) Two myogenic lineages within the developing somite. Development 114: 339–353
Pain B, Cochran M, Carck ME, Shen M, Etches R, Samarut J (1995) Identification et culture de cellules embryonnaires totipotentes aviaires. Abstract in Colloque Soc Biol Cell France, Lyon, p 40
Parakh KA, Kannan K (1993) Demonstration of a ubiquitin binding site on murine hemopoietic progenitor cells: implication of ubiquitin in homing and adhesion. Br J Haematol 84: 212–218
Parkinson EK (1992) Epidermal keratinocyte stem cells: their maintenance and regulation. Sem in Cell Biol 3: 435–444
Potten CS, Morris RJ (1988) Epithelial stem cells in vivo. J Cell Sci [Suppl] 10: 45–62
Primmett DRN (1988) Morphogenesis of the vertebral column in the chick. PhD thesis, University of Oxford
Primmett DRN, Stern CD, Keynes RJ (1988) Heat-shock causes repeated segmental anomalies in the chick embryo. Development 104: 331–339
Primmett DRN, Norris WE, Carlson GJ, Keynes RJ, Stern CD (1989) Periodic segmental anomalies induced by heat-shock in the chick embryo are associated with the cell cycle. Development 105: 119–130
Quinn LS, Holtzer H, Namerhoff M (1985) Generation of chick skeletal muscle cells in groups of 16 from stem cells. Nature 313: 692–694
Robertson EJ, Conlon FL, Barth KS, Costantini F, Lee JJ (1992) Use of embryonic stem cells to study mutations affecting postimplantation development in the mouse. Ciba Found Symp 165: 237–255
Rosendaal M, Green CR, Rahman A, Morgan D (1994) Up-regulation of the connexin-43 gap junction network in haemopoietic tissue before the growth of stem cells. J Cell Sci 107: 29–37
Schoenwolf GC (1992) Morphological and mapping studies of the paranodal and postnodal levels of the neural plate during chick neurulation. Anat Rec 233: 281–290
Selleck MAJ, Stern CD (1991) Fate mapping and cell lineage analysis of Hensen's node in the chick embryo. Development 112: 615–626
Selleck MAJ, Stern CD (1992a) Commitment of mesoderm cells in Hensen's node of the chick embryo to notochord and somite. Development 114: 403–415
Selleck MAJ, Stern CD (1992b) Evidence for stem cells in the mesoderm of Hensen's node and their role in embryonic pattern formation. In: Lash JW, Bellairs R, Sanders EJ (eds) Formation and differentiation of early embryonic mesoderm. Plenum, New York, pp 23–31
Senda T, Hirabayashi T, Fukazawa K, Fujita H (1992) Cytoskeletal architecture of the matrix cell and neuroblast in the neural tube of the chick embryo. Arch Histol Cytol 55: 45–55
Shuman RM (1991) Production of transgenic birds. Experientia 47: 897–905
Sigal SH, Brill S, Fiorino AS, Reid LM (1992) The liver as a stem cell and lineage system. Am J Physiol 263: G139–G148
Stemple DL, Anderson DJ (1992) Isolation of a stem cell for neurons and glia from the mammalian neural crest. Cell 71: 973–985
Stern CD, Bellairs R (1984) Mitotic activity during somite segmentation in the chick embryo. Anat Embryol 169: 97–102
Stern CD, Fraser SE, Keynes RJ, Primmett DRN (1988) A cell lineage analysis of segmentation in the chick embryo. Development 104 [Suppl]: 231–244
Stern CD, Hatada Y, Selleck MAJ, Storey KG (1992) Relationships between mesoderm induction and the embryonic axes in chick and frog embryos. Development 1992 [Suppl]: 151–156
Stern CD, Yu RT, Kakizuka A, Kintner CR, Mathews LS, Vale WW, Evans RM, Umesono K (1995) Activin and its receptors during gastrulation and the later phases of mesoderm development in the chick embryo. Dev Biol 172: 192–205
Tamayo E, Charbord P, Li J, Herve P (1993) A quantitative assay that evaluates the capacity of human stromal cells to support granulomonopoiesis in situ. Stem Cells 12: 304–315
Watanabe M, Kinutani M, Naito M, Ochi O, Takashima Y (1992) Distribution analysis of transferred donor cells in avian blastodermal chimeras. Development 114: 331–338
Xiang M, Zhou L, Peng YW, Eddy RL, Shows TB, Nathans J (1993) Brn-3b: a POU domain gene expressed in a subset of retinal ganglion cells. Neuron 11: 689–701
Zipori D (1992) The renewal and differentiation of hemopoietic stem cells. FASEB J 6: 2691–2697

Modulations of Cellular Interactions During Development of the Neural Crest: Role of Growth Factors and Adhesion Molecules

J.L. Duband, M. Delannet, F. Monier, S. Garret, and N. Desban

1 Introduction

The recent progress in the description of new growth, survival, differentiation and adhesion factors, the characterization of their biological activities, combined with the identification of new transcription factors have led to the emergence of new concepts that allow development of new strategies for elucidating the mechanisms that control embryonic development. The neural crest is certainly one of the embryonic systems that has benefited most from this burst of knowledge.

The neural crest is both a curiosity and a spectacular invention of vertebrates. It consists of a population of precursor cells endowed with striking migratory properties and possessing an extraordinary range of potentialities. The neural crest develops all along the embryonic axis, in the neural folds, at the boundary between the neural plate and the superficial ectoderm. However, neural crest cells do not remain at their site of origin after appearance. At a precise stage of embryonic development, e.g., soon after neural tube closure in birds, they venture out of the neural epithelium and disperse within the embryo along

Laboratoire de Biologie Cellulaire du Dévelopement, Institut Jacques Monod, Université Paris 7-Denis Diderot, 2 place Jussieu, 75251 Paris Cedex 05, France. Present address: Laboratoire de Biologie Moléculaire et Cellulaire du Développement, Université Peirre et Marie Curie, Bât. C-30, 9 quai Saint Bernard, 75005 Paris, France

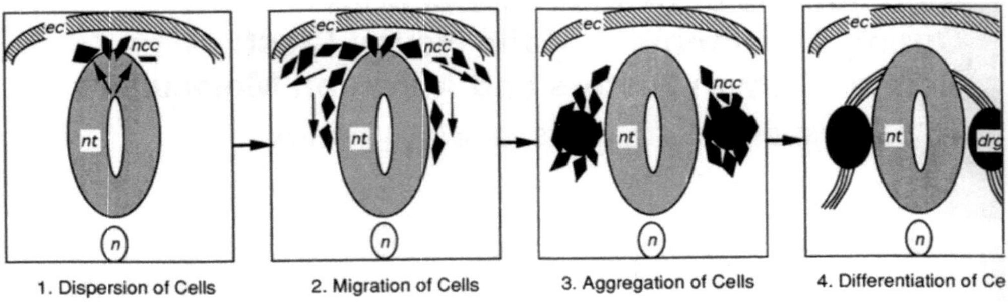

Fig. 1. The four major steps of neural crest development: (1) dispersion from the neural tube; (2) migration in defined migratory paths; (3) aggregation into the primordium of peripheral ganglia; and (4) cellular differentiation. *Lightly stippled area*: neural tube (*nt*); *hatched area*, ectoderm (*ec*); *heavily stippled area*, neural crest cells (*ncc*). *drg*, dorsal root ganglion; *n*, notochord

defined migratory pathways. After migration, neural crest cells settle in various regions of the embryo to generate a wide spectrum of derivatives ranging from cells that are truly neural (i.e., neurons of the peripheral nervous system) to others that are essentially mesodermal (Fig. 1; see for reviews, Le Douarin 1982; Le Douarin et al. 1993; Stemple and Anderson 1993).

The neural crest is regionalized such that cells originating from different axial levels undergo migration at different developmental stages, follow distinct migratory pathways, and produce defined subsets of derivatives (Fig. 2). In birds and mammals, four major regions can be identified and have been designated as cranial, vagal, truncal and lumbosacral. The cranial level corresponds to the forebrain, midbrain and anterior hindbrain. Cranial neural crest cells migrate laterally under the ectoderm to reach the ventral regions of the head where cells contribute to connective tissues and skeleton of the face (the so-called mesectoderm or ectomesenchyme) and to the ciliary and some cranial sensory ganglia. The vagal region corresponds to the posterior hindbrain up to the seventh somite, the term vagal deriving from the vagus nerve which emerges from the neural tube in this region. At this level, neural crest cells follow both a lateral pathway, to provide chiefly mesectodermal tissues and cranial sensory ganglia, and a ventral route, leading them to the periaortic region and the gut where they give rise primarily to sympathetic superior cervical ganglion and to the enteric nervous system. In the trunk region (from the eighth somite to the 28th somite), neural crest cells migrate essentially along a ventral route and differentiate into the spinal sensory ganglia, the sympathetic chain and the adrenal medulla. Finally, in the lumbosacral region (from the 29th somite to the caudal end of the animal), although the migratory pathways of neural crest cells are not known precisely, it is clear that they contribute, like their vagal counterparts, to the enteric nervous system as well as to spinal and sympathetic ganglia.

The diversity of cell types derived from the neural crest makes this cell population an extremely attractive system for investigating the processes of cell commitment and lineage diversification in vertebrates, and, more specifically, in

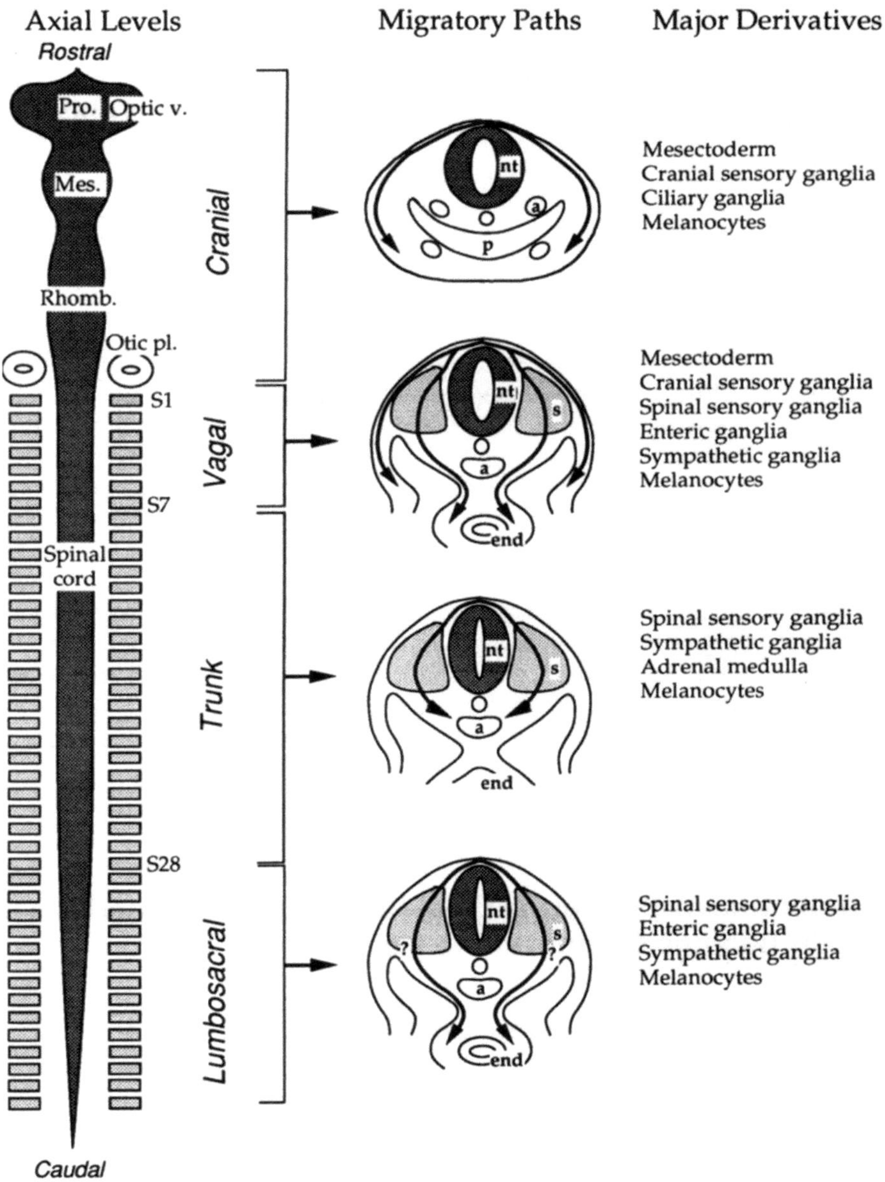

Fig. 2. The neural axis of an embryo showing the different portions of the central nervous system, i.e., prosencephalon (*Pro*), mesencephalon (*Mes.*), rhombencephalon (*Rhomb.*) and spinal cord, the otic placodes (*otic pl.*), the optic vesicles (*optic v.*), the paravertebral mesoderm (*somites S1*) and the corresponding neural crest regions: cranial, vagal, trunk and lumbosacral. The most representative derivatives of the neural crest populations originating from each region are indicated along with the principal migration routes of neural crest cells. *a*, arota; *end*, endoderm; *g*, gut; *nt*, neural tube; *p*, pharynx; *s*, somite

the nervous system (see article of N. Le Douarin, this volume). In this respect, because of the analogies between the neural crest system and hematopoietic cells, the term neuropoiesis has been proposed to account for the diverse range of neural cell types generated from this structure (Anderson 1989). However, due its unique migratory properties, the neural crest provides a provocative system for examining the mechanisms underlying cell locomotion. Finally, a nonnegligible interest of this system resides in the fact that neural crest cells can be isolated from the embryo and grown in culture under conditions that allowing investigators to mimic in vitro the different morphogenetic events that accompany the development of this structure in vivo, i.e., separation from the neural tube, migration, and differentiation.

Neural crest morphogenesis can be subdivided into four phases: (1) initiation of migration when cells segregate from the neural tube to become mesenchymal; (2) dispersion of cells through migratory pathways; (3) cessation of movement when cells come to reside at their final position; and (4) differentiation of cells (Fig. 1). The first phase is a typical example of an epithelium-to-mesenchyme transition. Indeed, presumptive neural crest cells are an integral part of the neural epithelium prior to migration and, once they are separated from it, they organize as a mesenchyme and locomote as individual cells. After migration, cessation of displacement is often accompanied by regroupment of neural crest cells into compact clusters of cells immediately followed by cellular differentiation. Thus, the migration phase starts with and results from an epithelium-to-mesenchyme transition and terminates with the reverse transition. Here, we shall review our present knowledge on the formation of the neural crest and on the mechanisms that cause modulations of cellular adhesiveness at the onset and cessation of displacement.

2 Definition of the Neural Crest: When Do Neurectodermal Cells Become Committed to Neural Crest Cells?

In amphibians, the neural crest population forms a transient ridge on the dorsal surface of the neural tube before migration, hence the term neural crest. In birds and mammals, however, neural crest cells do not form such a ridge and are not morphologically distinct from the rest of the neural tube prior to migration; consequently, they can be identified with certainty only at the onset of dispersion. This raises the intriguing question as to whether neural crest cells are segregated and committed prior to separation from the neural tube.

Numerous attempts have been made to find cellular markers that would specifically label neural crest cells. These markers include acetylcholine esterase activity (Cochard and Coltey 1983), wheat germ agglutinin (Chan and Tam 1988) and particular histological staining techniques (Nichols 1981). However, none of these procedures have allowed unambiguous discrimination of neural crest cells from their neighboring neural tube cells; thus it has not been possible to define infallibly the boundaries of the presumptive neural crest region. More recently, single cell lineage analyses have clearly established that, in all the species examined so far among amphibians, birds and mammals, cells situated in the dorsal aspect of the neural tube at trunk levels are able to give rise to descendants both in the neural tube and in the neural crest (Bronner-Fraser and Fraser 1989; Collazo et al. 1993; Serbedzija et al. 1994). This suggests that the neural crest is not a segregated population within the neural tube and that the fate of dorsal neural tube cells is not fixed before neural crest cell emigration.

Two series of experiments based on microsurgical approaches are clearly in favor of the absence of precocious segregation of neural crest cells from the neural tube and also from the ectoderm. In the axolotl, an amphibian urodele, neural crest cells can develop ectopically at the boundary between juxtaposed fragments of neural plate and ectoderm taken from regions that would not normally participate in formation of neural crest, indicating that local interactions, rather than classical neural induction, appear to be responsible for the formation of neural folds and neural crest (Moury and Jacobson 1989). Interestingly, in this system, both the ectoderm and the neural plate contribute to neural crest cells (Moury and Jacobson 1990; Selleck and Bronner-Fraser 1995). In the avian embryo, neural crest cells can reconstitute normally after ablation at cranial levels of considerable portions of both premigratory neural crest and neural tube either unilaterally or bilaterally until several hours after the normal onset of neural crest cell emigration. Furthermore, cell tracing using vital dye labeling clearly demonstrated that the reconstituted neural crest cells derive primarily from neural tube cells immediately ventral to the ablated region and not from adjacent neural crest cells (Scherson et al 1993). These results demonstrate that, at least at cranial levels of the avian embryo, the neural tube has a remarkable capacity to reform neural crest following either minor or major ablation and suggest that the dorsoventral commitment of the neural epithelium is not fixed precociously in this region. Therefore, the formation of the neural crest and its definition may reside essentially in the physical separation of this cell population from the neural tube. Whether this event is sufficient to modify irreversibly the commitment of neural crest cells with respect to neural tube and ectodermal cells remains, however, to be established. Nonetheless, although neural crest cells may not be totally committed before emigration, certain mechanisms of specification of these cells may operate precociously before initiation of migration, as indicated by the recent discovery of *dorsalin-1* and *slug* (see below; Basler et al. 1993; Nieto et al. 1994).

3 Neural Crest Cell Emigration out of the Neural Tube: A Problem of Hierarchy of Adhesion Molecules?

The proces of neural crest cell separation from the neural epithelium is precisely regulated, both spatially and temporally. As in vertebrates, the embryo develops progressively from anterior (or rostral) to posterior (or caudal), the neural crest also follows an overall rostrocaudal pattern of development. Cells separate from the neural epithelium in a wave that begins in the head, at the midbrain level, and extends essentially more caudally, in the hindbrain and then progressively in the trunk. In some species, e.g., in amphibians, separation of the neural crest from the neural tube appears at all axial levels to result directly from the apposition and subsequent fusion of the neural folds occurring at the closure of the neural tube (LÖFBERG and AHLFORS 1978; SPIETH and KELLER 1984). However, in most species, the emergence of neural crest cells is not in strict synchrony with the progressive closure of the neural tube along the rostrocaudal axis. Thus, separation from the neural epithelium may occur either long before neural folds come in contact, e.g., at the cranial level in mammals (NICHOLS 1981), or at the time when the folds are fusing, e.g., at the cranial level in birds (DUBAND and THIERY 1982; TOSNEY 1982), or finally long after neural fold fusion and separation of the overlying ectoderm from the neural tube, e.g., in the trunk both in mammals and in birds (TOSNEY 1978; ERICKSON and WESTON 1983). It is noteworthy that, in the trunk, the wave of emigration is approximately, but not exactly, parallel to the wave of segmentation of the axial mesoderm into somites (NEWGREEN and ERICKSON 1986). As it is apparently not consistently related to any of the morphogenetic events that take place during the same period of time, the pattern of initiation of neural crest cell migration has long appeared intriguing and somewhat difficult to interpret, particularly regarding the regulatory and triggering processes.

Morphological studies have permitted a detailed description of the main cellular events that accompany the onset of neural crest cell migration. A number of basic events can be recognized in the processes leading to the separation of the neural crest population, irrespective of their level of origin along the rostrocaudal axis. Typically, these events are similar to those occuring in any epithelium-to-mesenchyme transition, i.e., changes in cell shape, and alterations in cell-cell interactions and in the organization and composition of the extracellular matrix at the cell surface (Fig. 3). However, the time course of the occurrence of these changes may differ significantly in the cranial and truncal regions, indicating that neural crest cells originating from different axial levels may undergo morphogenesis in decidedly different manners (Fig. 4).

Changes in cell-cell adhesiveness among presumptive neural crest cells accompany the process of neural crest cell emigration and are supposed to allow release of the cells. At least in the avian embryo, cells of the dorsal neural tube initially extending to the luminal side possess adherens junctions (BANCROFT and BELLAIRS 1976; TOSNEY 1978; NEWGREEN and GIBBINS 1982; TOSNEY 1982; DUBAND et al. 1988). The cell adhesion molecule N-cadherin/A-CAM is concentrated in these junctions, but it is also found in the basolateral surface of the neural epithelial cells (HATTA et al. 1987; DUBAND et al. 1988; AKITAYA and BRONNER-FRASER 1992).

There is tantalizing evidence that, at the trunk level, loss of intercellular junctions and N-cadherin expression among the cells accompanies the detachment of neural crest cells from the neural tube, whereas at the cranial level both events occur well before separation of cells (NEWGREEN and GIBBINS 1982; HATTA

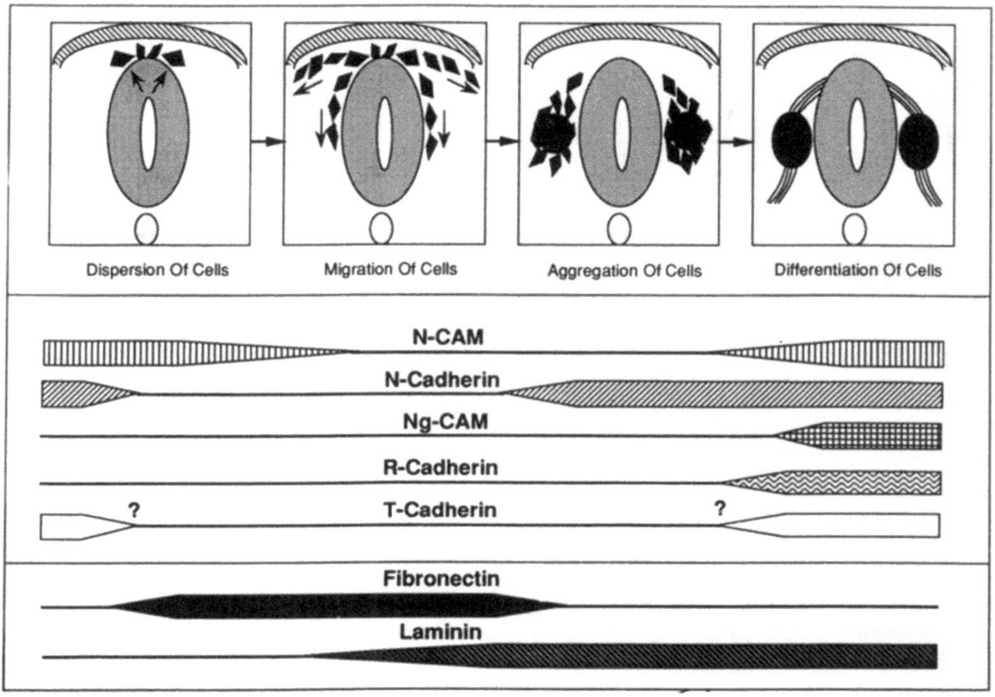

Fig. 3. The modulations in the expression of cell adhesion molecules and in the substratum adhesion properties of neural crest cells during the major steps of their development. Dispersion from the neural tube coincides with the disappearance of N-cadherin first, followed by that of N-CAM and with the acquisition of adhesion to fibronectin. Migration is correlated with low, if any, expression of cell adhesion molecules combined with a constantly strong adhesion to fibronectin and a progressively greater and greater adhesion to laminin. During aggregation into the primordium of peripheral ganglia, adhesion to fibronectin, but not to laminin, declines abruptly concommitant with the reexpression of N-cadherin. Finally, cellular differentiation is accompanied by the sequential expression of new cell adhesion molecules, Ng-CAM/L1 and R-cadherin. The time course of expression of T cadherin by neural crest cells has been tentatively indicated although it is not known precisely. *Lightly stippled area*: neural tube (*nt*); *hatched area*, ectoderm (*ec*); *heavily stippled area*, neural crest cells (*ncc*); *drg*, dorsal root ganglion; *n*, notochord

et al. 1987; DUBAND et al. 1988; AKITAYA and BRONNER-FRASER 1992). Beside N-cadherin, neural crest cells express other cell-adhesion molecules prior to migration, including N-CAM and T-cadherin (THIERY et al. 1982; DUBAND et al. 1985; RANSCHT and BRONNER-FRASER 1991; AKITAYA and BRONNER-FRASER 1992). While the available information about T-cadherin expression is only partial and does not allow precise determination of whether it disappears at the stage of emigration, it is well established that N-CAM is not readily lost at the onset of migration, but declines gradually as migration proceeds (THIERY et al. 1982; DUBAND et al. 1985; AKITAYA and BRONNER-FRASER 1992). Interestingly, neural crest cell emigration is delayed in the *splotch* mouse embryo, a mutant carrying mutations in the homeodomain of the Pax-3 gene and exhibiting failure of neural tube closure,

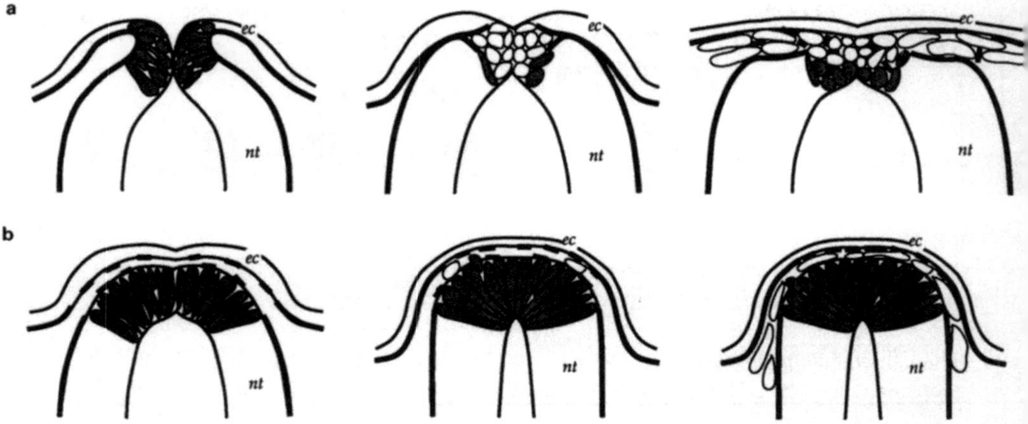

Fig. 4a,b. The major adhesive events occurring during separation of neural crest cells from the neural epithelium in the chick embryo at the cranial (**a**) and trunk (**b**) levels. The major steps of the neural tube closure and neural crest cell emigration are shown in each case. While at the cranial level, disappearance of N-cadherin from the surface of prospective neural crest cells occurs long before detectable changes in the organization of the basement membrane along neural crest cells, in the trunk region, there is no such continuous basement membrane along neural crest cells that could prevent their detachment and disappearance of N-cadherin occurs just upon cell release. Note that only the dorsalmost part of the neural tube is shown and that only prospective neural crest cells are represented. *Stippled areas*: cells expressing N-cadherin molecules; *thick lines*: basement membranes; *nt*, neural tube; *ec*, ectoderm

resulting in spina bifida and the lack of neural crest derivatives. This defect is correlated with abnormally high levels of N-CAM polypeptides as well as with an altered 140 kDa N-CAM isoform (Moase and Trasler 1990, 1991).

Electron microscopic studies have clearly established that the basal lamina lining the neural epithelium is interrupted, or even absent, in its dorsalmost aspect at the time of neural crest cell release (Bancroft and Bellairs 1976; Newgreen and Gibbins 1982; Tosney 1978, 1982; Erickson and Weston 1983). It is assumed that the lack of a continuous basement membrane clearly favors the release of neural crest cells as these cells are not able to penetrate intact basement membranes (Erickson 1987). However, while in the cranial region, the basal lamina is at first continuous under the neurectoderm and disappears progressively along the neural fold and ectoderm at the time of neural crest cell release (Tosney 1982; Duband and Thiery 1982a, 1987), at the trunk level, disruption does not correlate strictly with the onset of crest cell emigration as it occurs well before this event (Sternberg and Kimber 1986; Martins-Green and Erickson 1986, 1987). Therefore, in the trunk, the disappearance of the basal lamina along presumptive neural crest cells may not constitute the triggering event for emigration.

Parallel to the disorganization of the basal lamina, changes in the structure of the fibrillar extracellular matrix that constitute the migratory substratum are believed to take place during neural crest emigration. The matrix constituents identified so far as promoters of neural crest cell migration in vitro, such as fibronectin and collagens I, IV and VI, can be detected on the surface of the neural tube well before emigration commences and, as judged by immunofluorescence, their amount and spatial organization are apparently not substantially modified upon initiation of migration (Duband and Thiery 1982 a,b, 1987; Brauer et al. 1985; Perris et al. 1991a). Thus, changes in the organization and composition of the matrix, if they occur, may be subtle and are most likely restricted to minor components that have not been identified yet. There are, however, some documented examples in which changes in the matrix have been observed. For example, upon initiation of migration, neural crest cells secrete relatively high amounts of glycosaminoglycans that are supposedly able to expand intercellular spaces, thus favoring separation of cells (Pratt et al. 1975; Pintar 1978). However, fibulin, a matrix glycoprotein of M_r of about 100 kDa, is precisely deposited on the dorsal aspect of the neural tube and around prospective neural crest cells before migration at the mid- and hindbrain levels and is not found associated with neural crest cells at later stages of migration (Spence et al. 1992), thus raising the intriguing possibility that this molecule might be specifically involved in the process of epithelium-to-mesenchyme transition.

Morphological studies therefore suggest that disruption of cell-cell interactions might be the triggering event of the onset of cell dispersion at the trunk level, whereas changes in the organization of the extracellular matrix would be the deciding factor at the cranial level. Various experimental analyses, however, indicate that acquisition of cell-substratum adhesion may be a critical step in the process of neural crest cell emigration not only at the cranial level but in the trunk as well. When injected into the cranial mesenchyme lateral to the neural tube at the time of initiation of neural crest migration, antibodies to extracellular matrix components (fibronectin, a laminin-heparan sulfate proteoglycan, and tenascin), and antibodies to the β1-subunit of integrins, or RGD-containing peptides, all perturb neural crest cell migration and provoke neural tube anomalies associated with accumulation of neural crest cells into the lumen of the neural tube, suggesting that blockade of cell-matrix interactions affects not only locomotion along the migratory paths but emigration as well (Boucaut et al. 1984; Bronner-Fraser 1986; Poole and Thiery 1986; Bronner-Fraser and Lallier 1988). In contrast, antibodies to N-cadherin and N-CAM produce distorted neural tubes and ectopic neural crest cells but apparently do not interfere with the process of neural crest cell emigration itself (Bronner-Fraser et al. 1992a).

In 1985, Löfberg and coworkers described an original in vivo approach to test whether the onset of neural crest cell emigration might be stimulated by local changes in the composition of the extracellular matrix (Löfberg et al. 1985). The technique consists of adsorption of embryonic extracellular matrix material from an axolotl embryo onto a microcarrier consisting of a Nucleopore filter, which is then inserted into another embryo at a site close to premigratory neural crest cells. Using this technique, it was found that the dorsal epidermis and its associated extracellular matrix, derived from rostal regions at a time when neural crest cells are migrating, are able to provoke the precocious emigration of immature neural crest cells issued from more caudal regions. Consistent with the observation that extracellular fibrils increase in amount at the onset of neural crest cell emigration (Löfberg and Ahlfors 1978; Löfberg et al. 1980; Spieth and Keller 1984), this result suggests that presumptive neural crest cells need to be triggered from the environment, including the extracellular matrix, to start migration.

Moreover, by contrasting in vitro the migratory and adhesive properties of migrating and presumptive neural crest cells originating from rostral and caudal regions of the neural tube, respectively, it has been shown that presumptive neural crest cells initially exhibit reduced capacities of both adhesion and migration onto purified extracellular matrix molecules (Delannet and Duband 1992). The ability to respond to extracellular matrix material is acquired by the cells only gradually just prior to emigration, indicating that acquisition of functional, appropriate integrin receptors is a necessary step for migration (Delannet and Duband 1992). Most importantly, the integrity of the neural tube is necessary for the acquisition of substratum adhesion properties by the cells, indicating that intercellular communications are necessary for the maturation of cells.

Our knowledge on the repertoire of integrin receptors expressed by neural crest cells is still very sketchy and it is not possible at the present time to correlate the onset of neural crest cell emigration with the appearance of specific integrins on the surface of the cells. Both in vivo and in vitro studies have shown that neural crest cells express the $\alpha 1\beta 1$ integrin as a laminin receptor, but its expression is apparently not substantially modified upon initiation of migration (Duband et al. 1992; Lallier et al. 1994). In contrast, the $\alpha 6\beta 1$ integrin, another laminin receptor, is apparently not found on neural crest cells during early migration in vivo but, although the neural tube exhibits strong labeling for this integrin, the moment when neural crest cells lose expression of $\alpha 6\beta 1$ upon dissociation from the neural tube has not been determined precisely (Bronner-Fraser et al. 1992b). Recent studies have shown that αV integrins are abundant on neural crest cells cultured in vitro but their presence in vivo has not been documented yet (Delannet et al. 1994). Finally, there is good evidence that $\alpha 4\beta 1$, a receptor for fibronectin, becomes expressed by neural crest cells as soon as they segregate from the neural tube (Stepp et al. 1994). This is consistent with the finding that neural crest cells can interact in vitro with sequences on fibronectin molecules that are normally recognized by the $\alpha 4\beta 1$ integrin (Dufour et al. 1988; Humphries 1990).

4 Cessation of Neural Crest Cell Migration: The Opposite of Initiation of Migration?

The final step of neural crest cell migration involves the loss of migratory properties at defined sites of the embryo where cells terminate differentiation. In the case of the progenitors of the neurons and glial cells of the peripheral nervous system, cells that have shifted from a migratory to a stationary phenotype undergo condensation into compact clusters at the origin of the peripheral ganglia. Although relatively little is known at the present time on the extracellular signals that cause cells to lose their migratory capabilities and to coalesce, morphological studies indicate that cessation of migration occurs coincidently with profound alterations in the pattern of expression of cell-adhesion molecules and in the organization of the extracellular matrix associated with neural crest cells (Figs. 3,5). Whether these changes are directly responsible for cessation of movement or are instead secondary events aimed at stabilizing the nascent structure have not been addressed yet in experimental systems. In addition, it should be emphasized that, aside from a few neural crest-derived tissues, such as the mesenchyme of the branchial arches, for which farther migration is clearly prevented by the presence of a physical barrier ahead of the neural crest cell population, in most other cases, there is no obvious reason that explains the cessation of cell migration.

N-cadherin is the first cell adhesion molecule that is reexpressed on cells within the primordium of the peripheral ganglia (HATTA et al. 1987; DUBAND et al. 1988; AKITAYA and BRONNER-FRASER 1992). In contrast, N-CAM is found on neurons only later as they start growing neurites (THIERY et al. 1982a; DUBAND et al. 1985; AKITAYA and BRONNER-FRASER 1992)). Differentiation of neural crest cells into neurons is also accompanied by the de novo expression of a series of neuron-specific adhesion molecules, such as Ng-CAM/L1 and R-cadherin (THIERY et al. 1985; INUZUKA et al. 1991). The disappearance of N-cadherin from the cell surface at the onset of migration and its subsequent reexpression during cell aggregation suggest that, during migration, its expression is repressed in neural crest cells. Thus, in vitro, neural crest cells were found to be unable to retain stable cell-cell interactions during migration, even though they express functional N-cadherin molecules. In addition, stable N-cadherin-mediated junctions could be restored in neural crest cells upon treatment with kinase and phosphatase inhibitors, indicating that N-cadherin function is under the control of intracellular signals elicited by surface receptors (MONIER and DUBAND 1995).

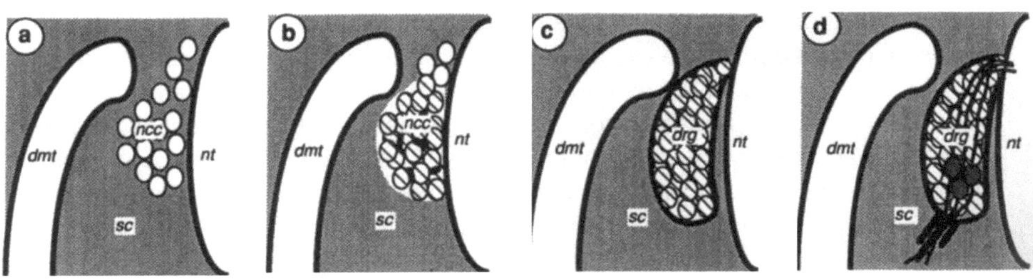

Fig. 5a–d. The major adhesive events occurring during aggregation of neural crest cells into dorsal root ganglia. Early accumulation of neural crest cells to form the prospective dorsal root ganglia coincides with the disappearance of fibronectin among the cell population and is accompanied by the reexpression of N-cadherin on the cell surface (**a, b**). A basement membrane is then gradually deposited around the ganglion (**b, c**) and is followed by the reexpression of N-CAM on the surface of cells undergoing neuronal differentiation (**d**). *Lightly stippled area*: regions containing fibronectin in the extracellular spaces; *hatched area*, cells expressing N-cadherin molecules; *heavily stippled area*, cells expressing both N-cadherin and N-CAM molecules; *thick lines*: basement membranes; *dmt*, dermamyotome; *drg*, dorsal root ganglion; *ncc*, neural crest cells; *nt*, neural tube; *sc*, sclerotome

The increase in cell-cell contacts among neural crest cells is associated with the disappearance of fibronectin and collagen type I from the mass of the ganglion and is followed by the progressive formation of a basal lamina surrounding the developing structure (Thiery et al. 1982b; Duband and Thiery 1987). Migration-promoting components of the extracellular matrix are replaced by other matrix molecules known to be inhibitory for neural crest cell movement in vitro. This is, for instance, the case for chondroitin sulfate and keratan sulfate proteoglycans and for decorin (Perris et al. 1991b). Changes in the repertoire of integrin receptors on the surface of neural crest cells have been noticed, but they appear to correlate essentially with the phenotypic expression of neuronal potentialities rather than with changes in motility behavior (Bronner-Fraser et al. 1992b; Duband et al. 1992).

5 Regulators of Adhesion Properties

Since the pioneering studies of Löfberg and colleagues using extracellular matrix-coated microcarriers (Löfberg et al. 1985), it has become clear that the extracellular matrix associated with neural crest cells contains some important regulatory cues that may govern both the adhesive properties and the differentiation program of neural crest cells (Kalcheim and Le Douarin 1986; Stemple and Anderson 1993). These instructive signals may reside in intrinsic constituents of the extracellular matrix and in nonmatrix molecules that are intimately associated with it.

5.1 Growth Factors/Cytokines

Among the various molecules that can be found in extracellular matrices, growth factors (or cytokines) are likely to play a pivotal role in the control of changes in cell adhesion during the development of the neural crest (see for example Nathan and Sporn 1991). Indeed, growth factors are often released in the extracellular matrix where they are trapped by some of the matrix constituents, such as heparin and various proteoglycans. Most importantly, the biological activity of many growth factors depends on their association with matrix elements which allow the proper presentation of the growth factor molecules to their receptors. Finally, growth factors are able to regulate the synthesis of extracellular matrix molecules, e.g. fibronectin, and their corresponding integrin receptors. Although considerable data have accumulated on the role of growth factors in governing the phenotypic expression of neural crest cells (see for example, Stemple and Anderson 1993), relatively little is known about the precise role of growth factors in controlling the adhesive properties of neural crest cells, with the possible exception of the transforming growth factors-β (TGF-β).

Because of their biological effects, members of the TGF-β superfamily appear as leading candidates among growth factors to participate in the control of the initiation of neural crest cell migration. Indeed, TGF-β is one of the factors that can be trapped by extracellular matrix materials and that has been implicated in a variety of biological events involving cell dispersion. In addition, it promotes the synthesis of fibronectin and collagens and causes marked alterations in the repertoire of integrins (see for a review, MASSAGUÉ 1990). The effects of TGF-β1 and TGF-β2 on avian neural crest cell emigration have been evaluated in vitro (DELANNET and DUBAND 1992). Both types were found to provoke a premature emigration of neural crest cells from neural tube explants taken from the caudal part of the embryo and to increase the substratum adhesion properties of cells. However, they appeared not to regulate the locomotory and intercellular adhesion properties of the cells, indicating that they do not control the entire emigration process. It was proposed that TGF-β may stimulate emigration of neural crest cells by modifying the expression pattern or the binding properties of integrins on the surface of these cells. Consistent with this finding, it has been shown that cranial neural crest cells synthesize and secrete a latent form of TGF-β and can activate it by proteolysis (BRAUER and YEE 1993). Which members of the TGF-β superfamily are involved in the control of cell migration in vivo remains, however, to be determined. Various reports have described the presence in the mouse embryo of different TGF-β both in migrating neural crest cells and in the neural tube at the time of neural crest cell emigration (FLANDERS et al. 1991; MILLAN et al. 1991; SCHMID et al. 1991). Recently, two novel members of the TGF-β family have been characterized as putative regulators of neural crest cell emigration. The first one, *dorsalin-1*, has been isolated in the chick embryo and its mRNA has been localized in the dorsal aspect of the trunk neural tube soon after closure (BASLER et al. 1993). Expression of *dorsalin-1* mRNAs in the prospective neural crest could be detected throughout the emigration process but persisted in the dorsal neural tube, though at lower levels, after complete separation of the neural crest. Most interestingly, neural tube fragments corresponding to the intermediate portion situated in between the dorsal and ventral thirds and which normally do not provide neural crest cells release cells exhibiting a neural crest phenotype when incubated in the presence of *dorsalin-1*, suggesting that *dorsalin-1* can act as motility-promoting factor. The second candidate, called *Radar*, has been isolated from the zebrafish embryo and is expressed in the prospective neural crest along the entire neural axis (RISSI et al. 1995). It is therefore plausible that various members of the TGF-β family may operate sequentially as morphogens, either in an autocrine or in a paracrine manner, to induce segregation and dispersion of neural crest cells from the neural tube.

Beside TGF-β, fibroblast growth factors (FGFs) are also candidates for participating in the cascade of events that control neural crest development as they, as well as their corresponding receptors are abundantly found in the neural tube in early developmental stages (see for a review BAIRD 1994). Although their exact role in the regulation of the adhesive properties of neural crest cells has not been directly addressed, it is likely that FGF, and more particularly FGF2 (also called basic FGF), are involved during certain steps of the ontogeny of the neural tube and neural crest. In this respect, it is noteworthy that FGF2 has been found to promote adhesion of neuroepithelial cells from avian trunk neural tube with extracellular matrix molecules (Kinoshita et al. 1993).

Recent studies in the mouse and *Xenopus* embryos indicate that platelet-derived growth factor (PDGF) may also govern more or less directly the ontogeny of neural crest cells, at least at the cranial level. Indeed, the *patch* mutation in mice, which is characterized by prominent defects in the head and neck, corresponds to a deletion of the gene encoding the PDGF receptor α subunit (PDGFRα). More specifically, pigment cells and the nonneuronal derivatives of the neural crest, but not the neuronal derivatives, are affected. This is accompanied by alterations in the organization of the matrix through which cells are supposed to migrate (Morrison-Graham et al. 1992). Interestingly, while PDGF mRNA is found in the ectoderm, PDGFRα is expressed in neural crest cells both prior to and during migration (Ho et al. 1994), raising the intriguing possibility that the ectoderm might control the development of the neural crest directly through PDGF.

It cannot be excluded that additional growth factors might be involved in the control of some of the events that accompany neural crest development. One is epidermal growth factor (EGF). Indeed, recent studies have demonstrated that cadherin-associated cytoskeletal components, catenins, are tyrosine phosphorylated in epithelial cells treated with EGF, resulting in the rapid deterioration of cadherin-mediated adherens junctions. Furthermore, the EGF receptor colocalizes with cadherin in adherens junctions in epithelial cell lines and is physically associated with β-catenin (Hoschuetzky et al. 1994), indicating that, upon stimulation by EGF, the EGF receptor may directly control the organization of cell-cell junctions. It would be therefore of great interest to determine whether the EGF receptor is expressed by neural crest cells and if it may also control activity of N-cadherin. Another growth factor of interest is scatter factor, also called hepatocyte growth factor (SF/HGF), as it has potent mitogenic, motogenic and morphogenic activities on epithelial cells in vitro (Weidner et al. 1990; Rosen et al. 1994). The expression patterns of SF/HGF and its corresponding receptor, *c-met*, are consistent with a role of this factor in epithelial-to-mesenchyme transitions during embryonic development (Sonnenberg et al. 1993). However, data about their possible implication in neural crest cell migration are not available yet. Finally, *Wnt* gene products, a family of putative growth factors, are believed to function as locally restricted morphogens during neural development (Nusse and Varmus 1992). Two of these genes, *Wnt-1* and *Wnt-3a*, are expressed in a discrete region in the dorsal aspect of the neural tube both at cranial levels and in the spinal cord. Dirsruption of the *Wnt-1* gene in the mouse either by homologous recombination or by antisense knockout leads to severe brain defects in the midbrain and hindbrain but leaves the spinal cord intact (Nusse and Varmus 1992; Augustine et al. 1993). In contrast, disruption of *Wnt-3a* targets all brain regions, but apparently less dramatically than for *Wnt-1*, as well as the spianl cord (Augustine et al. 1993). Although *Wnt* proteins are generally suspected to participate in signaling events during the establishment of boundaries between adjacent domains in the nervous system (NUSSE and VARMUS 1992), recent studies in amphibians revealed that some of them may possibly increase cell-cell adhesion and decrease cell migration (MOON et al. 1993). Consistent with these observations, expression of *Wnt-1* in cultured cell lines results in enhancement of plakoglobin, β-catenin and E-cadherin expression in the cells accompanied by increased stability of the β-catenin/ E-cadherin complex and greater cell-cell adhesion (HINCK et al. 1994). It can therefore be speculated that some of the *Wnt* genes may be implicated in stabilizing intercellular junctions at the dorsal midline during closure of the neural tube as well as after neural crest cell migration.

5.2 Adhesion Molecules

Recent studies have presented evidence that adhesion molecules themselves may be at the origin of cascades of intracellular events that control the expression and function of other adhesion molecules. For example, it has been shown that, in the epidermis, the down-regulation of integrin function that occurs during terminal differentiation of keratinocytes may be regulated, at least partly, by cadherins, as the selective loss of integrin mRNA and protein can be prevented by antibodies to cadherins (HODIVALA and WATT 1994). Likewise, the integrin $\alpha5\beta1$-mediated phagocytic activity of a macrophage cell line is totally abrogated upon ligation of the integrin $\alpha v\beta3$ with antibodies or with vitronectin. This effect is mediated by a signal transduction pathway involving protein kinase C (BLYSTONE et al. 1994). Thus, there is complex cross-talk among molecules of different families of adhesion molecules and among members of the same family.

The coordinated expression of cell-cell and cell-substratum adhesion molecules and of their corresponding ligands during the critical steps of neural crest development is also indicative of cross-talk between adhesion molecules in this system. In particular, during neural crest cell aggregation, the disappearance of fibronectin among cells immediately followed by expression of N-cadherin on the cell surface suggests that the loss of neural crest cell capability to interact with fibronectin may induce N-cadherin-mediated cell aggregation. When confronted with RGD peptides, antibodies to fibronectin or antibodies to integrins, neural crest cells cultured in vitro onto fibronectin substrates readily lose their migratory capabilities and form aggregates. This aggregation can be abolished by antibodies to N-cadherin. Furthermore, RGD peptide-induced aggregation of neural crest cells can be blocked by calcium ionophores, suggesting that, in locomoting neural crest cells, the function of N-cadherin may be under the negative control of integrins through activation of calcium channels. At the end of migration, inactivation of fibronectin-binding integrins would relieve the repression and allow cadherin molecules to become functional (MONIER, F., DELANNET, M., and DUBAND, J.-L., in preparation).

5.3 Transcription Factors

During the last decade, genes encoding transcription factors, and more particularly homeotic genes, have emerged as pivotal regulatory elements in the formation of the different compartments of the body, and especially of the nervous system (see for a review, RUBENSTEIN and PUELLES 1994). However, while a large body of data has accumulated on the spatiotemporal patterns of expression of homeotic genes and on their possible implication in the control of the regional specification and differentiation of the embryonic tissues, the nature of the effector genes that are directly under their control are still mostly unknown. Because homeotic genes control the whole spectrum of morphogenetic processes, the genes encoding adhesion receptors are excellent candidates for such downstream targets. Studies of the promoter of the *Ncam* gene have revealed the presence of clustered DNA sequences that constitute putative binding sites for homeotic genes (HIRSCH et al. 1990). Two mouse homeodomain proteins, Cux and Phox2, belonging to the Cut and Paired families, respectively, have been identified as transregulators of the activity of the *Ncam* gene promoter (VALARCHÉ et al. 1993). While Cux behaves as a repressor and is widely expressed during the embryonic development, Phox2 relieves this repression, although it is not an activator in itself, and it is restricted to the neural crest-derived autonomic nervous system which correlatively expresses high levels of N-CAM. *Ncam* may be under the control of other genes encoding transcription factors of the Paired family, including *Pax 8* (HOLST et al. 1994). Interestingly, two *Pax* genes, *Pax-3* and *Pax-7*, exhibit spatio temporal distributions during neural development consistent with a role in the formation of the neural crest (MANSOURI et al. 1994). As mentioned above, *Pax-3* is deficient in *splotch* mice, resulting in numerous severe defects including exencephaly, spina bifida and absence of many neural crest derivatives. Since N-CAM is at abnormal levels in *splotch* mutants (MOASE and TRASLER 1991), it is tempting to speculate that *Pax-3* controls *Ncam* gene expression, but this has not been proven yet. It is therefore quite plausible that *Ncam* might be the target of numerous *Pax* genes that may act either sequentially or in combination in neural crest cells.

In contrast to *Pax-3*, which is apparently not expressed by migrating neural crest cells, *AP2*, *ets-1*, and *fli* are transcription factors the expressions of which are found in both premigratory and migratory neural crest cells (VANDENBUNDER et al. 1989; MITCHELL et al. 1991; MEYER et al. 1993; MAROULAKOU et al. 1994). Numerous genes, including growth factors, proto-oncogenes as well as transcription factors, contain in their regulatory sequences binding sites for the *AP2* and *ets-1* transcription factors. Therefore, correlations between expression of these transcription factors and the occurence of cellular events during neural crest cell emigration are only speculative. It is, however, of interest to mention that binding sequences for *ets-1* and *Ap2* are both found in the promoter of the gene coding for the $\alpha 4$ integrin subunit, which is found on the surface of neural crest cells after their separation from the neural tube (ROSEN et al. 1991; STEPP et al. 1994). This would suggest that the $\alpha 4\beta 1$ integrin could be one of the targets of the *ets-1* and

AP2 transcription factors and that this event would be one of the different steps allowing the acquisition by cells of their substratum adhesion properties.

The last transcription factor that has been found in neural crest cells belongs to the *Snail* family. This factor, called *Slug*, is a zinc finger protein and exhibits a striking distribution pattern in mesodermal cells involuting in the primitive streak and in the neural folds and neural crest at all axial levels (Nieto et al. 1994). This distribution is suggestive of a role of *Slug* in the control of cellular events that accompany the dissociation and release of cells. The importance of *slug* in neural crest cell migration has been revealed in experiments in which early chick embryos were incubated in the presence of antisens oligonucleotides. Treated embryos exhibit anomalies in the mesoderm, which is partially lacking, and in the neural tube, which fails to close, and a deficit in neural crest cells. The presumptive neural crest remained integrated into the neural folds without any sign of epithelium-to-mesenchyme transition. Thus, at the present time, *slug* is undoubtedly the best candidate gene encoding a transcription factor for controlling the modulations of cellular adhesiveness during the ontogeny of the neural crest.

6 Concluding Remarks

Although considerable progress has been made during the last 10 years in the elucidation of the cellular mechanisms that are involved in the control of adhesiveness during the development of the neural crest, it is not possible at the present time to draw a tentative model taking into account all the observations that have been made so far and describing a coherent succession of the major cellular events. Clearly, migration of neural crest cells is a multistep process that is tightly regulated both spatially and temporally by both external signals and internal factors. There is a need for much more data in order to elucidate the intricate mechanisms of neural crest cell migration. In particular, although there is now ample evidence for the implication of growth factors, with a few exceptions, we still do not know at which steps they operate and which transcription factors are activated by the intracellular signals elicited by the various growth factors. Likewise, it will be important to determine which genes are regulated by these transcription factors and to evaluate how the triggered genes may in turn influence the response of individual cells and of the whole population.

Acknowledgments. J.-L.D. acknowledges financial support from the Centre National de la Recherche Scientifique (Programme ATIPE), the Ministère de la Recherche et de la Technologie (91.T.0011), the Association pour la Recherche contre le Cancer (ARC-6517), the Institut National de la Santé et de la Recherche Médicale (CRE 910705), the Association Française contre les Myopathies, the Ligue Nationale contre le Cancer, and the Fondation pour la Recherche Médicale. F.M. and S.G. are recipients of predoctoral fellowships of the Ministère de l'Enseignement Supérieur et de la Recherche.

References

Akitaya T, Bronner-Fraser M (1992). Expression of cell adhesion molecules during initiation of and cessation of neural crest cell migration. Dev Dynam 194: 12–20

Anderson DJ (1989). The neural crest cell lineage problem: neuropoiesis? Neuron 3: 1–12

Augustine K, Liu ET, Sadler TW (1993). Antisense attenuation of Wnt-1 and Wnt-3a expression in whole embryo culture reveals roles for these genes in craniofacial spinal cord, and cardiac morphogenesis. Dev Genet 14: 500–520

Baird A (1994). Fibroblast growth factors: activities and significance of non-neurotrophin neurtrophic growth factors. Curr Opin Neurobiol 4: 78–86

Bancroft M, Bellairs R (1976). The neural crest cells of the trunk region of the chick embryo studied by SEM and TEM. Zoon 4: 73–85

Basler K, Edlund T, Jessell TM, Yamada T (1993). Control of cell pattern in the neural tube: Regulation of cell differentiation by dorsalin-1, a novel TGFβ family member. Cell 73: 687–702

Blystone SD, Graham IL, Lindberg FP, Brown EJ (1994). Integrin αvβ3 differentially regulates adhesive and phagocytic functions of the fibronectin receptor α5β1 . J Cell Biol 127: 1129–1137

Boucaut J-C, Darribère T, Poole TJ, Aoyama H, Yamada KM, Thiery JP (1984). Biological active synthetic peptides as probes of embryonic development: a competitive peptide inhibitor of fibronectin function inhibits gastrulation in amphibian embryos and neural crest cell migration in avian embryos. J Cell Biol 99: 1822–1830

Brauer PR, Yee JA (1993). Cranial neural crest cells synthesize and secrete a latent form of transforming growth factor-β that can be activated by neural crest cell proteolysis. Dev Biol 155: 281–285

Brauer PR, Bolender DL, Markwald RR (1985). The distribution and spatial organization of the extracellular matrix encountered by mesencephalic neural crest cells. Anat Rect 211: 57–68

Bronner-Fraser M (1986). An antibody to receptor for fibronectin and laminin perturbs cranial neural crest development in vivo. Dev Biol 117: 528–536

Bronner-Fraser M, Lallier T (1988). A monoclonal antibody against a laminin-heparan sulfate proteoglycan complex perturbs cranial neural crest migration in vivo. J. Cell Biol 106: 1321–1329

Bronner-Fraser M, Fraser S (1989). Developmental potential of avian trunk neural crest cells in situ. Neuron 3: 755–766

Bronner-Fraser M, Wolf JJ, Murray BA (1992a). Effects of antibodies against N-cadherin and N-CAM on the cranial neural crest and neural tube. Dev Biol 153: 291–301

Bronner-Fraser M, Artinger M, Muschler J, Horwitz AF (1992b). Developmentally regulated expression of α6 integrin in avian embryos. Development 115: 197–211

Chan WY, Tam PPL (1988). A morphological and experimental study of the mesencephalic neural crest cells in the mouse embryo using wheat germ agglutinin-gold conjugate as the cell marker. Development 102: 427–442

Cochard P, Coltey P (1983). Cholinergic traits in the neural crest. acetylcholinesterase in crest cells of the chick embryo. Dev Biol 98: 221–228

Collazo A, Bronner-Fraser M, Fraser SE (1993). Vital dye labelling of Xenopus laevis trunk neural crest reveals multipotency and novel pathways of migration. Development 118: 363–376

Delannet M, Duband J.-L (1992). Transforming growth factor-β control of cell-substratum adhesion during avian neural crest cell emigration in vitro. Development 116: 275–287

Delannet M, Martin F, Bossy B, Cheresh DA, Reichardt LF, Duband J-L (1994). Specific roles of the αVβ1, αVβ3 and αVβ5 integrins in avian neural crest cell adhesion and migration on vitronectin. Development 120: 2687–2702

Duband J-L, Thiery JP (1982a). Distribution of fibronectin in the early phase of avian cephalic neural crest cell migration. Dev Biol 93: 308–323

Duband J-L, Thiery JP (1982b). Appearance and distribution of fibronectin during chick embryo gastrulation and neurulation. Dev Biol 94: 337–350

Duband J-L, Thiery JP (1987). Distribution of laminin and collagens during avian neural crest development. Development 101: 461–478

Duband J-L, Tucker GC, Poole TJ, Vincent M, Aoyama H, Thiery JP (1985). How do the migratory and adhesive properties of the neural crest govern ganglia formation in the avian peripheral nervous system? J Cell Biochem 27: 189–203

Duband J-L, Volberg T, Sabanay I, Thiery JP, Geiger B (1988). Spatial and temporal distribution of the adherens-junction-associated adhesion molecule A-CAM during avian embryogenesis. Development 103: 325–344

Duband JL, Belkin AM, Syfrig J, Thiery JP, Koteliansky VE (1992). Expression of α1 integrin, a laminin-collagen receptor, during myogenesis and neurogenesis in the avian embryo. Development 116: 585–600
Dufour S, Duband JL, Humphries MJ, Obara M, Yamada KM, Thiery JP (1988). Attachement, spreading and locomotion of avian neural crest cells are mediated by multiple adhesion sites on fibronectin molecules. EMBO J 7: 2661-2671
Erickson CA (1987). Behavior of neural crest cells on embryonic basal laminae. Dev Biol 120: 38–49
Erickson CA, Weston J (1983). An SEM analysis of neural crest cell migration in the mouse. J. Embryol Exp Morphol 74: 97–118
Flanders KC, Lüdecke G, Engels S, Cissel DS, Roberts AB, Kondaiah P, Lafyatis R, Sporn MB, Unsicker K (1991). Localization and actions of transforming growth factor-βs in the embryonic nervous system. Development 113: 183–191
Hatta K, Takagi S, Fujisawa H, Takeichi M (1987). Spatial and temporal expression pattern of N-cadherin cell adhesion molecules correlated with morphogenetic processes of chicken embryos. Dev Biol 120: 215–227
Hinck L, Nelson WJ, Papkoff J (1994). Wnt-1 modulates cell-cell adhesion in mammalian cells by stabilizing β-catenin binding to the cell adhesion protein cadherin. J Cell Biol 124: 729–741
Hirsch MR, Gaugler L, Deagostini-Bazin H, Bally-Cuif L, Goridis C (1990). Identification of positive and negative regulatory elements governing cell-type specific expression of the neural cell adhesion molecule gene. Mol Cell Biol 10: 1959–1968
Ho L, Symes K, Yordan C, Gudas LJ, Mercola M (1994). Localization of PDGF-A and PDGFRα mRNA in Xenopus embryos suggests signalling from neural ectoderm and pharyngeal endoderm to neural crest cells. Mech Dev 48: 165–174
Hodivala KJ, Watt FM (1994). Evidence that cadherins play a role in the downregulation of integrin expression that occurs during keratinocyte terminal differentiation J Cell Biol 124: 589–600.
Holst BD, Goomer RS, Wood IC, Edelman GM, Jones FS (1994). Binding and activation of the promoter for the neural cell adhesion molecule by Pax-8. J Biol Chem 269: 22245–22252
Hoschuetzky H, Aberle H, Kemler R (1994). β-catenin mediates the interaction of the cadherin-catenin complex with epidermal growth factor receptor. J Cell Biol 127: 1375–1380
Humphries MJ (1990). The molecular basis and specificity of integrin-ligand interactions J Cell Sci 97: 585-592
Inuzuka H, Redies C, Takeichi M (1991). Differential expression of R- and N- cadherin in neural and mesodermal tissues during early chicken development. Development 113: 959–967
Kalcheim C, Le Douarin NM (1986). Requirement of a neural tube signal for the differentiation of neural crest cells into dorsal root ganglia. Dev Biol 116: 451–466
Kinoshita Y, Kinoshita C, Heuer JG, Bothwell M (1993). Basic fibroblast growth factor promotes adhesive interactions of neuroepithelial cells from chick neural tube with extracellular matrix proteins in culture. Development 119: 943–956
Lallier T, Deutzmann R, Perris R, Bronner-Fraser M (1994). Neural crest cell interactions with laminin: structural requirements and localization of the binding site for α1β1 integrin. Dev Biol 162: 451–464
Le Douarin NM (1982). The neural crest. Cambridge University Press, Cambridge
Le Douarin NM, Ziller C, Couly GF (1993). Patterning of neural crest derivatives in the avian embryo: in vivo and in vitro studies. Dev Biol 159: 24–49
Löfberg J, Ahlfors K (1978). Extracellular matrix organization and early neural crest cell migration in the Axolotl embryo. Zoon 6: 87–101
Löfberg J, Ahlfors K, Fällström C (1980). Neural crest cell migration in relation to extracellular matrix organization in the embryonic Axolotl trunk. Dev Biol 75: 148–167
Löfberg J, Nynäs-McCoy A, Olsson C, Jonsson L, Perris R (1985). Stimulation of initial neural crest cell migration in the axolotl embryo by tissue grafts and extracellular matrix transplanted on micro-carriers. Dev Biol 107: 442–459
Mansouri A, Stoykova A, Gruss P (1994). Pax genes in development. J Cell Sci Suppl 18: 35–42
Maroulakou IG, Papas TS, Green JE (1994). Differential expression of ets-1 and ets-2 proto-oncogenes during murine embryogenesis. Oncogene 9: 1551–1565
Martins Green M, Erickson CA (1986). Development of neural tube basal lamina during neurulation and neural crest cell emigration in the trunk of the mouse embryo. J Embryol Exp Morphol 98: 219–236
Martins-Green M, Erickson CA (1987). Basal lamina is not a barrier to neural crest cell emigration: documentation by TEM and by immunofluorescent and immunogold labelling. Development 101: 517–533
Massagué J (1990). The transforming growth factor-β family. Annu Rev Cell Biol 6: 597–641

Meyer D, Wolff C-M, Stiegler P, Sénan F, Befort N, Befort JJ, Remy P (1993). X1-fli, the Xenopus homologue of the fli gene, is expressed during embryogenesis in a restricted pattern evocative of neural crest cell distribution. Mech Dev 44: 109–121
Millan FA, Denhez F, Kondaiah P, Akhurst P, Akhurst RJ, (1991). Embryonic gene expression patterns of TGF β1, β2, and β3 suggest different developmental functions in vivo Development 111: 131–144
Mitchell PJ, Timmons PM, Hébert JM, Rigby PWJ, Tjian R (1991). Transcription factor AP-2 is expressed in neural crest cell lineages during mouse embryogenesis. Genes Dev 5: 105–119
Moase CE, Trasler DG (1990). Delayed neural crest cell emigration from Sp and Sp^d mouse neural tube explants. Teratology 42: 171–182
Moase CE, Trasler DG (1991). N–CAM alterations in splotch neural tube defect mouse embryos. Development 113: 1049–1058
Monier F, Duband J.-L (1995). Control of N-cadherin-mediated intercellular adhesion in migrating neural crest cells in vitro (submitted)
Moon RT, DeMarais A, Olson DJ (1993). Response to Wnt signals in vertebrate embryos may involve changes in cell adhesion and cell movement. J Cell Sci [Suppl] 17: 183–188
Morrison-Graham K, Schatteman GC, Bork T, Bowen-Pope DF, Weston JA (1992). A PDGF receptor mutation in the mouse (Patch) perturbs the development of a non-neuronal subset of neural crest-derived cells. Development 115: 133–142
Moury JD, Jacobson AG (1989). Neural fold formation at newly created boundaries between neural plate and epidermis in the axolotl. Dev Biol 133: 44–57
Moury JD, Jacobson AG (1990). The origins of neural crest cells in the axolotl. Dev Biol 141: 243–253
Nathan C, Sporn M (1991). Cytokines in context. J Cell Biol 113: 981–986
Newgreen DF, Erickson CA (1986). The migration of neural crest cells. Int Rev Cytol 103: 89–145
Newgreen DF, Gibbins IL (1982). Factors controlling the time of onset of the migration of neural crest cells in the fowl embryo. Cell Tissue Res 224: 145–160
Nichols DH (1981). Neural crest formation in the head of the mouse embryo as observed using a new histological technique. J Embryol Exp Morphol 64: 105–120
Nieto MA, Sargent MG, Wilkinson DG, Cooke J (1994). Control of cell behavior during vertebrate development by Slug, a zinc finger gene. Science 264: 835–839
Nusse R, Varmus HE (1992). Wnt genes Cell 69: 1073–1087
Perris R, Krotoski D, Bronner-Fraser M (1991a). Collagens in avian neural crest development: Distribution in vivo and migration-promoting ability in vitro. Development 113: 969–984
Perris R, Krotoski D, Lallier T, Domingo C, Sorrell JM, Bronner-Fraser M (1991b). Spatial and temporal changes in the distribution of proteoglycans during avian neural crest development. Development 111: 583–599
Pintar JE (1978). Distribution and synthesis of glycosaminoglycans during quail neural crest morphogenesis. Dev Biol 67: 444–464
Poole TJ, Thiery JP (1986). Antibodies and a synthetic peptide that block cell-fibronectin adhesion arrest neural crest cell migration in vivo Prog Clin Biol Res 217B: 235–238
Pratt RM, Larsen MA, Johnston MC (1975). Migration of cranial neural crest cells in a cell-free hyaluronate-rich matrix. Dev Biol 44: 298–305
Ranscht B, Bronner-Fraser M (1991). T-cadherin expression alternates with migrating neural crest cells in the trunk of the avian embryo. Development 111: 15–22
Rissi M, Wittbrodt J, Délot E, Naegeli M, Rosa FM (1995). Zebrafish Radar: a new member of the TGF-β superfamily defines dorsal regions of the neural plate and the embryonic retina. Mech Dev 49: 223–234
Rosen GD, Brikenmeier TM, Dean DC (1991). Characterization of the α4 integrin gene promoter. Proc Natl Acad Sci USA 88: 4094–4098
Rosen EM, Nigam SK, Goldberg ID (1994). Scatter factor and the c-met receptor: a paradigm for mesenchymal/epithelial interaction. J Cell Biol 127: 1783–1787
Rubenstein JLR, Puelles L (1994). Homeobox gene expression during development of the vertebrate brain. Curr Top Dev Biol 29: 1–65
Scherson T, Serbedzija G, Fraser S, Bronner-Fraser M (1993). Regulative capacity of the cranial neural tube to form neural crest. Development 118: 1049–1061
Schmid P, Cox D, Bilbe G, Maier R, McMaster GK (1991). Differential expression of TGF β1, β2, β3 genes during mouse embryogenesis. Development 111: 117–130
Selleck MAJ, Bronner-Fraser M (1995). Origins of the avian neural crest: The role of neural plate-epidermal interactions. Development 121: 525–538

Serbedzija GN, Bronner-Fraser M, Fraser SE (1994). Developmental potential of trunk neural crest cells in the mouse. Development 120: 1709–1718

Sonnenberg E, Meyer D, Weidner KM, Birchmeier C (1993). Scatter factor/hepatocyte growth factor and its receptor the c-met tyrosine kinase, can mediate a signal exchange between mesenchyme and epithelia during mouse development. J Cell Biol 123: 223–235

Spence SG, Argraves WS, Walters L, Hungerford JE, Little CD (1992). Fibulin is localized at sites of epithelial-mesenchymal transitions in the early avian embryo. Dev Biol 151: 473–484

Spieth J, Keller RE (1984). Neural crest cell behavior in white and dark larvae of Ambystoma mexicanum: differences in cell morphology, arrangement and extracellular matrix as related to migration. J Exp Zool 229: 91–107

Stemple DL, Anderson DJ (1993). Lineage diversification of the neural crest: In vitro investigations Dev Biol 159: 12–23

Stepp MA, Urry LA, Hynes RO (1994). Expression of α4 integrin mRNA and protein and fibronectin in the early chicken embryo. Cell Adhes Commun 2: 359–375

Sternberg J, Kimber SJ (1986). The relationship between emerging neural crest cells and basement membranes in the trunk of the mouse embyo: a TEM and immunocytochemical study. J Embryol Exp Morphol 98: 251–268

Thiery JP, Duband J-L, Rutishauser U, Edelman GM (1982a). Cell adhesion molecules in early chicken embryogenesis. Proc Natl Acad Sci USA 79: 6737–6741

Thiery JP, Duband J-L, Delouvée A (1982b). Pathways and mechanism of avian trunk neural crest cell migration and localization. Dev Biol 93: 324–343

Thiery JP, Delouvée A, Grumet M, Edelman GM (1985). Appearance and distribution of the neuron glia cell adhesion molecule (Ng-CAM) in the chick embryo J Cell Biol 100: 442–456

Tosney KW (1978). The early migration of neural crest cells in the trunk region of the avian embryo: An electron microscopic study. Dev Biol 62: 317–333

Tosney KW (1982). The segregation and early migration of cranial neural crest cells in the avian embryo. Dev Biol 89: 13–24

Valarché I, Tissier-Seta J-P, Hirsch M-R, Martinez S, Goridis C, Brunet J-F (1993). The mouse homeodomain protein Phox2 regulates Ncam promoter activity in concert with Cux/CDP and is a putative determinant of neurotransmitter phenotype. Development 119: 881–896

Vandenbunder B, Pardanaud L, Jaffredo T, Mirabel MA, Stehelin D (1989). Complementary patterns of expression of c-ets 1, c-myb and c-myc in the blood-forming system of the chick embryo. Development 106: 265–274

Weidner KM, Behrens J, Vandekerckhove J, Birchmeier W (1990). Scatter factor: molecular characteristics and effect on the invasiveness of epithelial cells. J Cell Biol 111: 2097–2108

Establishment of Pathways in the Developing Neural System

M. CHIQUET*

1 Introduction: The Chick as a Model to Study Vertebrate Neurogenesis

For studying neural development, the chick embryo offers distinct advantages compared to other vertebrate models. The first is the accessibility of the embryos, which survive extensive manipulations in ovo. For example, it has been shown that axons in developing chick limb nerves generally follow stereotyped pathways which are predetermined by the limb mesenchyme (LEWIS et al. 1983; TOSNEY and LANDMESSER 1985). However, grafting experiments demonstrated that at certain crossroads growing axons can reorient their direction when they have been misguided (LANCE-JONES and LANDMESSER 1981; STIRLING and SUMMERBELL 1985). From such results, testable hypotheses were derived about the source, location, and mode of action of putative guidance cues along peripheral nerve pathways in vertebrates (TOSNEY and LANDMESSER 1985).

A second major advantage of the chick is the ease and low cost by which embryonic tissue is available for large scale biochemistry. Several cell-cell and

Department of Biophysical Chemistry, Biocenter of the University, Klingelbergstrasse 70, 4056 Basel, Switzerland
*Present Address: M.E. Müller-Institut für Biomechanik, Universität Bern, Murtenstrasse 35, 3010 Bern, Switzerland

cell-matrix adhesion molecules implicated in patterning of the vertebrate nervous system have first been purified from chick embryos. In recent exciting experiments, two distinct classes of chemo- or haptotactic molecules have been characterized which either inhibit or induce growth of specific vertebrate axons (Luo et al. 1993; Kennedy et al. 1994). In both cases, this was achieved by first fractionating extracts of thousands of embryonic chick brains according to activity in functional assays in vitro. In this brief review, only a few selected examples are presented in which the chick model has been extremely useful for studying the cellular and molecular mechanisms by which the vertebrate nervous system is wired during development. More extensive information is found in general reviews (Dodd and Jessel 1988; Reichardt and Tomaselli 1991; Hynes and Lander 1992).

2 Cellular Mechanisms of Peripheral Nerve Patterning

Newly born neurons are small cells of simple shape. To connect with their often distant targets, they have to grow axons and dendrites. Growth cones, the motile organelles at the tip of processes, trail these behind them and are responsible for pathfinding. Developing axons can bundle with other axons to form fiber tracts or peripheral nerves, or they can branch off from fascicles to explore new territories (see Dodd and Jessel 1988). How can the mechanisms be studied by which growing axons are guided and the nerve pattern is established? Before any guesses about the molecules involved can be made, the process needs to be described in detail during normal embryonic development. Then, it is useful to see how growing axons and nerves behave when their embryonic environment is experimentally altered. The chick embryo is very well suited for both description and manipulation, and its study has contributed most of what is known about the development of peripheral nerves in vertebrates.

Development of vertebrate segmental nerves starts with a massive outgrowth of motor axons from the ventral roots of the spinal cord (Tosney and Landmesser 1985). From the onset of growth, axons are accompanied by neural crest derived satellite cells, the precursors of Schwann cells. In each somite, axons and satellite cells are channeled through the anterior half of the sclerotome, hence giving rise to segmental nerves. Motor axons join with sensory fibers originating from the dorsal root ganglia. In the region of the limb buds, several segmental nerves bundle to form massive plexuses. From these, the individual limb nerves branch off to reach muscles and skin (Tosney and Landmesser 1985). The branching pattern of limb nerves is very stereotyped and practically identical for different embryos at the same developmental stage. It is different in wings and legs, but perfectly mirror symmetrical between left and right limbs of the same kind (Lewis et al. 1983). It should also be noted that the major nerve trunks are established in the limb before individual target muscles are formed.

Is the information for the pattern intrinsic to the growing nerves or provided by the peripheral environment? To address this question, several groups performed elegant transplantation experiments on chick embryos in ovo. To do so, a window is cut into the eggshell and the living embryo is operated on. The window is sealed, and the embryo is allowed to develop further before being analyzed. In this case, grafts were made at a time before the nerves had invaded the limb buds. In one set of experiments, limbs were rotated or replaced by limbs from the contralateral side. In a first approximation, the nerve pattern which formed corresponded to the one expected for the grafted limb, not for the host nerves at this site (STIRLING and SUMMERBELL 1985). Accordingly, many axons eventually innervated incorrect targets. Does this mean that peripheral axons do not know their target at the onset of their journey and that they simply follow public highways which are wholly predetermined by the environment? Closer analysis showed that this is not the case (see STIRLING and SUMMERBELL 1985). LANCE-JONES and LANDMESSER (1981) operated on young chick embryos and left the limb buds intact but reversed a small piece of the spinal cord along its axis at the level of the leg buds. After allowing an apparently normal nerve pattern to develop, they labeled sets of axons to follow their trajectory. If nerve pathways were wholly determined by the periphery, after spinal cord reversal anterior limb nerves should be populated by axons from originally posterior neurons and vice versa. However, this is not what happened. Although the misguided axons left the spinal cord via a wrong spinal nerve to enter the plexus, from there they still found the right exit nerve and innervated the correct target muscles. More extensive misplacements of the spinal cord led to innervation of limb muscles by incorrect spinal cord neurons, however. These experiments showed that, although peripheral axons do travel along common mesenchymal highways, from the beginning they "know" who they are and where to go. At intersections along the way (e.g., in the plexuses) they are able to make individual decisions, presumably by reading local "road signs." This works as long as axons reach the correct plexus. If misguided to a completely unfamiliar highway (as, e.g., in the case of a limb rotated by 180°), they are lost and, in default, start to innervate wrong targets.

The influence of targets on development of the nerve pattern was studied in muscle-free limbs (LEWIS et al. 1981). These can easily be generated in the chick embryo by irradiating, at an early stage, the myotomes which provide the precursor cells for all limb muscles. Surprisingly, although individual muscle nerve branches failed to develop in muscle-free limbs, the pattern of major nerve trunks appeared normal and hence developed independently of the presence of target muscles (LEWIS et al. 1981).

From these and similar experiments, several important conclusions can be drawn concerning the mechanisms of peripheral nerve patterning. First, growing limb nerves follow mesenchymal pathways which are permissive for growth, i.e., which allow local attachment and induce migration of growth cones. Pathways are delineated by nonpermissive areas, most notably the prechondrogenic areas of the posterior sclerotome and of the girdle regions (TOSNEY and LANDMESSER 1985). Second, local cues allow different types of axons to make choices at intersections along the highway (LANCE-JONES and LANDMESSER 1981; STIRLING and SUMMERBELL 1985). Third, in early peripheral nerve development there is no evidence for the involvement of long range signals secreted by target tissues, although such factors are clearly important for the establishment of neural pathways, e.g., in the spinal cord, as mentioned later (KENNEDY et al. 1994). Knowledge about the cellular interactions which lead to nerve pattern formation was a prerequisite to identify molecules involved. Among these are cell-matrix and cell-cell adhesion molecules for short range interactions (HYNES and LANDER 1992) and chemotactic or haptotactic molecules acting at a long range (DODD and JESSEL 1988). In the next paragraphs, examples for such molecules will be presented.

3 Tenascin: An Extracellular Matrix Protein Specifically Expressed Along Growing Peripheral Nerves

Only about half of the volume of embryonic mesenchyme is filled by cells; the rest is extracellular matrix (ECM). Cells interact with this matrix via specific receptors such as integrins, transmembrane proteins which inside the cell connect to the cytoskeleton. Since adhesion to a solid substrate is essential for axonal extension, it is not surprising that certain ECM proteins promote axonal growth via integrin receptors on the neuronal surface (HYNES and LANDER 1992). The best known example is laminin, which is extremely potent in promoting process formation by cultured peripheral and certain central neurons (REICHARDT and TOMASELLI 1991). Laminin is abundant in basement membranes, e.g., of Schwann cells, where it is thought to be important for regeneration of peripheral nerves after injury. Its function in peripheral nerve development is not clear, however, because the pathways of growing nerves do not follow laminin-rich basement membranes (TOSNEY and LANDMESSER 1985).

We became interested in another ECM protein, tenascin, because we found it to be highly expressed along developing nerves at phases of rapid growth (Fig. 1A; WEHRLE-HALLER et al. 1991). Tenascin is a multidomain protein with six large, extended and flexible subunits, giving it an insect-like appearance in the electron microscope. It was first purified from human and chick cell cultures and, coincidentally, from extracts of thousands of chick embryonic brains. Tenascin and related molecules turned out to be major ECM molecules in the developing and adult brain where they are found in defined regions and layers (for recent reviews, see TUCKER 1994).

Concerning the expression of tenascin in developing peripheral nerves, we found that its main source was not the mesenchyme along prospective nerve pathways, but instead the satellite cells (Schwann cell precursors) which comigrate with the growing axons. Hence, tenascin is not likely to be involved in early pathfinding but rather in later steps of nerve morphogenesis. To find out about its

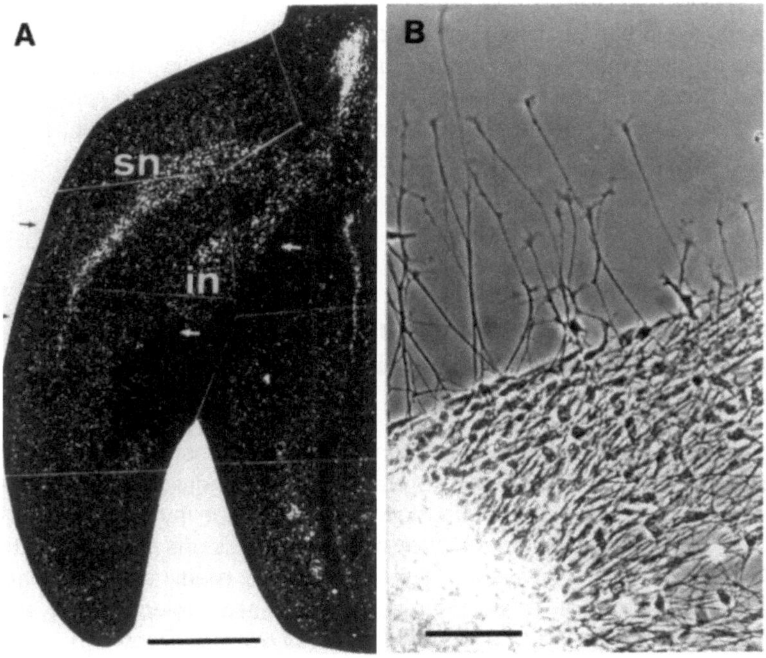

Fig. 1A,B. Expression of tenascin along developing peripheral nerves, and influence on neural cells as a substrate in vitro. **A** A cross-section through a 5 day old chick embryo on the level of the hindlimbs was hybridized with a radioactive probe specific for tenascin mRNA and autoradiographed; *sn*, superior brachialis longus nerve; *in*, inferior brachialis longus nerve. *Bar*, 400 μm. **B** A dorsal root ganglion explant from a 6 day old chick embryo (*lower left*) was plated on glass coated with a pattern of fibronectin (*bottom half*) vs tenascin (*top half*; for methods, see WEHRLE-HALLER and CHIQUET 1993). Note that sensory axons cross the border from fibronectin to tenascin whereas the spindle-shaped satellite cells are held back. *Bar*, 100 μm

function, we cultured neuronal explants on substrates covered with tenascin and found a strange dual mode of action on the adhesion of neurons vs satellite cells by this molecule; specific tenascin domains and cellular receptors are likely to be involved. Neuronal growth cones were able to attach and migrate on plain tenascin substrates, resulting in rapid process formation (WEHRLE and CHIQUET 1990). In contrast, satellite cells which extensively migrate on other ECM substrates were inhibited in their movements on tenascin and only spread along the surfaces of neurites (WEHRLE-HALLER and CHIQUET 1993). When confronted with a border between fibronectin (the major mesenchymal adhesive ECM protein) and tenascin, cultured neurites crossed the border whereas satellite cells piled up along it (Fig. 1B; WEHRLE-HALLER and CHIQUET 1993). Since the satellite cells themselves are the main source of tenascin in growing peripheral nerves, we believe that this protein helps them to set up boundaries between and around axon fascicles. The satellite cells seem to prevent themselves from migrating off into the surrounding adhesive mesenchyme and instead limit their movements to the surfaces of axons. In contrast, tenascin is permissive for the growth of axons which are free to explore the environment or sort out between fascicles. Thereby, tenascin might stabilize the developing nerve pattern while still allowing plasticity. This example shows that, on the molecular level, the "permissiveness" of an extracellular pathway might be subtle: a single ECM molecule can permit movements by one cell type and inhibit migration of another.

4 Collapsins/Semaphorins: Cell Surface-Associated Molecules Inhibiting the Growth of Specific Sets of Axons

As described earlier, growing peripheral axons bundle into fascicles, converge into large plexuses, and sort out again to branch off into individual nerves. At intersections along the way, sequential decisions are made by each axon of whether to stay with the fascicle or to branch off. These decisions are believed to depend on intimate contacts between neighboring axons, mediated by specific sets of cell adhesion molecules on their surfaces. These molecules are the cadherins and the CAMs of the Ig superfamily (for review, see Dodd and Jessel 1988; Hynes and Lander 1992). Several CAMs and cadherins have first been isolated from chick embryos; despite their importance for axonal growth and guidance, they will not be discussed further here.

Instead, we will focus on a recently discovered class of molecules associated with cell surfaces which, in contrast to the growth-promoting CAMs and cadherins, induce the instant collapse of advancing growth cones upon contact. In vertebrates, the first of these so called collapsins was again purified from the chick (Luo et al. 1993). The starting point was the observation by Kapfhammer and Raper (1987) that the growth cones of cultured sensory or sympathetic axons continued to migrate when meeting the axon of a like neuron. In contrast, the growth cone of a sensory neuron collapsed and its axon retracted upon contact with a sympathetic axon (or vice versa). Using their in vitro assay to screen for function, these authors then purified a similar collapsing activity from extracts of many embryonic chick brains (Raper and Kapfhammer 1990) and eventually cloned its cDNA (Luo et al. 1993). Most interestingly, the molecule collapsin turned out to be closely related to a protein in grasshopper and *Drosophila* called semaphorin I (or fasciclin IV), which is expressed in a transverse epithelial stripe in insect wing buds and apparently causes growing sensory axons to make a right angle turn (Kolodkin et al. 1993). Members of the semaphorin gene family are either transmembrane or extracellular proteins, indicating that some might function as short range cues whereas others act at a distance (Kolodkin et al. 1993). Biochemical isolation from the chick and an assay for function in vitro helped to elucidate the mode of action of these important molecules which guide growing axons by repulsion and which are shared by insects and vertebrates.

5 Netrins: Chemotactic/Haptotactic Molecules Which Attract Certain Classes of Growing Axons

So far, we have considered cell surface and substrate molecules which influence the steering of growth cones via short range interactions, as they predominate in peripheral nerve development. However, there are cases known in vertebrates in which a chemotactic gradient attracts growing axons at a long distance. For example, the commissural neurons are located dorsally in the developing spinal cord, and their axons travel towards the ventral floor plate. Experiments in vitro showed that a chemoattractant secreted by the floor plate is likely to guide the commissural axons. When an explant from the dorsal half of an embryonic spinal cord was embedded into a collagen gel in the vicinity of a piece of floor plate, commissural axons grew out from the neural explant in the direction of the floor plate (Fig. 2A; Tessier-Lavigne et al. 1988). Although first shown with explants from rat embryos, the proof that such a chemoattractant must also function in vivo again came from experiments on living chick embryos: When an isolated floor plate was transplanted laterally to the spinal cord, commissural axons were deviated from their normal path and started to grow towards the transplant (Placzek et al. 1990). In an effort to identify the responsible components, Tessier-Lavigne and colleagues used their in vitro assay to purify two similar activities from extracts of – once more – thousands of embryonic chick brains (Serafini et al.

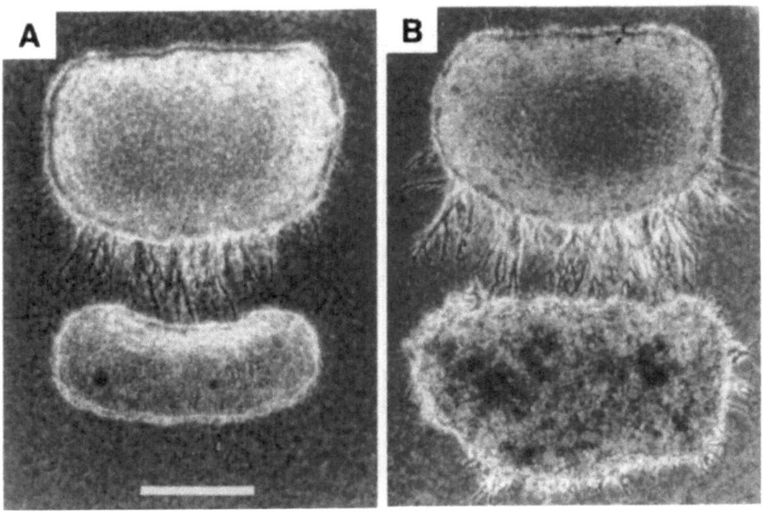

Fig. 2A,B. Attraction of growing commissural axons by cells secreting netrin-1 in vitro. **A** A dorsal spinal cord explant from an 11 day old rat embryo (*top*) was placed next to a floor plate explant (*bottom*) and cultured for 2 days. **B** Same as in **A**, except that a pellet of COS cells secreting recombinant chick netrin-1 (*bottom*) was placed next to the spinal cord explant (*top*). Note the axons growing from the spinal cord explant towards the source of netrin-1 in both cases. *Bar*, 200 μm. (From Kennedy et al. 1994)

1994). The factors were cloned and expressed in COS cells. When a pellet of the transfected COS cells was put at some distance from a dorsal spinal cord explant in a collagen gel, the commissural axons grew out towards the COS cell pellet which thus mimicked the floor plate (Fig. 2B; KENNEDY et al. 1994). This proved that the cloned proteins, which were called netrin I and II, were indeed chemoattractants for the growth cones of commissural axons. The structure of netrins turned out to be very interesting (SERAFINI et al. 1994). Half of their peptide sequence is homologous to the NH_2-terminal domain of the ECM protein laminin (which is known to be involved in laminin self-assembly; SCHITTNY and YURCHENCO 1990). The other half is very basic and probably responsible for the binding of netrins to heparin. These features of netrins make it quite likely that they are not freely diffusible chemotactic molecules, but that their diffusion is limited by binding to extracellular components. Thus, in the developing spinal cord the floor plate might set up substrate-bound, haptotactic netrin gradients. The expression patterns of netrin I and II are in agreement with this idea (KENNEDY et al. 1994).

Very exciting is the fact that also the netrins have close invertebrate relatives. A protein called UNC-6 is the structural homologue in *C. elegans* (ISHI et al. 1992; SERAFINI et al. 1994). There, the protein was identified genetically: In *unc*-6 mutant animals, the axons of certain ventral motor neurons are misguided and fail to enter a dorsal pathway (HEDGECOCK et al. 1990). Not only do nematode UNC-6 and vertebrate netrins share structural homology, but also their function as guidance cues for growing axons in the developing nervous system has been conserved during evolution.

6 Concluding Remarks

In the last two decades, much has been learned about the mechanisms and molecules involved in the wiring of the developing nervous system. Although the picture of this complex process is still very fragmentary, a few patterns emerged. The old question of whether growing axons are guided by permissive substrates, by specific short range cues, or by diffusibe long range factors became obsolete: These mechanisms are not mutually exclusive, but they complement each other. Their relative contribution depends on the neural pathway which is looked at, but there are proven examples for each of them, and the responsible molecules are being identified. By choosing a few examples, this brief article is meant to show the usefulness of the chick embryo as a model for elucidating the development of neural pathways. In my view, the history of the discoveries of collapsins and netrins is especially rewarding. In both cases, related molecules were identified in invertebrate models by mutational and functional analysis. In parallel, work on vertebrate embryos had shown that activities attracting or repulsing axons must exist in their developing nervous system. The chick allowed biochemical purification of active molecules by functional assays in vitro. Cloning of the cDNAs showed their homology to the corresponding molecules in invertebrates, and this is where the two completely independent approaches merged to yield a precise picture of how these molecules guide axons. Since it is increasingly clear that not only the mechanisms of axonal guidance but also the molecules involved are conserved between vertebrates and invertebrates, much synergy can be expected in the future by exploiting the specific virtues of various animal models, not the least the chick.

Acknowledgments. I thank Bernhard Wehrle-Haller and Manuel Koch for the pictures shown in Fig. 1, Marc Tessier-Lavigne for the permission to reproduce the photographs of Fig. 2, Ralph Brandenberger for critically reading the manuscript, and the Swiss National Fund for financial support.

References

Dodd J, Jessel TM (1988) Axon guidance and the patterning of neuronal projections in vertebrates. Science 242: 692–699

Hedgecock EM, Culotti JG, Hall DH (1990) The *unc-5*, *unc-6*, and *unc-40* genes guide circumferential migrations of pioneer axons and mesodermal cells on the epidermis in C. elegans. Neuron 4: 61–85

Hynes RO, Lander AD (1992) Contact and adhesive specificities in the associations, migrations, and targeting of cells and axons. Cell 68: 303–322

Ishi N, Wadsworth WG, Stern BD, Culottii JG, Hedgecock EM (1992) UNC-6, a laminin-related protein, guides cell and pioneer axon migrations in C. elegans. Neuron 9: 873–881

Kapfhammer JP, Raper JA (1987) Collapse of growth cone structure on contact with specific neurites in culture. J Neurosci 7: 201–212

Kennedy TE, Serafini T, de la Torre JR, Tessier-Lavigne M (1994) Netrins are diffusible chemotropic factors for commissural axons in the embryonic spinal cord. Cell 78: 425–435

Kolodkin AL, Matthes DJ, Goodman CS (1993) The semaphorin genes encode a family of transmembrane and secreted growth cone guidance molecules. Cell 75: 1289–1299

Lance-Jones C, Landmesser L (1981) Pathway selection by embryonic chick motoneurons in an experimentally altered environment. Proc R Soc Lond B 214: 19–52

Lewis J, Chevallier A, Kieny M, Wolpert L (1981) Muscle nerve branches do not develop in chick wings devoid of muscle. J Embryol Exp Morphol 64: 211–232

Lewis J, Al-Gaith L, Swanson G, Khan A (1983) The control of axon outgrowth in the developing chick wing. In: Fallon JF, Caplan AI (eds) Limb development and regeneration, part A. Liss, New York, pp 195–205

Luo Y, Raible D, Raper JA (1993) Collapsin: a protein in brain that induces the collapse and paralysis of neuronal growth cones. Cell 75: 217–227

Placzek M, Tessier-Lavigne M, Yamada T, Doff J, Jessell TM (1990) Guidance of developing axons by diffusible chemoattractants. Cold Spring Harbor Symp Quant Biol 55: 279–289

Raper JA, Kapfhammer JP (1990) The enrichment of a neuronal growth cone collapsing activity from embryonic chick brain. Neuron 4: 21–29

Reichardt LF, Tomaselli KJ (1991) Extracellular matrix molecules and their receptors: functions in neural development. Annu Rev Neurosci 14: 531–570

Schittny JC, Yurchenco PD (1990) Terminal short arm domains of the basement membrane protein laminin are critical for its self-assembly. J Cell Biol 110: 825–832

Serafini T, Kennedy TE, Galko MJ, Mirzayan C, Jessel TM, Tessier-Lavigne M (1994) The netrins define a family of axon outgrowth-promoting proteins homologous to C. elegans UNC-6. Cell 78: 409–424

Stirling RV, Summerbell D (1985) The behaviour of growing axons invading developing chick wing buds with dorsoventral or anteroposterior axis reversed. J Embryol Exp Morphol 85: 251–269

Tessier-Lavigne M, Placzek M, Lumsden AGS, Dodd J, Jessel TM (1988) Chemotropic guidance of developing axons in the mammalian central nervous system. Nature 336: 775–778

Tosney KW, Landmesser LT (1985) Development of the major pathways for neurite outgrowth in the chick hindlimb. Dev Biol 109: 193–214

Tucker R (ed) (1994) The function of tenascin: hypotheses and current knowledge. Perspectives on developmental neurobiology, vol 2, 1. Gordon and Breach, New York

Wehrle B, Chiquet M (1990) Tenascin is accumulated along developing peripheral nerves and allows neurite outgrowth in vitro. Development 110: 401–415

Wehrle-Haller B, Chiquet M (1993) Dual function of tenascin: simultaneous promotion of neurite growth and inhibition of glial migration. J Cell Sci 106: 597–610

Wehrle-Haller B, Koch M, Baumgartner S, Spring J, Chiquet M (1991) Nerve-dependent and -independent tenascin expression in the developing chick limb bud. Development 112: 627–637

Limb Development

P. FRANCIS-WEST* and C. TICKLE

1 Introduction

A fundamental problem in chick limb development is to understand the mechanisms that ensure that structures form in their correct positions with appropriate cell and tissue arrangements. The ideas of positional information have been particularly influential in analysis of this problem (WOLPERT 1971). According to these ideas, cells in the limb bud will be exposed to positional cues and then will interpret this positional information in order to participate in forming structures appropriate to that position. It has been useful to think of specification of cell

---

Departments of Anatomy and Developmental Biology, University College and Middlesex School of Medicine, University College London, Windeyer Building, Cleveland St., London W1P 6DB, UK
*Present address: Department of Craniofacial Development, UMDS, Guy's Tower, Floor 28, London Bridge, London SE1 7RT, UK

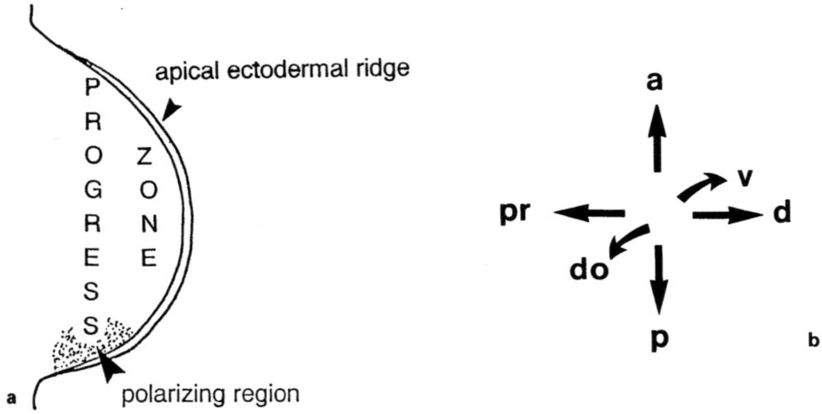

Fig. 1a,b. a Early chick limb bud with two localized signaling regions, the thickened apical ectodermal ridge rimming the bud tip and the polarizing region. Signals from these regions and from the ectoderm covering the limb bud operate in undifferentiated mesenchyme of the progress zone. **b** Orientation of proximo-distal (*pr-d*), antero-posterior(*a-p*) and dorsal-ventral (*do-v*) axes

position in a 3-dimensional coordinate system with respect to proximo-distal, antero-posterior and dorso-ventral axes (Fig. 1). Classical embryological experiments have defined three major interactions that lead to patterning along each of the axes. More recently, candidates have emerged for molecules that signal and mediate responses and the molecular networks have begun to be dissected. There is probably a limited range of molecular networks that control patterns of cell differentiation and the same systems are used at different times and in different sites in embryonic development. Therefore, the study of the chick limb may illuminate general principles.

2 Outline of Chick Limb Development

Limbs begin developing in chick embryos at around 2 days of incubation. During the next 2 days the pattern of tissues is set up. By 10 days of incubation, tissue differentiation has laid down the skeleton as a cartilage model and the musculature is defined.

The first external sign of limb development is the formation of paired thickenings in the body wall opposite somites 15–20, which will develop into wings; and opposite somites 26–33, which will give rise to legs. These thickenings soon become discrete buds, which, during the next 24 h, elongate rapidly. Since outgrowth occurs predominantly from the posterior part of the bud, the elongated buds do not project straight out from the body wall but are angled posteriorly. Next, an indentation appears about halfway along the bud. This indentation marks the future elbow/knee region. At the flat and broadened tip of the bud, thickened ridges (rays) become apparent and are forerunners of the digital skeleton. Throughout these changes in shape, the limb bud rudiment continues to grow, particularly in length. Separation of individual digits occurs as soft tissue between the digital rays regresses.

The early limb bud consists of a core of undifferentiated mesenchyme cells encased in ectoderm. Despite the apparent homogeneity of the undifferentiated mesenchyme cell population, the cells have two separate origins. Cells from the lateral ventral part of the somite migrate into the early limb forming region and give rise to the myogenic cells of the muscles (ORDAHL and LE DOUARIN 1992). By interchanging somites it has been shown that muscle cells from somites from any axial level are able to supply muscle cells to the developing limb (BUTLER et al. 1988). Therefore, the muscle cell precursors must respond to environmental signals to form the appropriate structures in the correct pattern. The other population of mesenchyme cells in the limb bud is derived from lateral plate mesoderm and will give rise to the various connective tissues including cartilage, tendon, perichondrium, and muscle connective tissues. The cell lineage of the various connective tissues is not known. An unanswered question is whether there is a multipotential precursor cell that gives rise to all connective tissue cell types or whether there is a special precursor cell for each type of connective tissue.

Initiation of budding of the limb involves local maintenance of cell proliferation, whereas proliferation of cells in regions on either side of the buds drops (SEARLS and JANNERS 1971). Initially, the ectodermal covering of the limb bud is two cells thick, as elsewhere over the body, and consists of a loosely packed layer of cuboidal cells, overlain by flattened periderm cells, sitting on the basement membrane. Later, as the bud is forming, the ectoderm at the tip of the limb becomes pseudostratified and known as the apical ectodermal ridge. Cells in the apical ridge are elongated, tightly packed and linked by extensive gap junctions (FALLON and KELLEY 1977). The ridge arises in the chick embryo at a very early stage of limb budding, while in mouse embryos the ridge develops later when the bud is quite pronounced. When a quail ridge was grafted in place of a chick ridge, the quail cells remained in the ridge during subsequent development. Thus the ridge appears to constitute a separate self-perpetuating cell population distinct from nonridge ectoderm (SAUNDERS et al. 1976), and cells do not appear to be recruited into the ridge from nonridge ectoderm during bud outgrowth.

Immediately below the ridge is a zone of undifferentiated mesenchyme cells. This zone remains at the bud tip during outgrowth, while at the bud base cells begin to differentiate. The first signs of cartilage differentiation are condensations of mesenchyme cells in the center of the limb bud. These condensations may be due to aggregation of mesenchyme cells or failure of cells to move apart following cell division (reviewed WOLPERT 1982). As the condensations begin to expand due to cartilage cell differentiation and secretion of extracellular matrix, surrounding cells become oriented around the condensation to form a perichondrium. Above and below the condensations, accumulations of potentially myogenic cells, known as muscle masses, arise and these undergo a series of splittings to give individual muscles encased in connective tissue sheaths (Shellswell and Wolpert 1977; Robson et al. 1994). The muscle pattern is dictated by the pattern of connective tissue differentiation. Both nerve processes and neural crest cells, which give rise to pigment cells and Schwann cells, enter the limb bud once cartilage and muscle differentiation is underway.

The limb bud is vascularised from an early stage in development. A central artery supplies a capillary bed which drains into a submarginal vein. The submarginal vein in the chick limb bud lies some distance beneath the ectoderm and thus there is an avascular rim to the bud, whereas in the mouse limb bud the vessels are much closer to the ectoderm. As the bud grows, so the capillary network expands. This expansion is fueled by proliferation of the endothelial cell population in preexisting vessels rather than involving cell recruitment from undifferentiated limb mesenchyme (Wilson 1983).

3 Cell Death

During limb development, there are several regions of cell death which have been identified by staining pycnotic cells with vital dyes. These are found at the posterior and anterior margins of the chick limb bud (stage 24/25); in between the forming ulna and radius (the opaque patch, stage 24/25) and in the interdigital spaces (Saunders and Gasseling 1962). These localized regions of cell death may be important in sculpting the limb. Whilst the importance of interdigital cell death is clear in the fact that it separates the digits, the role of anterior and posterior necrotic zones and the opaque patch are uncertain as mouse limb buds appear to lack all these necrotic zones (Martin 1990).

4 Cell Interactions That Lead to Patterning

The three major patterning interactions are: (1) an epithelial-mesenchymal interaction between the apical ectodermal ridge and underlying mesenchyme which governs outgrowth and proximo-distal pattern; (2) an interaction between posterior mesenchyme cells and mesenchyme cells at the tip of the bud that governs antero-posterior pattern; and (3) an interaction between ectoderm and mesenchyme governing dorso-ventral pattern (Fig. 1).

Outgrowth of the bud is mediated by a signal from the apical ectodermal ridge. When the apical ridge is removed from a limb bud, truncated limbs develop (Fig. 2c. Saunders 1948). The extent of the truncation depends on the stage of limb development at which the ridge was removed (Summerbell 1974a). When the ridge is removed early, the limbs are severely truncated, whereas when the ridge is removed later the limbs are more complete and lack only distal structures.

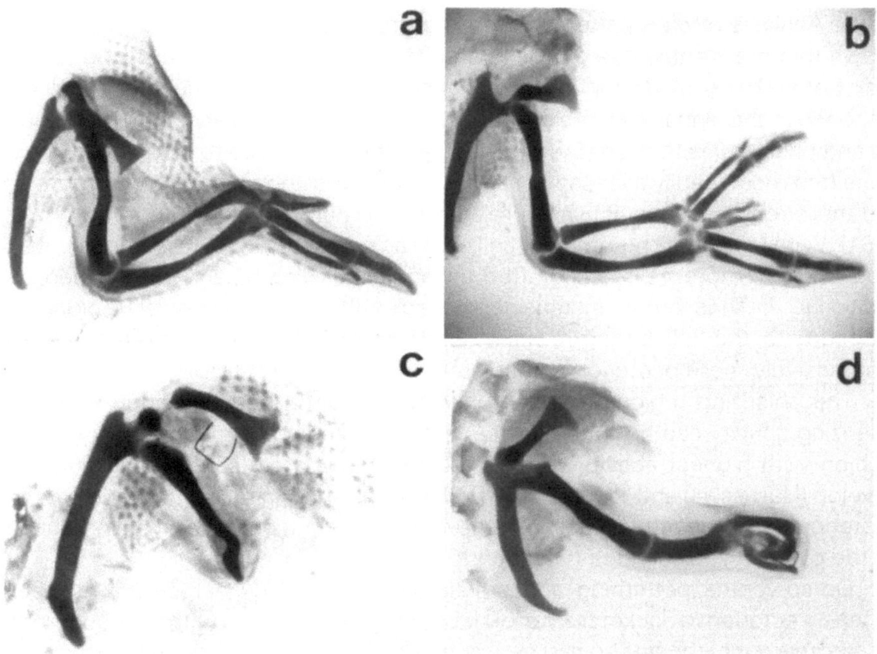

Fig. 2a–d. Whole mounts of chick wings stained with alcian green to show skeletal pattern. **a** Normal wing, digit pattern *234*; **b** wing following application of retinoic acid to anterior margin, digit pattern *4 3 2 2 3 4*; **c** wing following removal of apical ridge from early bud. Only part of the humerus has developed. **d** Wing following removal of apical ridge at early bud stage and application of FGF-4

Specification of cells to participate in forming an appropriate structure along the proximo-distal axis appears to depend on a timing mechanism. The apical ectodermal ridge signal maintains the zone of undifferentiated mesenchyme cells at the tip of the limb bud as the bud elongates. This zone of cells is known as the progress zone (SUMMERBELL et al. 1973). The length of time cells spend in the progress zone determines which structures they will form. Cells that leave the progress zone early form proximal structures whereas cells that leave later form distal ones.

Patterning across the antero-posterior axis of the limb is specified by a signal from the polarizing region, a small group of mesenchyme cells at the posterior margin of the bud (SAUNDERS and GASSELING 1968). The signal from the polarizing region acts on cells in the progress zone. When the polarizing region is grafted to the anterior margin of a limb bud, additional digits are induced in mirror-image symmetry with the normal set. Thus a polarizing region graft to a wing bud results in six digits arranged in the pattern *4 3 2 2 3 4* instead of three digits in the normal pattern *2 3 4* (reading anterior to posterior; Fig. 2b). Forearm pattern can also be affected if the graft has time to act before this region of the limb has been laid down. Analysis of digit patterns when the graft is placed in different positions shows that the identity of a digit is related to distance from the polarizing region (Tickle et al. 1975). The additional digits arise from the host tissue sequentially in response to the signal from the grafted polarizing region, first an additional *2*, then a *3* and finally a *4* is formed (Smith 1980). The ability of tissue to induce additional digits (the most readily assessed change in pattern) is known as polarizing activity and the strength of this activity is related to the number of cells grafted (Tickle 1981). With very small numbers of cells only an additional digit *2* is induced but as progressively more cells are grafted an additional digit *3* arises and then an additional *4*. Thus either increasing the strength of the signal or prolonging exposure to the full strength signal leads to specification of digits that are progressively more posterior in character.

The polarizing region is present in very early limb buds although potential polarizing activity can be found in the flank, prior to limb formation, in a graded fashion with highest activity where the posterior margin of the wing bud will develop (Hornbruch and Wolpert 1991). The polarizing region is maintained at the posterior margin near the distal tip of the elongating limb bud by a signal produced by the overlying apical ridge (Vogel and Tickle 1993).

Dorso-ventral patterning appears to be regulated by ectodermal signals. When an ectodermal jacket from a left limb bud is recombined with a mesodermal core from a right limb bud so that ventral ectoderm now covers dorsal mesoderm, the structures that arise distally are reversed dorso-ventrally to conform with the polarity of the ectoderm (MacCabe et al. 1974). In addition, when an apical ridge is grafted either dorsally or ventrally promoting outgrowth, then a second limb axis is induced with either a double dorsal or a double ventral pattern.

5 Signaling Molecules in the Developing Chick Limb Bud

The signaling cascades in limb bud outgrowth and patterning include both growth factors and retinoids. Growth factors known to be important in limb bud development include fibroblast growth factor (FGF) and transforming growth factor-β (TGF β) families, and *Hedgehog* and *Wnt* gene families. Other growth factors expressed in the developing limb include scatter factor (hepatocyte growth factor; Myokai et al. 1995) and insulin-like growth factors (IGFs; Ralphs et al. 1990). Scatter factor is expressed throughout the mesenchyme of the early limb bud and it later becomes localized to the distal tip of the bud suggesting that it may be involved in epithelial-mesenchymal interactions in the progress zone (Myokai et al. 1995). IGF is expressed in the progress zone and at the periphery of the limb bud. Expression of platelet-derived growth factor (PDGF) receptors in the developing limb bud has been detected by PCR (Potts and Carrington 1993).

6 Fibroblast Growth Factors and Their Receptors

The FGF family comprises nine FGFs (FGF-1 to FGF-9)(Miyamoto et al. 1993). The genes encoding some of these, for example, *Fgf-2* and *Fgf-4*, have potent oncogenic properties (reviewed MASON 1994). FGFs are small peptides that bind to heparin and exert their effects on cells via cell surface FGF receptors. FGF receptors operate as dimers and binding of the growth factor is transduced by an intracellular tyrosine kinase domain. There are four different receptors (FGFR1-4). An alternatively spliced variant of FGFR-2 is the product of the *bek* oncogene. In vitro studies show that many FGF family members can act as ligands for both FGFR-1 and FGFR-2, FGF-7 is the ligand for the *bek* product.

Of the FGFs, FGF-2, FGF-4 and FGF-8 are expressed in early vertebrate limb buds. FGF-2 protein is fairly widely distributed in both ectoderm and mesoderm of chick embryos as judged by immunohistochemistry (DONO and ZELLER 1994), and *Fgf-4* transcripts are localized to the posterior part of the ridge of chick embryos (Fig. 3a; NISWANDER et al. 1994). In mouse embryos, *Fgf-8* transcripts are found throughout the ridge and can be detected earlier than *Fgf-4* transcripts (HEIKINHEIMO et al. 1994; CROSSLEY and MARTIN 1995). With respect to the receptors, *Fgr-1* is expressed in mesenchymal cells and *Fgfr-2* and *bek* in overlying ectoderm including apical ridge in mouse embryos.

Both FGF-2 and FGF-4 have been shown to substitute for the apical ectodermal ridge in chick limb buds. It is not yet known whether FGF-8 has the same property. When the apical ridge is removed from early limb buds, limb truncations can result and only shoulder girdle and upper arm may develop. However, application of two beads soaked in either FGF-4 or FGF-2 can lead to

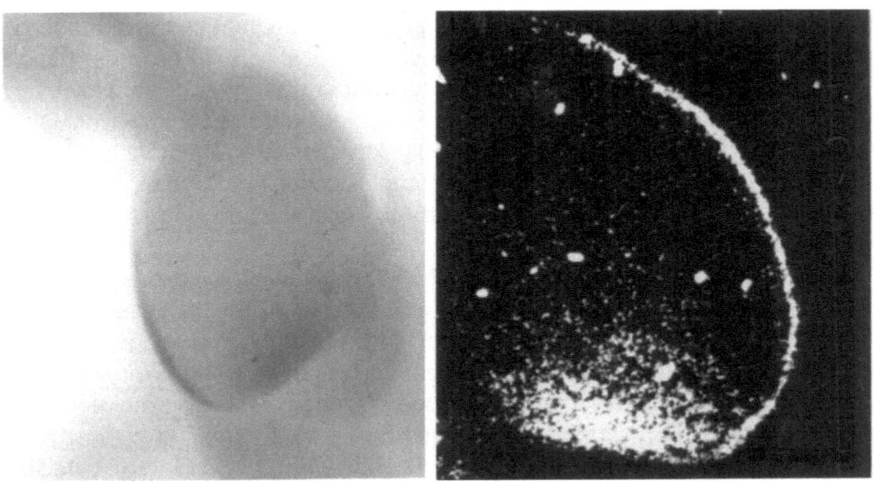

Fig. 3a,b. Two signaling molecules expressed in normal chick limb buds: **a** *Fgf-4* transcripts in posterior apical ridge; **b** *Bmp-2* transcripts in posterior mesenchyme

development of a complete set of limb structures in the absence of the ridge (Niswander et al. 1993; Fallon et al. 1994).

Posterior application of FGF-4 has been shown to also maintain the polarizing region after ridge removal (Vogel and Tickle 1993). This maintenance of polarizing activity in posterior mesenchyme is correlated with the formation of distal limb structures in limb buds stripped of a ridge (Niswander et al. 1993; Fig. 2d). Application of FGF-4 can stimulate proliferation of anterior mesenchyme cells but application in this position does not generate distal structures, just a very thick humerus. Thus, both an FGF and a polarizing region signal appear to be required to maintain the progress zone.

An FGF signal has also been recently implicated in initiating limb development. In mouse embryos which develop from chimeras containing cells that constitutively express *Fgf-4*, ectopic limb bud-like structures develop in the flank between forelimbs and hindlimbs (Abud et al., in preparation). When an FGF bead is placed in the flank of a chick embryo before limb buds develop, an ectopic bud is produced which establishes its own polarizing region and apical ectodermal ridge and can then go on to develop into a complete limb (Cohn et al. 1995).

7 The Transforming Growth Factor-β Family and Receptors

The TGFβ family consists of secreted proteins which are thought to have both autocrine and paracrine actions in development (reviewed Massague 1990). The family consists of a wide range of related proteins including TGFβ itself (types 1,2,3), activins, and the bone morphogenetic proteins (BMP 2–8). These proteins act as dimers and can form heterodimers or homodimers, within the subgroups, which may have different biological activities (Hseuh et al. 1987). The identification of TGFβ and BMP receptors in the chick has not yet been reported. However, some of these receptors have been cloned in several different animals including humans and *C. elegans* (Massague et al. 1994). Human TGFβ receptors have been analyzed in detail and the TGFβ receptor complex consists of two proteins, type 1 and type 11, both of which are required for signal transduction (Wrana et al. 1992, 1994).

Of the TGFβ family, BMP-2, BMP-4, BMP-7 are known to be expressed in the early chick limb bud (Francis et al. 1994; Francis-West et al. 1995a). In addition, the activin 11A receptor is expressed in chick limb mesenchyme suggesting that activin or activin related proteins are present (Ohuchi et al. 1992). *Bmp-2, Bmp-4* and *Bmp-7* are expressed in the apical ectodermal ridge and in discrete regions in the mesenchyme. This suggests that these BMPs have roles in mesenchymal signaling and epithelial-mesenchymal interactions. Of particular interest are the mesenchymal expression domains of *Bmp-2* and *Bmp-7* which are associated with the polarizing region (Fig. 3b; Francis et al. 1994; Francis-West et al. 1995a).

In later limb buds, Bmp *2, 4, 7* and also Tgfβ *1, 2, 3* are expressed in regions of cartilage differentiation (THORP et al. 1992; FRANCIS et al. 1994; P.H. FRANCIS-WEST et al., unpublished data). TGFβ-1 and β-2 and the BMPs (BMP-2, -3, -4) have all been shown to promote chondrogenesis when added to chick mesenchymal cells in vitro (SCHOFIELD and WOLPERT 1990; CARRINGTON et al. 1991; CHEN et al. 1991; ROARK and GREER 1994). This is consistent with their expression pattern in older limbs and suggests that they have a role in skeletal development.

8 The *Hedgehog* Gene Family

The vertebrate hedgehog family consists of three members, *sonic hedgehog, desert hedgehog* and *Indian hedgehog* which are homologues of the *Drosophila* segment polarity gene, *hedgehog* (LEE et al. 1992). The *hedgehog* gene family encodes proteins which have a hydrophobic leader sequence and therefore are thought to act as secreted molecules (FIETZ et al. 1994). The primary protein product of verterbrate *sonic hedgehog* gives rise to two protein species by internal autoproteolytic cleavage (CHANG et al. 1994; LEE et al. 1994). In *Drosophila* there is some evidence that suggests that the product of the gene *patched* may act as a receptor for *Hedgehog* (see FIETZ et al. 1994).

In developing chick limbs, sonic hedgehog (shh) is expressed in the polarizing region of the limb bud (RIDDLE et al. 1993). The *shh* gene is also expressed in other regions of the embryo such as Hensen's node and the floor plate which are involved in patterning embryonic structures (RIDDLE et al. 1993). Ectopic expression of *shh* at the anterior margin of a chick limb bud induces digit duplications showing that *shh* can provide a polarizing signal (RIDDLE et al. 1993).

9 The *Wnt* Gene Family

The *Wnt* gene family consists of up to ten or more closely related genes (reviewed NUSSE and VARMUS 1992) which encode glycoproteins. *Wnt-1* is the homologue of *wingless* in *Drosophila*, which is required for the correct patterning of each segment and other structures including the wing imaginal disc and was also identified as an oncogene by insertion of the mouse mammary tumor virus. The receptors which mediate *Wnt* gene family signaling are at present unknown.

At least two of members of the *Wnt* gene family are expressed in the developing limb. In chick limb buds, *Wnt-5a* is expressed in the apical ectodermal ridge and in the mesenchyme. In the mesenchyme, highest levels of transcripts are found in the progress zone and progressively lower levels in more proximal mesenchyme (DEALY et al. 1993). It has been suggested that the levels of *Wnt-5a* transcripts may determine the fate of distinct regions along the proximo-distal axis such that they form hand plate, radius/ulna and humerus but there is no direct evidence to support this.

Wnt-7a transcripts are localized to the dorsal ectoderm (Parr et al. 1993; Dealy et al. 1993). In mice, knockout of *Wnt-7a* function results in ventralization of distal limb structures suggesting that *Wnt-7a* is a signal that helps to specify the dorso-ventral axis of the limb bud (Parr and McMahon 1995). In *Drosophila, wingless* organizes the dorso-ventral axis of the leg. Thus, some of the signaling interactions that determine patterning of limb structures have been conserved through evolution. In *Drosophila, wingless* appears to act locally over a few cell diameters (reviewed Burrus 1994). It is not yet known whether *Wnt-7a* acts locally in the vertebrate limb.

10 Retinoids and Retinoic Acid Receptors

Derivatives of vitamin A are collectively known as retinoids. The main form of vitamin A in animal tissues is retinol, which can be metabolized irreversibly into retinoic acid. There are several naturally occurring isomers of retinoic acid, of which all-*trans*-retinoic acid is the main isomer but other isomers such as 9-*cis*-retinoic acid are also found. In a parallel metabolic pathway, another related retinoid, all-*trans*-3, 4-didehydro-retinoic acid can be generated from 3, 4-didehydroretinol (Thaller and Eichele 1990). Metabolism of retinoic acid itself gives a complicated set of polar and nonpolar products.

Retinoic acid is a small lipid-soluble molecule and can interact with binding proteins and nuclear receptors in target cells. Cellular retinoic acid binding proteins (CRABP I and II) may regulate the amount of retinoic acid that reaches the nucleus to interact with nuclear retinoic acid receptors, either by acting as a shuttle between cytoplasm and nucleus or sequestering retinoic acid in the cytoplasm. There are two main classes of nuclear retinoid receptor, retinoic acid receptors (RARs) and retinoid-X-receptors (RXRs). These receptors have the same structural plan as steroid hormone receptors and contain a DNA binding domain and a ligand binding domain. The ligands for RARs are all-*trans* and 9-*cis*-retinoic acid whereas only 9-*cis* binds to RXRs. In mice, three different RARs have been identified and three RXRs, designated, in each case α, β, γ. In addition, there are up to seven different isoforms of each RAR (reviewed Mendelsohn et al. 1992). It seems likely that the chicken genome contains genes that code for the same number of RARs and RXRs.

Retinoids can be extracted from chick limb buds. All-*trans*-retinoic acid is estimated to be present at nanomolar concentrations whereas retinol, the precursor of retinoic acid, is much more abundant. In an experiment in which over 300 chick limb buds were dissected, it was found that relatively more retinoic acid could be extracted from posterior tissue, which contains the polarizing region, than from anterior tissue (THALLER and EICHELE 1987). In addition, 3, 4-didehydro-retinoic acid has been shown to be present endogenously. Limb bud cells can metabolize retinol to retinoic acid and all-*trans*-retinoic acid has a half-life of less than 30 min in limb bud tissue (EICHELE et al. 1985). Therefore retinoic acid could act as a transient signaling molecule in the limb bud.

CRABP protein can be detected by antibody labeling in the progress zone of chick limb buds with a gradation in labeling intensity suggesting higher protein levels anteriorly (MADEN et al. 1988). Chick limb bud cells also express transcripts of RARs and RXRs although the most complete expression maps for these genes have been made in mouse embryos (MENDELSOHN et al. 1992). In both chickens and mice, RAR β is expressed in proximal limb bud cells although an isoform of RARβ has also been reported in distal chick limb cells (ROWE et al. 1991a; SMITH and EICHELE 1991). It seems likely that, in the chicken as in the mouse, RAR α and RAR γ will be expressed throughout the limb bud. Both epithelium and mesenchyme of early chick limb buds express RXR α (SELEIRO et al. 1995). In contrast, RXR γ is only expressed in neural crest cells migrating into the limb bud (ROWE et al. 1991b).

Over 10 years ago all-*trans*-retinoic acid was found to mimic signaling of the polarizing region (TICKLE et al. 1982). When a bead soaked in retinoic acid is implanted at the anterior margin of a chick limb bud a mirror-image pattern of digits is induced which is essentially similar to that produced by a polarizing region graft (TICKLE et al. 1985; Fig. 2b). Local application of retinoic acid produces dose-dependent changes in wing pattern; low doses specifying just an additional digit *2* and progressively higher doses an additional *3* and then a *4*. The same sequential addition of digits is produced by altering the length of exposure to the signal, with the most anterior digit, digit *2*, forming first, then an additional *3* and finally an additional *4*. These time/dose relationships are similar to those found for polarizing region cells. When increasing numbers of polarizing cells are grafted or when a grafted polarizing region is removed after various lengths of time, the same sequential specification of digits is seen (reviewed TICKLE 1991). Both 3, 4-didehydro-retinoic acid and 9-*cis*-retinoic acid also lead to digit duplications with 9-*cis* retionic acid being considerably more potent than all-*trans*-retinoic acid (THALLER et al. 1993). Other embryonic signaling regions such as floor plate and Hensen's node, "organizers" of the neural tube and early body axis, respectively, can provoke digit duplications in chick wings and also have been shown to generate retinoic acid (HOGAN et al. 1992; WAGNER et al. 1990).

One possibility is that retinoic acid provides a long range positional signal across the antero-posterior axis of the limb bud. The more abundant CRABP protein in anterior cells could sequester retinoic acid and this would steepen the gradient of free retinoids across the limb; thus enhancing the two fold difference between levels of retinoic acid found in the posterior and anterior mesenchyme. However, in other systems, it has been shown that even a two fold difference in concentration of a signaling molecule is sufficient to specify different cell fates (GREEN and SMITH 1990). Which of the receptors mediate the duplicating effects of retinoic acid in the developing limb is not known. RAR β expression can itself be induced by retinoic acid and when a bead soaked in retinoic acid is implanted at the anterior margin of a chick wing bud, RAR β transcripts are induced in anterior mesenchyme (Noji et al. 1991). This change in receptor expression is not seen when a polarizing region is grafted, which has led to the conclusion that the duplicating effect of the graft is not due to production of retinoic acid (Brockes 1991). Retinoic acid could act directly on both mesenchymal and ectodermal cells of the limb and different receptors could operate in each tissue. Recombination experiments between mesenchyme and ectoderm of retinoid-treated and normal limbs show that changes in the mesenchyme are irreversible whereas any epithelial changes are reversible (Tickle et al. 1989).

11 Gap Junctional Communication

Gap junctions are communication channels between cells which allow the passage of small molecules and are, therefore, one mechanism of intercellular signaling. Freeze fracture studies have shown that there are extensive and numerous gap junctions between cells in the apical ectodermal ridge (Fallon and Kelley 1977). In limb bud mesenchyme, there are differences in the number of gap junctions between cells distally and proximally with many more junctions between distal mesenchyme cells (Kelley and Fallon 1983). Immunohistochemistry with antibodies to connexin43, a component of chick gap junctions, also shows much more extensive labeling between mesenchyme cells at the tip of the chick limb bud than between cells more proximally (Green et al. 1994). Removal of the apical ridge leads to a reduction in mesenchymal connexin43 labeling. More direct evidence that gap junctions are instrumental in patterning interactions is that blocking communication between polarizing region and anterior chick limb mesenchyme cells with a gap junction antibody reduces the ability of a mixed graft of polarizing region and anterior cells to induce digit duplications (Allen et al. 1990). It is interesting that two classes of signaling molecules implicated in limb bud signaling, retinoids and the products of *Wnt* genes, have been shown in other systems to affect cell-cell communication via gap junctions (Mehta et al. 1989; Olson et al. 1991).

12 Homeobox Genes and Limb Patterning

Homeobox genes encode transcription factors and were originally identified in *Drosophila* by mutations which led to replacement of body structures with others structures that normally form elsewhere. In vertebrates, there are four gene complexes (Hox A, B, C and D) which appear to be involved in determining cell fate in different regions of the body (reviewed by Hunt and Krumlauf 1992;

McGinnis and Krumlauf 1992). Genes at the 3' end of each cluster are expressed earlier in development than the more 5' members, and successive genes (from 3' to 5') are expressed generally in sequentially restricted domains within each other. In the chick limb, four genes of the Hox A complex (Hox a-9, -10, -11, & -13) and five genes of the HoxD complex (Hoxd-9 to -13) are activated sequentially during development (Izpisua-Belmonte et al. 1991; Yokouchi et al. 1991). During early limb development, these members of the HoxA and HoxD complexes are expressed within nested domains along the p-d and a-p axes, respectively (Yokouchi et al. 1991; Izpisua-Belmonte et al. 1991; Nohno et al. 1991). This has led to the suggestion that cell fate is determined by the combination of these genes that are expressed (Izpisua-Belmonte et al. 1991; Yokouchi et al. 1991). Along the p-d axis, the most distal structures would be formed by cells expressing all the HoxA complex whereas more proximal structures would arise from cells which express only the more 3' members of the HoxA gene complex. Along the a-p axis, the most posterior digit, digit *4*, would arise from cells which express all the HoxD cluster whereas the most anterior digit would form from cells expressing only Hoxd-9.

The model that expression of the Hoxd complex can specify cell fate has been supported by studies in which one member of the HoxD complex, Hoxd-11, was overexpressed in limb buds such that cells which normally only express Hoxd-9 and Hoxd-10 also express Hoxd-11. In these studies a transformation of leg digit *2* to one with a more posterior phenotype was observed in 30% of the cases (Morgan et al. 1992). Furthermore, by studying the induction of these genes by polarizing signals, changes in patterning are linked to the induction of the HoxD complex (Izpisua-Belmonte et al. 1991; Nohno et al. 1991). Tissue with low polarizing activity such as the neural tube only induces expression of more 3'members of the HoxD complex, whereas tissues with high polarizing activity induce 5' members of the complex. Therefore, there is a correlation between gene members of the HoxD complex induced and the fate of cells (Izpisua-Belmonte et al. 1992a). However, the ability of Hox genes to directly specify cell fate within the limb has recently been challenged by studies in mice in which the functional gene for one member of the HoxD complex, Hoxd-13, has been knocked out (Dollé et al. 1993). In these mice, there appears to be growth retardation and slight changes in skeletal phenotype but not in a-p patterning. Thus, the predicted phenotype, that of loss of posterior structures, is not seen. It is possible that other Hox genes such as the paralogue Hoxa-13 could compensate for the lack of expression of Hoxd-13.

Another Hox gene appears to establish cell position in the limb as it forms. Hoxb-8 is normally expressed in the posterior part of the early forelimb bud of mouse embryos (Charitie et al. 1994). However, studies in mice have shown that, if it is also expressed further anteriorly, digit duplications are induced. Thus, Hoxb-8 expression appears to be involved in the induction of polarizing signals in the forelimb bud.

13 Signaling Cascades in the Limb Bud

A-p, p-d and d-v patterning pathways comprise intercellular signaling molecules, such as growth factors, and transcription factors encoded by genes, such as those of Hox complexes. All of these signaling pathways appear to interact, probably through the progress zone (SUMMERBELL 1974b), to determine precise patterning and morphogenesis along all three axes of the limb. For example, expression of *shh* is regulated by signals from the apical ectodermal ridge and dorsal ectoderm (NISWANDER et al. 1994; LAUFER et al. 1994; PARR and MCMAHON 1995; YANG and NISWANDER 1995).

Two factors can mimic polarizing activity and induce digit duplications when added to anterior mesenchyme. These are retinoic acid and expression of the *sonic hedgehog* gene. Both of these signals are found at the posterior margin of the limb bud and in other regions of the embryo with polarizing activity such as Hensen's node and floor plate. Anterior mesenchyme cells distal to a bead soaked in retinoic acid acquire polarizing activity (WANEK et al. 1991) and this has led to the idea that retinoic acid induces digit duplications by inducing a new polarizing region in anterior mesenchyme. More recently it has been shown that retinoic acid can activate expression of *shh* suggesting that retinoic acid patterns the mesenchyme through *shh* (RIDDLE et al. 1993). However, polarizing activity of cells does not appear to be totally related to the expression of *shh* suggesting that other factors contribute to polarizing activity. These factors may be retinoic acid or other, as yet unidentified, factors. In limb buds of the polydactylous chicken mutant *talpid*^{3}, whose digits are all morphologically similar, the ability of cells to produce a polarizing signal was found to be graded across the antero-posterior axis being weak anteriorly but high posteriorly. Yet transcripts encoding *shh* were only found in posterior mesenchyme (FRANCIS-WEST et al. 1995a). The molecular basis for the weak polarizing ability of anterior mesenchyme in *talpid* limb buds is unknown. The relationship between *shh* expression and polarizing activity may be even more complex since other experiments suggest that cells may express *shh* but do not have polarizing activity (HELMS et al. 1994). This again demonstrates that the expression of shh is not always related to the presence of polarizing activity.

Bmp-2 transcripts map to the polarizing region during early limb development (Fig. 3b). In the *Drosophila* wing imaginal disc, *hedgehog* induces the expression of other genes including *decapentaplegic* (*dpp*) that encodes a protein which is homologous to BMP-2 and BMP-4. Furthermore, *dpp* itself can pattern the wing and act as a morphogen specifying different cell fates (DIAZ-BENJUMEA et al. 1994). In *Drosophila*, one of the functions of *hedgehog* appears to be to activate *dpp* expression. If the chick limb bud is analogous, *shh* and *Bmp-2* should be sequential signals. Recently it has been shown that ectopic expression of *shh* does indeed activate *Bmp-2* expression in chick wing buds (LAUFER et al. 1994). However, application of recombinant BMP-2 on beads to the anterior margin of chick limb buds does not provide a polarizing signal and no digit duplications are induced (Francis et al. 1994). One possibility is that BMP-2 may require other cooperative factors such as BMP-7 which is also expressed at the mRNA level in posterior mesenchyme and which is known to form heterodimers with BMP-2 in vivo (Sampath et al. 1990).

The production of polarizing signals and the response to them requires cooperation with a signal from the apical ridge. When a polarizing region graft or a bead soaked in retinoic acid is placed anteriorly and the ridge is removed, genes encoding polarizing signals and Hox-d genes are not activated and no digit duplications are induced (Izpisua-Belmonte et al. 1992b; Niswander et al. 1994; Koyama et al. 1993). FGF-2 and FGF-4 are known to substitute for the apical ectodermal ridge. FGF-4 (and possibly FGF-2) probably signals from ridge to mesenchyme continuously throughout the process that leads to specification of additional digits. Both retinoic acid and *shh* can lead to induction of *Fgf-4* expression in anterior ridge (Niswander et al. 1994; Laufer et al. 1994). During limb outgrowth, a continuous interplay between ridge and mesenchyme is established and is based on a positive feedback loop between *shh* expression in the mesenchyme and *Fgf-4* expression in the ridge. However, the intervening steps in the loop are not known. In the *talpid*[3] mutant, transcripts of genes encoding BMPs are expressed across the entire antero-posterior axis in association with overlying *Fgf-4* expression in the apical ridge, whereas *shh* expression is restricted posteriorly (Francis-West et al. 1995a). It is therefore possible that BMPs act as a signal to the ridge.

Two sets of signals are known to regulate the expression of *shh* in the posterior mesenchyme. One is from the dorsal ectoderm and the other is the apical ectodermal ridge. Removal of either of these results in the down-regulation of *shh* expression (Niswander et al. 1994; Laufer et al. 1994; Parr and McMahon 1995; Yang and Niswander 1995). In mice, knockout of *Wnt 7a* results in the down-regulation of *shh* expression, and thus the product of *Wnt 7a* appears to be one of the signals needed for expression of *shh* (Parr and McMahon 1995). The signal from the ridge which maintains expression of *shh* as well as the HoxD complex and *Bmp-2* in the posterior mesenchyme is FGF (Niswander et al. 1994; Laufer et al. 1994).

FGF acts on cells in the progress zone both to maintain *shh* expression and to promote outgrowth. Genes associated with the progress zone include the homeobox genes Msx-1 and Msx-2 (Robert et al. 1991) and the gene encoding the growth factor BMP-4 (Francis et al. 1994). Expression of Msx-1 has been shown to be dependent on signals from the apical ridge. When tissue from the proximal part of a mouse limb bud is grafted distally beneath the ridge in a chick wing bud, Msx-1 expression is induced rapidly within 5 h (Davidson et al. 1991). Furthermore, levels of Msx-1 expression are related to distance of cells from the ridge with lower levels of expression in cells farther away from the ridge (Davidson et al. 1991; Brown et al. 1993). FGF-2 can maintain Msx-1 expression in the absence of the apical ecotodermal ridge (Fallon et al. 1994) but this could be a direct or indirect effect. In tooth development, *Msx-1* expression is induced by BMP-4 (Vainio et al. 1993). However, in early limb buds, BMP-4 application on beads or overexpression with a retrovirus does not maintain or induce *Msx-1* expression (P.H. Francis-West and M.K. Richardson, unpublished data).

14 Cell Differentiation

Once the pattern of tissues has been established by the signaling networks, tissues must then undergo appropriate differentiation and become arranged in precise arrays. There is at present a considerable gap in understanding how distinct expression patterns of say, homeobox genes, could lead to formation of individual bones and muscles. However, there has been considerable progress in understanding signals that are involved in controlling cartilage and muscle differentiation.

During skeletal development, key signals that control cartilage differentiation include BMPs and the related factors, growth differentiation factors, (GDFs). In mice, two of these factors have been shown to be essential for the development of specific skeletal elements. Mutation in the gene that encodes BMP-5 results in changes in the shape of the ear and an absence of certain ribs (reviewed Kingsley 1994). In contrast, mutation of the gene that encodes GDF-5 does not affect development of the axial skeleton but affects appendicular skeleton, i.e., limbs are shorter and thinner (Storm et al. 1994). It is not yet clear whether the changes in morphogenesis of the bones is related to changes in the initial founding cell population or in later growth and differentiation.

Cartilage growth and differentiation involves a series of steps from initial cell condensation to formation of an established cartilage element which, in limbs, consists of an ordered array of rounded, flattened and hypertrophic chondrocytes (Rooney et al. 1984). In mice, expression of different members of the BMP family have been mapped to different events during cartilage growth and morphogenesis. For example, *Bmp-2* and *Bmp-5* expression is associated with initial condensations whereas expression of *Bmp-3*, *Bmp-7* and *Bmp-6* is associated with hypertrophic (mature) chondrocytes (Lyons et al. 1989; Wozney et al. 1993). The expression of these related molecules during distinct phases of chondrocyte differentiation has led to the suggestion that each BMP may have a distinct role (Lyons et al. 1989; reviewed Kingsley 1994). Overexpression studies in the developing chick limb have shown that BMPs can alter cartilage growth, morphogenesis and chondrocyte differentiation. Overexpression of BMP-4, which is expressed in the perichondrium of cartilage elements, results in gross thickening of cartilage elements and prevents hypertrophy of the chondrocytes (Francis-West et al. 1995b).

In contrast to cartilage differentiation, little is known about the signaling molecules which control differentiation and growth of myogenic cells. However, several of the muscle-specific transcription factors are known (see Buckingham 1994) including MyoD, Myogenin, MRF-4 and MYF5. In development of limb muscles in the mouse, these genes are expressed sequentially with *Myf-5* being expressed first, myogenin and MyoD slightly later and MRF-4 much later (OTT et al. 1991; SASSOON et al. 1989; BOBER et al. 1991). In mice, knockout of *Myf5* or MyoD results in almost normal muscle development except in the *Myf5* knockout, muscle development is delayed (BRAUN et al. 1992; RUDNICKI et al. 1992). In the MyoD knockout, *Myf5* RNA levels are increased suggesting that *Myf5* expression can compensate for absence of MyoD. Consistent with this, knockout of both *Myf5* and MyoD prevents any skeletal muscle development demonstrating that at least one of these factors is essential for the development of skeletal muscle (RUDNICKI et al. 1993). Little is known about growth factors involved in muscle development. *Fgf-4*, *-5* and *Bmp-2* are expressed in muscle cells (FLORINI et al. 1991). However, in vitro, FGF-2 prevents differentiation of myogenic cells in chick limb cultures (SCHOFIELD and WOLPERT 1990) and BMP-2 prevents myogenic differentiation of a cell line (YAMAGUCHI et al. 1991). The relevance of these in vitro studies to the normal role of these factors duing muscle development is at present unclear. Furthermore, it is probable that these factors can have different effects depending on the stage of muscle development.

In addition to cartilage and muscle, other connective tissue must be specified and differentiate appropriately. These include tendons, ligaments and adipose tissue. The signals involved in specifying and controlling the development of these different tissues are not yet known.

References

Allen F, Tickle C, Warner A (1990) The role of gap junctions in the patterning of the chick limb bud. Development 108: 623–634

Bober E, Lyons GE, Braun T, Cossu G, Buckingham M, Arnold H-H (1991) The muscle regulatory gene Myf-6 has a biphasic pattern of expression during early mouse development. J Cell Biol 113: 1255–1265

Braun T, Rudnicki MA, Arnold H-H, Jaenish R (1992) Targeted inactivation of the muscle regulatory gene Myf-5 results in abnormal rib development and perinatal death. Cell 71: 369–382

Brockes JP (1991) We may not have a morphogen. Nature 350: 15

Brown JM, Wedden SE, Millburn GH, Robson LG, Hill RE, Davidson DR, Tickle C (1993) Experimental analysis of the control of expression of the homeoboxgene *Msx-1* in the developing limb and face. Development 119: 41–48

Buckingham M (1994) Molecular biology of muscle development. Cell 78: 15–21

Burrus LW (1994) *Wnt-1* as a short-range signalling molecule. Bioessays 16: 155–157

Butler J, Cosmos E, Cauwnbergs PC (1988) Positional signals: evidence for a possible role in muscle fibre-type patterning of the embryonic avian limb. Development 102: 763–772

Carrington JL, Chen P, Yanagishita M, Reddi AH (1991) Osteogenin (Bone Morphogenetic Protein-3) stimulates cartilage formation by chick limb bud cells in vitro. Dev Biol 146: 406–415

Chang DT, Lopez A, von Kessler DP, Chiang C, Simandl BK, Zhao R, Seldin MF, Fallon JF, Beachey PA (1994) Products, genetic linkage and limb patterning activity of a murine *hedgehog* gene. Development 120: 3339–3353

Charitie J, De Graaf W, Shen S, Deschamps J (1994) Ectopic expression of Hoxb-8 causes duplication of the ZPA in the forelimb and homeotic transformation of axial structures. Cell 78: 589–601

Chen P, Carrington JL, Hammonds RG, Reddi AH (1991) Chick limb bud mesodermal cell chondrogenesis: inhibition of isoforms of platelet-derived growth factor and reversal by recombinant bone morphogenetic protein 2B (BMP-2B) and modulation by transforming growth factor B1 and B2. Exp Cell Res 195: 509–515

Cohn MJ, Izpisua-Belmonte J-C, Abud H, Heath JK, Tickle C (1995) Fibroblast growth factors induce additional limb development from the flank of chick embryos. Cell 80: 239–246

Crossley PH, Martin GR (1995) The mouse *Fgf-8* gene encodes a family of polypeptides and is expressed in regions that direct outgrowth and patterning in the developing embryo. Development 121: 439–451

Davidson DR, Crawley A, Hill RE, Tickel C (1991) Position-dependent expression of two related homoeobox genes in developing vertebrate limbs. Nature 352: 429–431

Dealy CN, Roth A, Ferrari D, Brown AMC, Kosher RA (1993) *Wnt-5a* and *Wnt-7a* are expressed in the developing chick limb bud in a manner suggesting roles in pattern formation along the proximo-distal and dorsoventral axes. Mech Dev 43: 175–186

Diaz-Benjumea FJ, Cohen B, Cohen SM (1994) Cell interactions between compartments establishes the proximal-distal axis of *Drosophila* legs. Nature 372: 175–179

Dollé P, Dierich A, LeMeur M, Schimmang T, Schuhbauer B et al (1993) Disruption of the *Hoxd-13* gene induces localized heterochrony leading to mice with neotenic limbs. Cell 75: 431–441

Dono R, Zeller R (1994) Cell-type-specific nuclear translocation of fibroblast growth factor-2 isoforms during chicken kidney and limb morphogenesis. Dev Biol 163: 316–330

Eichele G, Tickle C, Alberts B (1985) Studies on the mechanism of retinoid-induced pattern duplications in the early chick limb bud: temporal and spatial aspects. J Cell Biol 101: 1913–1920

Fallon JF, Kelley RO (1977) Ultrastructural analysis of the apical ectodermal ridge during vertebrate limb morphogenesis. II. Gap junctions as distinctive ridge structures common to birds and mammals. J Embryol Exp Morphol 41: 223–32

Fallon JF, Lopez A, Ros MA, Savage MP, Olwin BB, Simandl K (1994) FGF-2: apical ectodermal ridge outgrowth signal for chick limb development. Science 264: 104–107

Fietz MJ, Concordet J-P, Barbosa R, Johnson R, Krauss S, McMahon AP, Tabin C, Ingham PW (1994) The *hedgehog* gene family in *Drosophila* and vertebrate development. Development supplement: 43–51

Florini JR, Eurton DZ, Magri KA (1991) Hormones, growth factors and myogenic cell differentiation. Annu Rev Physiol 53: 201–216

Francis PH, Richardson MK, Brickell PM, Tickle C (1994) Bone morphogenetic proteins and a signalling pathway that controls patterning in the developing chick limb. Development 120: 209–218

Francis-West PH, Robertson KE, Ede DA, Rodriguez C, Izpisua- Belmonte J-C, Houston B, Burt DW, Gribbin C, Brickell PM, Tickle C (1995a) Expression of genes encoding Bone Morphogenetic proteins and sonic hedgehog in talpid (ta3) limb buds: their relationships in the signalling cascade involved in limb patterning. Dev Dyn 203: 187–197

Francis-West PH, Bell E, Richardson MK, Archer CW, Wolpert L, Brickell PM (1995b) Overexpression of BMP-4 using a retroviral vector alters the size and shape of developing skeletal elements (submitted)

Green CR, Bowles L, Crawley A, Tickle C (1994) Expression of the connexin 43 gap junctional protein in tissues at the tip of the chick limb bud is related to the epithelial-mesenchymal interactions that mediate morphogenesis. Dev Biol 161: 12–21

Green JB, Smith JC (1990) Graded change in dose of Xenopus activin homolog elicits stepwise transition of embryonic cell fate. Nature 347: 391–394

Heikinheimo M, Lawshe A, Shackleford GM, Wilson DB, MacArthur CA (1994) *Fgf-8* expression in the post-gastrulation mouse suggests roles in the development of the face, limbs and central nervous system. Mech Dev 48: 129–138

Helms J, Thaller C, Eichele G (1994) Relationship between retinoic acid and *sonic hedgehog*, two polarizing signals in the chick wing bud. Development 120: 3267–3274

Hogan BLM, Thaller C, Eichele G (1992) Evidence that Hensen's node is a site of retinoic acid synthesis. Nature 359: 237–241

Hornbruch A, Wolpert L (1991) The spatial and temporal distribution of polarizing activity in the flank of the pre-limb-bud stages in the chick embryo. Development 111: 725–731

Hseuh AJW, Dahl KD, Vaughan J, Tucker E, Rivier J, Bardin CW, Vale W (1987) Heterodimers and homodimers of inhibin subunits have different paracrine action in the modulation of luteinizing hormone-stimulated androgen biosynthesis. Proc Natn Acad Sci USA 84: 5082–5086

Hunt P, Krumlauf R (1992) Hox codes and positional specification in vertebrate embryonic axes. Annu Rev Cell Biol 8: 227–256

Izpisúa-Belmonte J-C, Tickle C, Dollé P, Wolpert L, Duboule D (1991) Expression of the homeobox Hox-4 genes and the specification of position in chick wing development. Nature 350: 585–589

Izpisúa-Belmonte J-C, Brown JM, Crawley A, Duboule D, Tickle C (1992a) Hox4 gene expression in mouse/chicken heterospecific grafts of signalling regions to limb buds reveals similarities in patterning mechanisms. Development 115: 553–560

Izpisúa-Belmonte J-C, Brown JM, Duboule D, Tickle C (1992b) Expression of Hox-4 genes in the chick wing links pattern formation to the epithelial-mesenchymal interactions that mediate growth. EMBO J 11: 1951–1957

Kelley RO, Fallon JF (1983) A freeze fracture and morphometric analysis of gap junctions of limb bud cells: initial studies on a possible mechanism for morphogenetic signalling during development. In: Fallon JF, Caplan AI (ed) Limb development and regeneration part A. Liss, New York, pp 119–130

Kingsley DM (1994) What do BMPs do in mammals? Clues from the mouse short-ear mutation. Trends Genet 10: 16–21

Koyama E, Noji S, Nohno T, Myokai F, Ono K, Nishijima K, Kuriowa A, Ide H, Taniguchi S, Saito T (1993) Cooperative activation of *Hox*-D homeobox genes by factors from the polarizing region and the apical ridge in chick limb morphogenesis. Dev Growth Differ 35: 189–198

Laufer E, Nelson CE, Johnson RI, Morgan BA, Tabin C (1994) *Sonic hedgehog* and *Fgf-4* act through a signalling cascade and feedback loop to integrate growth and patterning of the developing limb bud. Cell 79: 993–1003

Lee JJ, von Kessler DP, Parks S, Beachey PA (1992) Secretion and localized transcription suggest a role in positional signalling for products of the segmentation gene *hedgehog*. Cell 71: 33–50

Lee JJ, Ekker SC, von Kessler DP, Porter JA, Sun BI, Beachy PA (1994) Autoproteolysis in *hedgehog* protein biogenesis. Science 266: 1528–1535

Lyons KM, Pelton RW, Hogan BLM (1989) Patterns of expression of murine Vgr-1 and BMP-2a RNA suggest that transforming growth factor-b-like genes coordinately regulate aspects of embryonic development. Genes Dev 3: 1657–1668

MacCabe JA, Errick J, Saunders JW (1974) Ectodermal control of the dorsoventral axis in the leg bud of the chick embryo. Dev Biol 39: 69–82

Maden M, Ong DE, Summerbell D, Chytil F (1988) Spatial distribution of cellular protein binding to retinoic acid in chick limb bud. Nature 335; 733–735

Martin PM (1990) Tissue patterning in the mouse limb. Int J Dev Biol 34: 323–336

Massague J (1990) The Transforming Growth factor-b family. Annu Rev Cell Biol 6: 597–647

Massague J, Attisano L, Wrana JL (1994) The TGF-beta family and its composite receptors. Trends Cell Biol 4: 172–178

Mason IJ (1994) The ins and outs of fibroblast growth factors. Cell 78: 547–552

McGinnis W, Krumlauf R (1992) Homoebox genes and axial patterning. Cell 68: 283–302

Mendelsohn C, Ruberte E, Chambon P (1992) Retinoid receptors in vertebrate limb development. Dev Biol 152: 50–61

Mehta PP, Bertram JS, Loewenstein WR (1989) The action of retinoids on cellular growth correlate with their actions on gap junctional communication. J Cell Biol 108: 1053–1065

Miyamoto M, Narou K, Seko C, Marsumoto S, Kondo T, Kurokawa T (1993) Molecular cloning of a novel cytokine cDNA encoding the ninth member of the fibroblast growth factor family, which has a unique secretion property. Mol Cell Biol 13: 4251–59

Morgan BA, Izpisúa-Belmonte J-C, Duboule D, Tabin CJ (1992) Targeted misexpression of Hox-4.6 in the avian limb causes apparent homeotic transformations. Nature 358: 236–239

Myokai F, Washio N, Asahara Y, Yamaai T, Tanda N, Ishikawa T, Aoki S, Kurihara H, Murayama Y, Saito T, Matsumoto K, Nakamura T, Noji S, Nohno T (1995) Expression of the Hepatocyte Growth Factor gene during chick limb development. Dev Dyn 202: 80–90

Niswander L, Tickle C, Vogel A, Booth I, Martin G (1993) FGF-4 replaces the apical ectodermal ridge and directs outgrowth and patterning of the limb. Cell 75: 579-587

Niswander L, Jeffrey S, Martin G, Tickle C (1994) Positive feedback loop in signalling in vertebrate limb development. Nature 372: 609–612

Nohno T, Noji S, Koyama E, Ohyama K, Myokai F, Kuriowa A, Saito T, Taniguchi S (1991) Involvement of the *Chox-4* chicken homeobox genes in determination of antero-posterior axial polarity during limb development. Cell 64: 1197–205

Noji S, Nohno T, Koyama E, Muto K, Ohyama K et al (1991) Retinoic acid induces polarizing activity but is unlikely to be a morphogen in the chick limb bud. Nature 350: 83–86

Nusse R, Varmus HE (1992) *Wnt* genes. Cell 69: 1073–1087

Ohuchi H, Noji S, Koyama E, Myokai F, Nishikawa K, Nohno T, Tashiro K, Shiokawa K, Matsuo N, Taniguchi S (1992) Expression pattern of the activin receptor 11A gene during differentiation of chick neural tissues, muscle and skin. FEBS Lett 303: 185–189

Olson DJ, Christian JL, Moon RT (1991) Effect of *Wnt-1* related proteins on gap junctional communication in *Xenopus* embryos. Science 252: 1173–1176

Ordahl CP, Le Douarin NM (1992) Two myogenic lineages within the developing somite. Development 114: 339–353

Ott MO, Bober E, Lyons G, Arnold H-H, Buckingham M (1991) Early expression of the myogenic regulatory genes, Myf-5, in precursor cells of skeletal muscle in the mouse embryo. Development 111: 1097–1107

Parr BA, McMahon AP (1995) Dorsalizing signal *Wnt-7a* required for normal polarity of D-V and A-P axes of mouse limb. Nature 374: 350–353

Parr BA, Shea MJ, Vassileva G, McMahon AP (1993) Mouse *Wnt* genes exhibit discrete domains of expression in the early embryonic CNS and limb buds. Development 119: 247–261

Potts JD, Carrington JL (1993) Selective expression of chick plateletderived-growth-factor-alpha (PDGF-alpha) receptor during limb bud development. Dev Dyn 198: 14–21

Ralphs JR, Wylie L, Hill DJ (1990) Distribution of insulin-like growth factor peptides in the developing chick embryo. Development 109: 51–58

Riddle RD, Johnson RL, Laufer E, Tabin C (1993) *Sonic hedgehog* mediates the polarizing activity of the ZPA. Cell 75: 1401–1416

Roark EF, Greer K (1994) Transforming growth factor-b and Bone morphogenetic protein-2 act by distinct mechanisms to promote chick limb cartilage differentiation in vitro. Dev Dyn 200: 103–116

Robert B, Lyons G, Simandl BK, Kuriowa A, Buckingham M (1991) The apical ectodermal ridge regulates *Hox-7* and *Hox-8* gene expression in developing chick limb buds. Genes Dev 5: 2363–2374

Robson LG, Kara T, Crawley A, Tickle C (1994) Tissue and cellular patterning of the musculature in chick wings. Development 120: 1265–1276

Rooney P, Archer C, Wolpert L (1984) Morphogenesis of cartilaginous long bone rudiments. In: Trelstad RL (ed) The role of extracellular matrix in development. Liss, New York, pp 305–322

Rowe A, Richman JM, Ochanda JO, Brickell PM (1991a) Retinoic acid alters the pattern of retinoic acid receptor beta expression in the embryonic chick limb. In: Hinchliffe JR, Hurle J, Summerbell D (eds) Developmental patterning of the vertebrate limb. Plenum, New York, pp 97–101

Rowe A, Eager N, Brickell PM (1991b) A member of the RXR nuclear receptor family is expressed in neural-crest-derived cells of the developing chick peripheral nervous system. Development 111: 771–778

Rudnicki MA, Braun T, Hinuma S, Jaenisch R (1992) Inactivation of MyoD in mice leads to up-regulation of the myogenic HLH gene myf-5 and results in apparently normal muscle development. Cell 71: 383–390

Rudnicki MA, Schnegelsberg PNJ, Stead RH, Braun T, Arnold H-H, Jaenish R (1993) MyoD or Myf-5 is required for the formation of skeletal muscle. Cell 75: 1351–1359

Sampath TK, Coughlin JE, Whetstone RM, Banach D, Corbett C, Ridge RJ, Ozkaynak E, Oppermann H, Rueger DC (1990) Bovine osteogenic protein is composed of OP-1 and BMP-2A, two members of the transforming growth factor-b superfamily. J Biol Chem 265: 13198–13205

Sassoon D, Lyons G, Wright WE, Lin V, Lassar A, Weintraub H, Buckingham M (1989) Expression of two myogenic regulatory factors, myogenin and MyoD1 during mouse embryogenesis. Nature 341: 303–307

Saunders JW (1948) The proximo-distal sequence of origin of limb parts of the chick wing and the role of the ectoderm. J Exp Zool 108: 363–404

Saunders JW, Gasseling MT (1962) Cellular death in morphogenesis of the avian wing. Dev Biol 5: 147–178

Saunders JW, Gasseling MT (1968) Ectodermal-mesenchymal interactions in the origin of limb symmetry. In: Fleishmajer R, Billingham RE (eds) Epithelial-mesenchymal interactions. Williams and Wilkins, Baltimore, pp 78–97

Saunders JW, Gasseling MT, Errick J (1976) Inductive activity and enduring cellular constitution of a supernumerary apical ectodermal ridge grafted to the limb bud of the chick embryo. Dev Biol 50: 16–25

Schofield JN, Wolpert I (1990) Effect of TGF-betal, TGF-beta2 and bFGF on chick cartilage and muscle cell differentiation. Exp Cell Res 191: 144–148

Searls RL, Janners MY (1971) The initiation of limb bud outgrowth in the embryonic chick. Dev Biol 24: 198–213

Seleiro EAP, Rowe A, Brickell PM (1995) The chicken retinoid-X receptor alpha (RXRa) gene and its expression in the developing limb. Wilh Roux Arch 204: 244–249

Shellswell GB, Wolpert L (1977) The pattern of muscle and tendon development in the chick wing. In: Ede DA, Hinchliffe JR, Balls M (eds) Vertebrate limb and somite morphogenesis. Cabridge University Press, Cambridge, pp 71–86

Smith JC (1980) The time required for positional signalling in the chick wing bud. J Embryol Exp Morphol 60: 321–28
Smith SM, Eichele G (1991) Temporal and regional differences in the expression pattern of distinct retinoic acid receptor-beta transcripts in the chick embryo. Development 111: 245–52
Storm EE, Huynh TV, Copeland NG, Jenkins NA, Kingsley DM, Lee S-J (1994) Limb alterations in brachypodism mice due to mutations in a new member of the TGF-beta superfamily. Nature 368: 639–643
Summerbell D (1974a) A quantitative analysis of the effect of excision of the AER from the chick limb bud. J Embryol Exp Morphol 32: 651–660
Summerbell D (1974b) Interaction between the proximo-distal and anteroposterior co-ordinates of positional values during specification of positional information in the early development of the chick limb bud. J Embryol Exp Morphol 32: 227–237
Summerbell D, Lewis J, Wolpert L (1973) Positional information in chick limb morphogenesis. Nature 224: 492–496
Thaller C, Eichele G (1987) Identification and spatial distribution of retinoids in the developing chick limb bud. Nature 327: 625–628
Thaller C, Eichele G (1990) Isolation of 3,4-didehydroretinoic acid, a novel morphogenetic signal in the chick wing. Nature 345: 815–819
Thaller C, Hofmann C, Eichele G (1993) 9-*cis*-retinoic acid, a potent inducer of digit pattern duplications in the chick wing bud. Development 118: 957–965
Thorp BH, Anderson I, Jakowlew SB (1992) Transforming Growth factors b1, -b2, and -b3 in cartilage and bone cells during endochondral ossification in the chick. Development 114: 907–911
Tickle C (1981) The number of polarizing region cells required to specify additional digits in the developing chick wing. Nature 289: 295–298
Tickle C (1991) Retinoic acid and chick limb bud development. Development [Suppl] 1: 113–121
Tickle C, Summerbell D, Wolpert L (1975) Positional signaling and specification of digits in chick limb morphogenesis. Nature 254: 199–202
Tickle C, Alberts B, Wolpert L, Lee J (1982) Local application of retinoic acid to the limb bud mimics the action of the polarizing region. Nature 296: 564–566
Tickle C, Lee J, Eichele G (1985) A quantitative analysis of the effect of all-*trans*-retinoic acid on the pattern of chick wing development. Dev Biol 109: 92–95
Tickle C, Crawley A, Farrar J (1989) Retinoic acid application to chick wing buds leads to a dose-dependent reorganization of the apical ectodermal ridge that is mediated by the mesenchyme. Development 106: 691–705
Vainio S, karavanova I, Jowett A, Thesleff I (1993) Induction of BMP-4 as a signal mediating secondary induction between epithelial and mesenchymal tissues during tooth development. Cell 75: 45–58
Vogel A, Tickle C (1993) FGF-4 maintains polarizing activity of posterior limb bud cells in vivo and in vitro. Development 119: 199–206
Wagner M, Thaller C, Jessell T, Eichele G (1990) Polarizing activity and retinoid synthesis in the floor plate of the neural tube. Nature 345: 819–822
Wanek N, Gardiner DM, Muneoka K, Bryant SV (1991) Conversion by retinoic acid of anterior cells into ZPA cells in the chick wing bud. Nature 350: 81–83
Wilson DB (1983) The origin of the endothelium in the developing marginal vein of the chick wing bud. Cell Differ 13: 63–67
Wolpert L (1971) Positional information and pattern formation. Curr Top Dev Biol 6: 183–224
Wolpert L (1982) Cartilage morphogenesis in the limb. In: Bellairs R, Curtis ASG, Dunn G (eds) Cell Behaviour. Cambridge University Press, Cambridge, pp 359–373
Wozney JM, Capparella J, Rosen V (1993) The bone morphogenetic proteins in cartilage and bone development. In: Bernfield M, Molecular basis of morphogenesis. Wiley-Liss, New York, pp. 221–230
Wrana JL, Attisano L, Carcarno J, Zentrella A, Doody J, Laiho M, Wang X-F, Massague J (1992) TGF beta signals through a heteromeric protein kinase receptor complex. Cell 71: 1003–1014
Wrana JL, Attisano L, Wieser R, Ventura F, Massague J (1994) Mechanism of activation of the TGF-beta receptor. Nature 370: 341–347
Yamaguchi A, Katagiri T, Ikeda T, Wozney JM, Rosen V, Wang EA, Kahn AJ, Suda T, Yoshiki S (1991) Recombinant human bone morphogentic protein-2 stimulates osteoblastic maturation and inhibits myogenic differentiation in vitro. J Cell Biol 113: 681–687
Yang Y, Niswander L (1995) Interaction between the signaling molecules WNT and SHH during vertebrate limb development: dorsal signals regulate anteroposterior patterning. Cell 80: 939–947
Yokouchi Y, Sasaki H, Kuriowa A (1991) Homeobox gene-expression correlated with the bifurcation process of limb cartilage development. Nature 353: 443–446

Appendix

Monoclonal Antibodies Against Chicken Leukocyte Antigens

Name of antibody	Isotype	Recognized molecule and/or MW	Tissue distribution	Reference
Dr. Richard Boyd, Dept. of Pathology and Immunology, Monash Medical School, Commercial Road, Prahran 3181, Victoria, Australia (Tel: 61-3-520 2738; Fax: 61-3-529 6484)				
MUI36	IgG2a	Chicken T and B cell antigen-1 (CTB-1)/ 68 kDa	Pan B-cell, subpopulation of thymocytes and macrophages	BOYD et al. 1992 Dev Immunol 2: 51–66 BEAN 1993 PhD thesis
MUI83	IgG1	Chicken thymus cell antigen-1 (CTCA-1)/85 and 40–60 kDa	Primarily on immature thymocytes, however also present on a subset of mature thymocytes and peripheral T cells	BOYD et al. 1992 Dev Immunol 2: 51–66 BEAN 1993 PhD thesis
MUI78	IgG2a	MHC class II	Subpopulations of cells in numerous tissues	BOYD et al. 1992 Dev Immunol 2: 51–66
MUI79	IgM	Unknown	60%–70% of blood monocytes and isolated subpopulations of macrophages in the spleen, bursa and thymus	BOYD et al. 1992 Dev Immunol 2: 51–66
Dr. Chen-Lo Chen, Wallace Tumor Institute 378, University of Alabama at Birmingham 1824 6th Avenue, South, Birmingham, AL 35294–3300, USA (Tel: 1-205-934 3370; Fax: 1-205-934 1875)				
CLA-1	IgG1	CD45R1; 195 and 180 kDa restricted isoform	Thymic and peripheral T cells, embryo bursa, some Mø	CHEN and COOPER 1987 In: Avian immunology CRC Boca Raton, pp 137–154
CLA-2	IgG1	chL2; 180 and 100 kDa	All leukocytes	CHEN et al. 1991 In: Avian cellular immunology. CRC, Boca Raton, pp 1–22
CLA-3	IgG1	chL3; 90–100 kDa	All leukocytes	CHEN et al. 1991 In: Avian cellular immunology. CRC, Boca Raton
TCR1	IgG1	TCRγδ	All T γδ cells	SOWDER et al. 1988 J Exp Med 167: 315–322
TCR3	IgG1	TCRVβ2	All Vβ2 T cells	CHAR et al. 1990 J Immunol 145: 3547–3555

Name of antibody	Isotype	Recognized molecule and/or MW	Tissue distribution	Reference
M1	IgG2b	Igμ chain	IgM⁺ cells	CHEN et al. 1982 J Immunol 129: 2580–2585
M4	IgM	Ig μ chain	IgM⁺ cells	CHEN et al. 1982 J Immunol 129: 2580–2585
G1	IgG1	Ig γ chain	IgG⁺ cells	CHEN et al. 1982 J Immunol 129: 2580–2585
A1	IgG2b	Ig α chain	IgA⁺ cells	CHEN et al. 1982 J Immunol 129: 2580–2585
L1	IgG1	Ig λ chain	IgM⁺, Ig G⁺ and IgA⁺ cells	CHEN et al. 1982 J Immunol 129: 2580–2585
Cld1	IgM	Ig (anti-N-acetylglucosamine)	<1% B cells	CHANH et al. 1982 J Immunol 129: 2541–2547
Cld2	IgM	Ig (anti-*p*-aminobenzoic acid)	<1% B cells	CHANH et al. 1982 J Immunol. 129: 2541–2547
Cld3	IgG1	Ig	<1% B cells	WRIGHT et al. 1987 In: Avian immunology. Liss, ⊠, pp 81–85
CV_H1	IgG1	IgH chains	A subset of B cells	WRIGHT et al. 1987 In: Avian immunology. Liss, ⊠, pp 81–85
CT1 CT1a	IgG1 IgG3	chT1; 65 and 45 kDa	Most thymocytes, few peripheral T cells	CHEN et al. 1984 Eur J Immunol 14: 385–391
CT3	IgG1	CD3; Multi-chains 20,19 17 and 2 x 16 kDa	All αβ and γδ T cells	CHEN et al. 1986 J Exp Med 164: 375–380
CT4	IgG1	CD4; 64 kDa	Thymic and peripheral T cell subset	CHAN et al. 1988 J Immunol 140: 2133–2138
CT8	IgG1	CD8α; 2 x 34 kDa	Thymic and peripheral T cell subset	CHAN et al. 1988 J Immunol 140: 2133–2138
S3	IgG1	CD6 candidate; 110 kDa	Some thymocytes, peripheral T cells	GÖBEL et al. 1994 12th Eur Immunol Meeting Abstract W05/18

Name of antibody	Isotype	Recognized molecule and/or MW	Tissue distribution	Reference
A19	IgG1	$\alpha^E\beta_7$ candidate; 120,90 and 28 kDa	Intestinal T cells	HAURY et al. 1993 Eur J Immunol 23: 313–319
CB1	IgG1	cHB1; 2 x 50 kDa	Bursal B cells and low intensity on peripheral B cells	CHEN and COOPER 1987 In: Avian immunology. CRC, Boca Raton, pp 137–154
CB2	IgM	chB2; 80–125 kDa	Bursa	CHEN and COOPER 1987 In: Avian immunology. CRC, Boca Raton, pp 137–154
CB3	IgG1	chB3; 50 kDa	Bursa	PICKEL et al. 1990 Immunogenetics 32: 1–7
CB4	IgG1	chB4; 107, (53 and 39 kDa)	Bursal and peripheral B cells	CHEN and COOPER 1987 In: Avian immunology, CRC, Boca Raton pp 137–154
CB5	IgM	chB5; 167 kDa	Bursa and peripheral B	CHEN and COOPER 1987 In: Avian immunology, CRC, Boca Raton, pp 137–154
Cla	IgM	MHC class II	B cells, monocytes, avtivated T cells	EWERT et al. 1984 J Immunol 132: 2524–2530

Dr. Catherine Corbel, Institut d'Embryologie du CNRS 49 bis, ave. de la Belle Gabrielle, 94736 Nogent-sur-Marne, France (Tel: 33-1-48-73 6090; Fax: 33-1-48-73 4377)

BEN	IgG1	95 – 110 kDa, member of Ig superfamily	Myeloid and erythroid, progenitors, immature T cells, activated T cells, subpopulations of neurons and epithelial cells	CORBEL et al. 1992 Exp Cell Res 203: 91–99 POURQUITÉ et al. 1992 PNAS 89: 5261–5265
TAP1	IgG2a	MHC class II		GUILLEMOT et al. 1984 J Exp Med 160: 1803–1819

Name of antibody	Isotype	Recognized molecule and/or MW	Tissue distribution	Reference
Dr. Fred Davison, AFRC-Institute for Animal Health, Compton Laboratory, Compton, Nr. Newbury, Berkshire, RG16 oNN, UK (Tel: 44-635-578 411; Fax: 44-635-578 844)				
AV5	IgG_1	Common leukocyte	Lymphocytes and macrophages antigen 100 kDa	Unpublished
AV6	IgG_1	Common leukocyte	Lymphocytes and macrophages antigen 100 kDa	Unpublished
AV7	IgG_1	CD28, 40 kDa	T cells	YOUNG et al. 1994 J Immunol 152: 3848–3851
AV12	IgG_1	CD8α, 2 x 34 kDa	T cell subpopulation	TREGASKES et al. 1995 J Immunol 154: 4485–4494
AV13	IgG_1	CD8α	T cell subpopulation	TREGASKES et al. 1995 J Immunol 154 : 4485–4494
AV14	IgG2b	CD8α	T cell subpopulation	TREGASKES et al. 1995 J Immunol 154: 4485–4494
AV29	IgG2b	CD4	T cell subpopulation	Unpublished
AV30	IgG1	CD4	T cell subpopulation	Unpublished
AV36	IgG1	CD3	All T cells	Unpublished
AV20	IgG1	ChB8 (Bu-1)	Recognises both allotypes, Bu-la and Bu-lb, on B cells and a subpopulation of macrophages	Unpublished
Dr. Françoise Dieterlen-Lievre, Institut de d'Embryologie Cellulaire et Moleculaire du CNRS 49 bis, Avenue de la Belle Gabrielle, 94736 Nogent-sur-Marne, France (Tel: 33-1-48-73 6090; Fax: 33-1-48-73 4377)				
QH_1/MB_1	IgM	α-macroglobulin	All hemopoietic and blood cells, except erythrocytes and endothelial cells	PARDANAUD et al. 1987 Development 100: 339–349 PÉAULT et al. 1983 PNAS 80: 2976–2980 LABASTIE 1989 Cell Diff Dev 27:151–162

Name of antibody	Isotype	Recognized molecule and/or MW	Tissue distribution	Reference
Dr. Karel Hála, Inst. for General and Exp. Pathology, University of Innsbruck, Fritz-Pregl-Str. 3, 6020 Innsbruck, Austria (Tel: 43-512-507 3103; Fax: 43-512-507 2867)				
INN-CH-113	IgM	MHC class II		HALA et al. 1984 Dev Comp Immunol 8: 673 WICK et al. 1984 Mol Immunol 21:259
INN-CH-16	IgM	IL-2α chain, 50kDa	Activated T cells	HALA et al. 1986 Eur J Immunol 16: 1331 SCHAUENSTEIN et al. 1988 Dev Comp Immunol 12: 823
Dr. Suzan H.M. Jeurissen, Central Veterinary Institute, Immunology, POB 65, 8200 AB Lelystad, Netherlands (Tel: 31-3200-73 253; Fax: 31-3200-73 473)				
HIS-C7	IgG2a	CD45, 180 kDa	All leucocytes	JEURISSEN et al. 1988 Vet Immunol Immunopathol 19: 225–238 PARAMITHIOTIS et al. 1991 J Immunol 147: 3710
HIS-C1	IgG1	Bu-1, 70 kDa	B cells, bursa cells	JEURISSEN et al. 1988 Vet Immunol Immunopathol 19: 225–238
HIS -C12	IgG1	Ig μ chain	IgM	JEURISSEN et al. 1988 Vet Immunol Immunopathol 19: 225–238
59.7	IgG1	Ig μ chain	IgM	JEURISSEN et al. 1988 Vet Immunol Immunopathol 19: 225–238 BIANCHI et al. 1990 Vet Immunol Immunopathol 24: 125–134

Name of antibody	Isotype	Recognized molecule and/or MW	Tissue distribution	Reference
47.3	IgG1	Ig γ chain	IgG	JEURISSEN et al. 1988 Vet Immunol Immunopathol 19: 225–238 BIANCHI et al. 1990 Vet Immunol Immunopathol 24: 125–134
46.5	IgG1	Ig α chain	IgA	JEURISSEN et al. 1988 Vet Immunol Immunopathol 19: 225–238
47.5	IgG2b	Ig λ chain	Ig light chain	BIANCHI et al. 1990 Vet Immunol Immunopathol 24: 125–134
74.1	IgG1	CD8α	T cell subpopulation	NOTEBORN et al. 1991 J Virol 65: 3131–3139
68.1	IgG1	87 kDa 87 kDa/30 kDa	Monocytes, interdigitating cells, macrophages, B cells	JEURISSEN et al. 1988 Dev Comp Immunol 128: 55–864
74.2	IgG1		Mature macrophages	JEURISSEN et al. 1992 Immunology 77: 75–80
68.2	IgG1	45 kDa	Ellipsoid reticulocytes	JEURISSEN et al. 1989 Vet Immunol Immunopathol 22: 123–133
74.3	IgG1		Follicular dendritic cells FDC precursors	JEURISSEN et al. 1992 Immunology 77: 75–80 JEURISSEN et al. 1994 Processes during lymphopoiesis and immunopoiesis. Plenum, New York, pp 237–241

Name of antibody	Isotype	Recognized molecule and/or MW	Tissue distribution	Reference
118.1	IgG1		Endothelial cells, basal membrane	JEURISSEN et al. 1994 Poultry Science Rev 5: 183–207

Dr. Claus Koch, Statens Seruminstitut, Artillerivej 5, 2300 Copenhagen S, Denmark (Tel: 45-32-68 32 68; Fax: 45-32-68 38 68)

HYB 12-1		Chicken complement component C3		
HYP 14-1		Chicken Ig		
HYP 208-1		Chicken Ig		
HYP 22-1		Chicken Factor B		
HYP 135-1		Denatured Chicken fact. B		
HYP 182-1		Chicken MBP		

Dr. Uli Lösch, Institute for Animal Physiology, Ludwig-Maximilians-Universität, Veterinärstrasse 13, 80539 Munich, Germany (Tel: 49-89-2180 2190; Fax: 49-89-344 937)

TCR2	IgG1	TCR Vβ1	All Vβ1 T cells	CIHAK et al. 1988 Eur J Immunol 18: 533–538
				CHEN et al. 1988 Eur J Immunol 18: 539–543
CL-1		CD45	All leukocytes	HOUSSAINT et al. 1987 Eur J Immunol 17: 287–290

Dr. Marcia Miller, Beckman Research Institute of the City of Hope, 1450 E. Duarte Road, Duarte CA 91010, USA (Tel: 1-818-301 8264; Fax: 1-818-301 8280)

maAE1	IgG2a	B-G antigens	Many tissues in addition to erythrocytes	MILLER et al. 1982 Dev Biol 94: 400–412
				MILLER et al. 1982 Immunogenetics 29: 127–132
maAE2	IgG2b	B-G antigens	Same as maAE1	MILLER et al. 1982 Dev Biol 94: 400–407
C6B12		B-F antigens	Precipitates MHC class I and recognises B-Fα chain in immunoblots	SHAMANSKY et al. Fasab J 2: 482 (abstr.)
2D5		B-L antigens	Precipitates MHC class II	Unpublished
maEE1	IgG2b	Chicken embryonic erythrocyte antigen. Also called HT7 by others.	Many tissues	MILLER and PINK 1986 Exp Cell Res 167: 295–310 *See also:*

Name of antibody	Isotype	Recognized molecule and/or MW	Tissue distribution	Reference
				MILLER et al. 1982 Dev Biol 94: 400–412

Dr. Kelly McNagny, EMBL, Postfach 10.2209, Meyerhofstrasse 1, 69117 Heidelberg, Germany (Tel: 49-6221-387 417; Fax: 49-6221-387 7516)

Name of antibody	Isotype	Recognized molecule and/or MW	Tissue distribution	Reference
MEP17	IgG1,γ	125/150 kDa (NR) α2 (β1) integrin VLA-2, CD49b	All leukocytes except late erythrycytes and granulocytes	MCNAGNY et al. 1992 Leukemia 6: 975–983
JS8		Transferrin receptor	All proliferating cells	SCHMIDT et al. 1986 Leuk Res 10: 257–272
MEP21	IgG1	150 kDa phosphoprotein	T-MEP and thrombocytes	MCNAGNY et al. 1992 Leukemia 6: 975–983
11C3		αIIbβ3 integrin, CD41/CD61	T-MEP and thrombocytes	LACOSTE-ELEAUME et al. 1994 Exp Cell Res 213: 198–209
MEP26	IgG1	47–60 kDa homodimer	T,E-MEP, thrombocytes, some early erythrycytes and eosinophils	MCNAGNY et al. 1992 Leukemia 6: 975–983
LIBS6		human αIIbβ3 integrin, CD41/CD61 (activated form)	Activated thrombocytes	FRELINGER et al. 1991 J Biol Chem 266: 17106–17111
K1		135 kDA	Thrombocytes and macrophages	KASPERS et al. 1993 Vet Immunol Immunopathol 36: 333–346
LM609		β3 integrin, CD61	T-MEP, thrombocytes and macrophages	CHERESH and SHAPIRO 1987 J Biol Chem 262: 17703–17711
MYL51/2		170 kDa	Myeloblasts, macrophages and granulocytes	KORNFELD et al. 1983 Exp Cell Res 43: 383–394
EOS47	IgG1	Melanotransferrin	Eosinophils	MCNAGNY et al. 1992 Leukemia 6: 975–983
JS3		Glycophorin?	Erythrocytes	SCHMIDT et al. 1986 Leuk Res 10: 257–272

Name of antibody	Isotype	Recognized molecule and/or MW	Tissue distribution	Reference
JS4			E-MEP, erythrocytes	SCHMIDT et al. 1986 Leuk Res 10: 257–272

Dr. Michael J.H. Ratcliffe, Dept. of Microbiology and Immunology, McGill University, 3775 University St., Montreal P.Q. H3A 2B3, Canada (Tel:1-514-398 3934; Fax: 1-514-398 7052)

Name of antibody	Isotype	Recognized molecule and/or MW	Tissue distribution	Reference
EP96	IgM	CD4	T cell subpopulation	MARMOR et al. 1993 J Exp Med 177: 647
EP72	IgG2b	CD8α	T cell subpopulation	TREGASKES et al. 1995 J Immunol 154: 4485–4494
EP42	IgG2a	CD8β	T cell subpopulation	TREGASKES et al. 1995 J Immunol 154: 4485–4494
EP25		MHC class II		
LT40	IgM	CD45	All leukocytes	PARAMITHIOTIS et al. 1991 J Immunol 147: 3710
LT2		B cell differentiation antigen		this volume
EP57	IgG1	Lymphocytes, except thymic CD3-4-8 cells		PARAMITHIOTIS et al. 1991 J Immunol 147: 3710
HY30	IgG1	ChB12 (Bu-2) (pan-B)	B cells	HUFFNAGLE et al. 1989 Hybridoma 8: 589
LT28	IgG1	Pan B cell	B cells	BENATAR et al. 1991 Eur J Immunol 21: 2529–2536
LT7	IgG1	Pan B cell	B cells	BENATAR et al. 1991 Eur J Immunol 21: 2529–2536
LT10	IgM	Immature B cell		BENATAR et al. 1991 Eur J Immunol 21: 2529–2536
LT14	IgM	Immature B cell		BENATAR et al. 1991 Eur J Immunol 21: 2529–2536

Name of antibody	Isotype	Recognized molecule and/or MW	Tissue distribution	Reference
LT19	IgM	Immature B cell		BENATAR et al. 1991 Eur J Immunol 21: 2529–2536

Dr. Jan Salomonsen, Department of Virology and Immunology, The Royal Veterinary and Agricultural University, Bulowsvej, 1870 Frederiksberg C, Denmark (Tel: 45-35-283128 ; Fax: 45-35-282742)

2G11	IgG1	MHC class II β chain	B-cells, monocytes	KAUFMAN et al. 1990 J Immunol 144: 2258–2272
F21-21	IgG1	MHC class I	All cells	CRONE and SIMONSEN 1987 In: Avian immunology. vol II: CRC, Boca Raton, pp 25–41
F21-2	IgG1	β2 microglobulin	All cells	SKJØDT et al. 1986 Mol Immunol 23: 1301–1309
A4G2	IgG2a	β2 microglobulin	Only free β2 microglobulin	DUNON et al. 1990 EMBO J 9: 3315–3322
5-6C8	IgG2a	B-G antigen	B6 and B13 erythrocytes, thrombocytes	SALOMONSEN et al. 1987 Immunogenetics 25: 373–382
5–7D7	IgG3	B-G antigen	Several haplotypes, erythrocytes thrombocytes	SALOMONSEN et al. 1987 Immunogenetics 25: 373–382
mabG1	IgG2a	B-G antigen	Erythrocytes and thrombocytes polymorphic	SALOMONSEN et al. 1991 PNAS 88: 1359–1363
mabG3	IgM	B-G antigen?	B21 erythrocytes and B cells on all haplotypes	SALOMONSEN et al. 1991 PNAS 88: 1359–1363
mabG4	IgM	B-G antigen?	Erythrocytes, T and B cells on all haplotypes	SALOMONSEN et al. 1991 PNAS 88: 1359–1363

Dr. Claudio D.Stern, Dept. of Genetics and Development, College of Physicians and Surgeons, Columbia 701 West, 168th Street, New York, NY 10032, USA (Tel:1-212-305 7915 ; Fax: 1-212-923 2090)

WEN 3H7	IgG_1	Tenascin/J1/cytotactin		Unpublished
WEN 1E9	IgG_1	N-cadherin-like antigen	Somite, notochord (chick)	Unpublished

Name of antibody	Isotype	Recognized molecule and/or MW	Tissue distribution	Reference
α PnA-R	Rabbit (polyclonal antibody)	48 kDa, 55 kDa (red) 200 kDa and other (non-red)	Posterior half of somite	Unpublished
α Hyaluromidase	Rabbit (polyclonal antibody)	48 kDa, 55 kDa (red)		STERN 1984 Cell Biol Int Rep 8: 703–717

Dr. Olli Vainio, Department of Medical Microbiology, Turku University, Kiina myllynkatu 13, 20520 Turku, Finland (Tel: 35-821-3337400; Fax: 35-821-2330008)

37C.18	IgG1	MHC class I	All cells, recognizes a polymorphic determinant	PINK et al. 1985 Immunogenetics 21: 293–297
21-1A6	IgG1	MHC class II		VEROMAA et al. 1988 Eur J Immunol 18: 225–230
L22	IgG1	Bu-1a	B cells and a subpopulation of macrophages	PINK and RIJNBECK 1983 Hybridoma 2: 287–296
21-1A4	IgG1	Bu-1a	B cells and a subpopulation of macrophages	VEROMAA et al. 1989 Hybridoma 7: 41–53
5-11G2	IgG1	Bu-1b	B cells and a subpopulation of macrophages	VEROMAA et al. 1989 Hybridoma 7: 41–53
11A9	IgM	ov antigen, chL12	Most embryonic hemopoietic cells, T cells, subpopulation of B cells	HOUSSAINT et al. 1991 J Exp Med 174: 397–406
RR5-89	IgG2b	64 kDa	Thymocytes, few peripheral T cells (like CT1)	KATEVUO et al. in preparation
T10A6	IgM	64 kDa	Thymocytes, few peripheral T cells (like CT1)	HOUSSAINT et al. 1987 Eur J Immunol 17: 287–290
2-6	IgG1	CD4	T cell subpopulation	LUHTALA et al. 1993 Hybridoma 12: 633–646
2-35	IgG2b	CD4	T cell subpopulation	LUHTALA et al. 1993 Hybridoma 12: 633–646

Name of antibody	Isotype	Recognized molecule and/or MW	Tissue distribution	Reference
11-39	IgG1	CD8α	T cell subpopulation	LUHTALA et al. 1995 Scand J Immunol 42: 171–174
3-298	IgG2b	CD8α	T cell subpopulation, recognises polymorphic determinant	Unpublished
11-13	IgG1	CD8α, allelic	T cell subpopulation, recognises polymorphic determinant	LUHTALA et al. 1995 Scand J Immunol 42: 171–174
11-38	IgM	CD8α, allelic	T cell subpopulation	LUHTALA et al. 1995 Scand J Immunol 42: 171–174
2-191	IgG1	CD5, 64 kDa	T and B cells	KOSKINEN et al. in preparation
3-58	IgG1	CD5, 64 kDa	T and B cells	Koskinen et al. in Preparation
2-4	IgG2a	CD28, 40 kDa	T cells, except most of γδ cells in periphery	VAINIO et al. 1991 J Immunol 147: 1593–1599 YOUNG et al. 1994 J Immunol 152: 3848–3851
2-264	IgG2b	MUCRel, 100 kDa, member of Ig superfamily	Hemopoietic precursors, thymocytes, few peripheral T cells, endothelial cells, muscle, subpopulation of neurons	VAINIO et al. in preparation

Lymphocyte Markers: An Overview

Molecule	mAb	Contact Person
CD3	CT3	C.-L. H. Chen
	AV36	F. Davison
CD4	CT4	C.-L. H.Chen
	2–6, 2–35	O. Vainio
	AV29, Av30	F. Davison
	EP96	M.J.H. Ratcliffe
CD5	2–191,3–58	O. Vainio
CD6(putative)	S3	C.-L.H. Chen
CD8α	CT8	C.-L.H. Chen
	AV12, AV13, AV14	F.Davison
	74.1	S.Jeurissen
	EP72	M.J.H. Ratcliffe
	11–39, 11–38, 11–13, 3–298	O. Vainio
CD8β	Ep42	M.J.H. Ratcliffe
CD28	2–4	O. Vainio
	AV7	F.Davison
Bu-1a	L22, 21–1A4	O.Vainio
Bu-1b	5–11G2	O.Vainio
Bu-1a/b (both allotypes)	AV20	F.Davison
OV, ChL12	11A9	O.Vainio
Bu-2	HY30	M.J.H. Ratcliffe
CD45 (putative)	LT40	M.J.H. Ratcliffe
	CLA-1	C.-L.H. Chen
	HIS-C7	S. Jeurissen
	CL-1	U.Lösch
TCRγδ	TCR1	C.-L.H. Chen
TCRVβ1	TCR2	U.Lösch
TCRVβ2	TCR3	C.-L.H. Chen

Subject Index

Current Topics in Microbiology and Immunology

Volumes published since 1989 (and still available)

Vol. 170: **Compans, Richard W. (Ed.):** Protein Traffic in Eukaryotic Cells. Selected Reviews. 1991. 14 figs. X, 186 pp. ISBN 3-540-53631-0

Vol. 171: **Kung, Hsing-Jien; Vogt, Peter K. (Eds.):** Retroviral Insertion and Oncogene Activation. 1991. 18 figs. X, 179 pp. ISBN 3-540-53857-7

Vol. 172: **Chesebro, Bruce W. (Ed.):** Transmissible Spongiform Encephalopathies. 1991. 48 figs. X, 288 pp. ISBN 3-540-53883-6

Vol. 173: **Pfeffer, Klaus; Heeg, Klaus; Wagner, Hermann; Riethmüller, Gert (Eds.):** Function and Specificity of γ/δ TCells. 1991. 41 figs. XII, 296 pp. ISBN 3-540-53781-3

Vol. 174: **Fleischer, Bernhard; Sjögren, Hans Olov (Eds.):** Superantigens. 1991. 13 figs. IX, 137 pp. ISBN 3-540-54205-1

Vol. 175: **Aktories, Klaus (Ed.):** ADP-Ribosylating Toxins. 1992. 23 figs. IX, 148 pp. ISBN 3-540-54598-0

Vol. 176: **Holland, John J. (Ed.):** Genetic Diversity of RNA Viruses. 1992. 34 figs. IX, 226 pp. ISBN 3-540-54652-9

Vol. 177: **Müller-Sieburg, Christa; Torok-Storb, Beverly; Visser, Jan; Storb, Rainer (Eds.):** Hematopoietic Stem Cells. 1992. 18 figs. XIII, 143 pp. ISBN 3-540-54531-X

Vol. 178: **Parker, Charles J. (Ed.):** Membrane Defenses Against Attack by Complement and Perforins. 1992. 26 figs. VIII, 188 pp. ISBN 3-540-54653-7

Vol. 179: **Rouse, Barry T. (Ed.):** Herpes Simplex Virus. 1992. 9 figs. X, 180 pp. ISBN 3-540-55066-6

Vol. 180: **Sansonetti, P. J. (Ed.):** Pathogenesis of Shigellosis. 1992. 15 figs. X, 143 pp. ISBN 3-540-55058-5

Vol. 181: **Russell, Stephen W.; Gordon, Siamon (Eds.):** Macrophage Biology and Activation. 1992. 42 figs. IX, 299 pp. ISBN 3-540-55293-6

Vol. 182: **Potter, Michael; Melchers, Fritz (Eds.):** Mechanisms in B-Cell Neoplasia. 1992. 188 figs. XX, 499 pp. ISBN 3-540-55658-3

Vol. 183: **Dimmock, Nigel J.:** Neutralization of Animal Viruses. 1993. 10 figs. VII, 149 pp. ISBN 3-540-56030-0

Vol. 184: **Dunon, Dominique; Mackay, Charles R.; Imhof, Beat A. (Eds.):** Adhesion in Leukocyte Homing and Differentiation. 1993. 37 figs. IX, 260 pp. ISBN 3-540-56756-9

Vol. 185: **Ramig, Robert F. (Ed.):** Rotaviruses. 1994. 37 figs. X, 380 pp. ISBN 3-540-56761-5

Vol. 186: **zur Hausen, Harald (Ed.):** Human Pathogenic Papillomaviruses. 1994. 37 figs. XIII, 274 pp. ISBN 3-540-57193-0

Vol. 187: **Rupprecht, Charles E.; Dietzschold, Bernhard; Koprowski, Hilary (Eds.):** Lyssaviruses. 1994. 50 figs. IX, 352 pp. ISBN 3-540-57194-9

Vol. 188: **Letvin, Norman L.; Desrosiers, Ronald C. (Eds.):** Simian Immunodeficiency Virus. 1994. 37 figs. X, 240 pp. ISBN 3-540-57274-0

Vol. 189: **Oldstone, Michael B. A. (Ed.):** Cytotoxic T-Lymphocytes in Human Viral and Malaria Infections. 1994. 37 figs. IX, 210 pp. ISBN 3-540-57259-7

Vol. 190: **Koprowski, Hilary; Lipkin, W. Ian (Eds.):** Borna Disease. 1995. 33 figs. IX, 134 pp. ISBN 3-540-57388-7

Vol. 191: **ter Meulen, Volker; Billeter, Martin A. (Eds.):** Measles Virus. 1995. 23 figs. IX, 196 pp. ISBN 3-540-57389-5

Vol. 192: **Dangl, Jeffrey L. (Ed.):** Bacterial Pathogenesis of Plants and Animals. 1994. 41 figs. IX, 343 pp. ISBN 3-540-57391-7

Vol. 193: **Chen, Irvin S. Y.; Koprowski, Hilary; Srinivasan, Alagarsamy; Vogt, Peter K. (Eds.):** Transacting Functions of Human Retroviruses. 1995. 49 figs. IX, 240 pp. ISBN 3-540-57901-X

Vol. 194: **Potter, Michael; Melchers, Fritz (Eds.):** Mechanisms in B-cell Neoplasia. 1995. 152 figs. XXV, 458 pp. ISBN 3-540-58447-1

Vol. 195: **Montecucco, Cesare (Ed.):** Clostridial Neurotoxins. 1995. 28 figs. XI., 278 pp. ISBN 3-540-58452-8

Vol. 196: **Koprowski, Hilary; Maeda, Hiroshi (Eds.):** The Role of Nitric Oxide in Physiology and Pathophysiology. 1995. 21 figs. IX, 90 pp. ISBN 3-540-58214-2

Vol. 197: **Meyer, Peter (Ed.):** Gene Silencing in Higher Plants and Related Phenomena in Other Eukaryotes. 1995. 17 figs. IX, 232 pp. ISBN 3-540-58236-3

Vol. 198: **Griffiths, Gillian M.; Tschopp, Jürg (Eds.):** Pathways for Cytolysis. 1995. 45 figs. IX, 224 pp. ISBN 3-540-58725-X

Vol. 199/I: **Doerfler, Walter; Böhm, Petra (Eds.):** The Molecular Repertoire of Adenoviruses I. 1995. 51 figs. XIII, 280 pp. ISBN 3-540-58828-0

Vol. 199/II: **Doerfler, Walter; Böhm, Petra (Eds.):** The Molecular Repertoire of Adenoviruses II. 1995. 36 figs. XIII, 278 pp. ISBN 3-540-58829-9

Vol. 199/III: **Doerfler, Walter; Böhm, Petra (Eds.):** The Molecular Repertoire of Adenoviruses III. 1995. 51 figs. XIII, 310 pp. ISBN 3-540-58987-2

Vol. 200: **Kroemer, Guido; Martinez-A., Carlos (Eds.):** Apoptosis in Immunology. 1995. 14 figs. XI, 242 pp. ISBN 3-540-58756-X

Vol. 201: **Kosco-Vilbois, Marie H. (Ed.):** An Antigen Depository of the Immune System: Follicular Dendritic Cells. 1995. 39 figs. IX, 209 pp. ISBN 3-540-59013-7

Vol. 202: **Oldstone, Michael B. A.; Vitković, Ljubiša (Eds.):** HIV and Dementia. 1995. 40 figs. XIII, 279 pp. ISBN 3-540-59117-6

Vol. 203: **Sarnow, Peter (Ed.):** Cap-Independent Translation. 1995. 31 figs. XI, 183 pp. ISBN 3-540-59121-4

Vol. 204: **Saedler, Heinz; Gierl, Alfons (Eds.):** Transposable Elements. 1995. 42 figs. IX, 234 pp. ISBN 3-540-59342-X

Vol. 205: **Littman, Dan. R. (Ed.):** The CD4 Molecule. 1995. 29 figs. XIII, 182 pp. ISBN 3-540-59344-6

Vol. 206: **Chisari, Francis V.; Oldstone, Michael B. A. (Eds.):** Transgenic Models of Human Viral and Immunological Disease. 1995. 53 figs. XI, 345 pp. ISBN 3-540-59341-1

Vol. 207: **Prusiner, Stanley B. (Ed.):** Prions Prions Prions. 1995. 42 figs. VII, 163 pp. ISBN 3-540-59343-8

Vol. 208: **Farnham, Peggy J. (Ed.):** Transcriptional Control of Cell Growth. 1995. 17 figs. IX, 141 pp. ISBN 3-540-60113-9

Vol. 209: **Miller, Virginia L. (Ed.):** Bacterial Invasiveness. 1996. 16 figs. IX, 115 pp. ISBN 3-540-60065-5

Vol. 210: **Potter, M.; Rose, N. R. (Eds.):** Immunology of Silicones. 1996. 136 figs. XX, 430 pp. ISBN 3-540-60272-0

Vol. 211: **Wolff L.; Perkins A. S. (Eds.):** Molecular Aspects of Myeloid Stem Cell Development. 1996. 98 figs. XIV, 298 pp. ISBN 3-540-60414-6

Springer-Verlag and the Environment

We at Springer-Verlag firmly believe that an international science publisher has a special obligation to the environment, and our corporate policies consistently reflect this conviction.

We also expect our business partners – paper mills, printers, packaging manufacturers, etc. – to commit themselves to using environmentally friendly materials and production processes.

The paper in this book is made from low- or no-chlorine pulp and is acid free, in conformance with international standards for paper permanency.

MIX
Papier aus verantwortungsvollen Quellen
Paper from responsible sources
FSC® C105338

If you have any concerns about our products,
you can contact us on
ProductSafety@springernature.com

In case Publisher is established outside the EU,
the EU authorized representative is:
Springer Nature Customer Service Center GmbH
Europaplatz 3, 69115 Heidelberg, Germany

Printed by Libri Plureos GmbH
in Hamburg, Germany